THE SCIENCE OPPORTUNITIES FOR THE WARM SPITZER MISSION WORKSHOP

To learn more about AIP Conference Proceedings, including the Conference Proceedings Series, please visit the webpage
http://proceedings.aip.org/proceedings

THE SCIENCE OPPORTUNITIES FOR THE WARM SPITZER MISSION WORKSHOP

Pasadena, California 4 – 5 June 2007

EDITORS

Lisa J. Storrie-Lombardi
Nancy A. Silbermann
Spitzer Science Center
California Institute of Technology
Pasadena, California

SPONSORING ORGANIZATION
Spitzer Science Center

Melville, New York, 2007
AIP CONFERENCE PROCEEDINGS ■ VOLUME 943

Editors:

Lisa J. Storrie-Lombardi
Nancy A. Silbermann

Spitzer Science Center
California Institute of Technology
1200 E. California Boulevard, MS 314-6
Pasadena, CA 91125
U.S.A.

E-mail: lisa@ipac.caltech.edu
nancys@ipac.caltech.edu

L.C. Catalog Card No. 2007937126
ISBN 978-0-7354-0457-1
ISSN 0094-243X
Printed in the United States of America

CONTENTS

5. - CONTRIBUTED WHITE PAPERS

Preface

When the Spitzer Space Telescope cryogen is exhausted the Observatory can operate with IRAC bands 1 and 2 (3.6 and 4.5 μm) at full sensitivity. The other channels of all instruments will not operate at the elevated temperatures (25-30K). A major element of planning for this phase is developing the science operations philosophy for the Warm Spitzer mission. The need to substantially reduce the operations cost for the Warm Mission and the much simplified demands of operating a one instrument mission offer the opportunity to execute science projects that are substantially larger in scope than those that can be contemplated in the Cryogenic Spitzer era. During this phase of the Spitzer mission, a vigorous archival research program will be conducted that will exploit the entire Spitzer archival dataset.

In September 2006 the Spitzer Science Center (SSC) Director's office started planning a community workshop to explore in detail the science drivers for the warm Spitzer mission and help the SSC develop the science operations philosophy to for the Warm Mission. The Director recruited a steering committee to generate white papers outlining science drivers for the warm mission. The Steering Committee members are:

Chair: **Pat McCarthy** (OCIW)
Nearby Universe: **Daniela Calzetti** (U. Mass)
Extrasolar Planets: **Drake Deming** (GSFC)
Stars, Brown Dwarfs: **Jill Knapp** (Princeton)
Solar System: **Carey Lisse** (JHU-APL)
Galactic Structure and ISM: **Mike Skrutskie** (U. Virginia)
Star Formation: **Steve Strom** (NOAO)
Distant Universe: **Pieter van Dokkum** (Yale)

Each steering committee recruited their own subject area subcommittee from the community to write the white papers and delivered the white papers to the SSC to post for the community prior to the workshop, which was held June 4-5, 2007, at the Westin Hotel in Pasadena, CA. The SSC prepared one general white paper and one archive white paper. In addition the SSC received five contributed white papers are that are also included in this volume. About 90 people attended the two-day workshop. We discussed science goals and how to operate the mission with the reduced resources available, both using (a) the shortest two IRAC channels, and (b) archival research with the rich Spitzer archive. Pat McCarthy summarizes the highlights from the meeting in first paper in this volume.

The SSC would like to again thank the Steering Committee members, their subcommittees, the authors of the contributed white papers, and all the workshop participants for their time and efforts in helping us make plans for a scientifically productive and successful warm mission.

Lisa Storrie-Lombardi, John Stauffer, Luisa Rebull and Nancy Silbermann
Spitzer Science Center, California Institute of Technology

1. – WORKSHOP SUMMARY

Spitzer Science in the Post Cryogen Era

Patrick J. McCarthy

Carnegie Observatories, 813 Santa Barbara St. Pasadena, CA, 91101, USA

Abstract. I review the broad scientific opportunities offered by the Spitzer warm mission and attempt to put them into context with other priorities and future and ongoing programs in astronomy and astrophysics. The warm mission offers a number of unique opportunities for scientific investigations that are beyond the reach of other facilities or are not practical with Spitzer operating in its current mode. Some possible approaches to making the best use of the facility during a 3-5 year duration are considered. These include undertaking large and ambitious surveys while preserving small programs in fields where they are most effective.

Keywords: Spitzer Space Telescope, infrared astronomical observations
PACS: 95.85.Hp

1. INTRODUCTION

Once the cryogen is depleted on Spitzer it will rapidly lose its great sensitivity at wavelengths longer than about 5μm. The observatory will, however, remain uniquely powerful in the 3-5μm region of the spectrum by providing imaging with sensitivity that will not be matched until JWST flies. In its warm state Spitzer can provide both cutting edge science and prepare a foundation on which JWST, ALMA and other next generation astronomical facilities can build. The goal of this workshop was to identify and articulate these scientific opportunities and consider how to best use the observatory in an efficient manner. As documented throughout this volume, the scientific potential of the warm mission is impressive. In this contribution I will attempt to pull together some of the disparate science goals into a more or less coherent whole and consider some options of how the Spitzer Science Center and the user community might approach the challenge of maximizing the unique opportunities offered by the warm mission.

2. THE POWER OF SPITZER IN ITS WARM STATE

As the temperature in the telescope and instruments rises the spectrometers and Multi-band Imaging Photometer (MIPS) will cease to function, as internal backgrounds will swamp any astronomical signals and saturate the detectors in even the shortest exposures. The two long wavelength channels (5.8 and 8μm bands) of the Infrared Array Camera (IRAC) will also be rendered inoperable. The short wavelength 3.5 and 4.5μm channels will, however, not only remain operable, their sensitivities will be uncompromised. We can put the IRAC channels 1 & 2 sensitivities in some

CP943, *The Science Opportunities for the Warm Spitzer Mission Workshop*,
edited by L. J. Storrie-Lombardi and N. A. Silbermann

perspective by comparing with ground-based 8m telescopes. In 100 seconds IRAC has a 5σ point-source sensitivity of ~ 3μJy. An 8m telescope working on the ground at K (2.2μm) can achieve this same sensitivity in about 200 seconds, so the 85cm Spitzer telescope is about a factor of ~ 2 faster at L than a ground-based 8m telescope operating at K. At L-band Spitzer is 120 times faster than a ground-based 8m in good conditions, at M-band the speed gain is closer to a factor of 1000. While adaptive optics will improve the comparison with the ground somewhat, the orders of magnitude reduction in background from a cryogenic telescope operating in space produces a gain in sensitivity and speed that will likely never be matched on the ground.

The power of the IRAC instrument on Spitzer arises not just from its sensitivity, but also from its impressive mapping speed. The large pixels of IRAC (1.2″ x 1.2″) provide a field of view just over 5′ x 5′. This large field cannot be easily matched on the ground as the background prohibits such large pixels. The resulting mapping speed for IRAC, while below that of all-sky mapping instruments like WISE, provides a powerful survey capability. Most of the programs discussed at this workshop make use of this unique combination of sensitivity and mapping speed to carry out surveys that are beyond the reach of any currently planned ground- or space-based facilities.

In its warm state the lifetime of Spitzer is not limited by on-board consumables. Rather the Earth-trailing orbit of the spacecraft will ultimately put it beyond the reach of the ground-system needed for control and data transfer. Thus the lifetime of the warm mission is likely to be set by programmatic priorities with NASA and the US astronomical community. For the purpose of this workshop we are considering a lifetime in the 3-5 year range. Spitzer is a highly efficient observatory, much more so than satellites in low Earth orbits, such as HST. A 5-year Spitzer warm mission will yield roughly as much observing time as Hubble has in its 16 years of operations to date. Allowing for the large difference in sensitivity and observing efficiency, the Spitzer warm mission will out strip *3000 years of 3.5μm and 4.5μm observing on a ground-based 8 to 10m-class telescope*. This is a powerful capability that should not be taken lightly.

The power of the warm mission to carry out large surveys is illustrated in Figure 1. I plot the total areal coverage possible using IRAC with exposure times ranging from 30 seconds to several hours in programs with durations that range from 500 hours to the full 5 year lifetime of the mission. I take a year of observing as equal to 7000 hours and assume that one continues to gain in depth like $\sqrt{t}$, which is probably optimistic for the longest exposure times. I've not plotted below 100nJy as this is likely near the confusion limit and no brighter than 5μJy as programs shallower than this are highly inefficient. The program duration for any combination of depth and area can be inferred from the contours in Figure 1. A useful benchmark is provided by the observation that in the 5-year lifetime of the warm mission one could map 1000 square degrees to 1μJy.

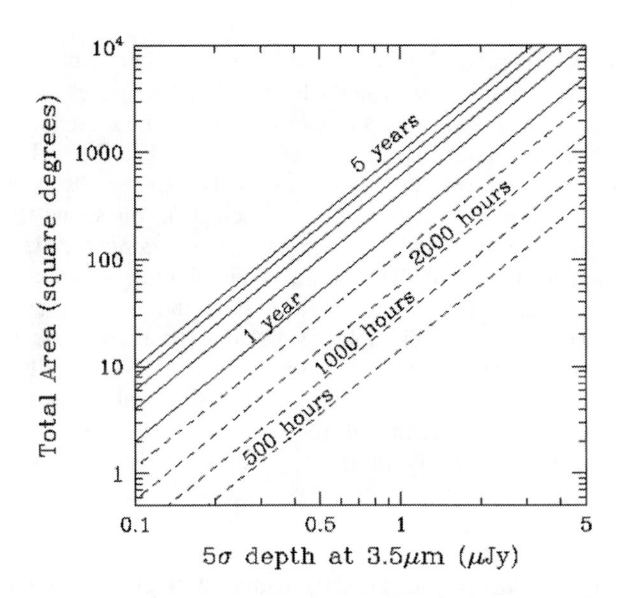

FIGURE 1. Area vs. depth contours for IRAC mapping programs with durations ranging from 500 hours to the life of the mission. A year of observing is taken as 7000 hours of on-source time. Each contour corresponds to exposure times ranging from 30 seconds to several hours. The duration of a program with a particular combination of area and depth at 3.5μm can be read from the graph. A benchmark is provided by the fact that over the entire 5-year mission one could image 1000 square degrees to a 5σ depth of 1μJy.

The NASA explorer program provides an interesting benchmark to put the warm mission in context. The cost cap for the *SMEX* and *MIDEX* missions are currently $105M and $180M, far smaller than the cost of a *Great Observatory* mission, but in the same range as the incremental cost of the warm mission. The *Explorer* class astronomy missions have had lifetimes that range from as short as one year to nearly 20 years, but a 3-5 year lifetime is quite typical for *SMEX* program. These missions tend to have a narrower science focus than a Great Observatory program, but many contain a mix of core survey and guest investigator programs. The Spitzer warm mission thus fits many of the criteria of an *Explorer* class mission with one critical exception – the space craft is in orbit simply waiting for the science mission to be defined by the user community. This is a scientific opportunity that does not often arise – the chance to carry out a cutting edge space astronomy program of the *Explorer* class without the years of suffering that normally accompanies such an endeavor.

3. THE WARM MISSION AND NATIONAL SCIENTIFIC PRIORITIES

While astronomers who are active users of Spitzer, and IRAC in particular, might consider it a foregone conclusion that the warm mission should be supported, the broader community might rightly ask where this fits into our national scientific priorities. There are a number or places where national priorities are considered and tabulated, but the National Academy's Decadal Survey is the most authoritative source for long-range planning in astronomy and astrophysics. The Bachall report from the 1991 survey played a key role in the genesis of the Spitzer observatory. The most recent decadal survey identified a number of scientific questions that are both of fundamental importance and ripe for progress in the coming decade. These include advancing our understanding of the nature and distribution of matter and energy on large scales, identifying the first generation of stars and galaxies, understanding the formation and evolution of black holes, stars and planetary systems and, finally, exploring the impact of the astronomical environment on the Earth. These rather broad goals encompass much of contemporary astrophysics and Spitzer has played an important role in addressing many of these goals. In its warm state Spitzer can continue to have large impacts in most of these areas. As discussed in other contributions to this volume, the warm mission can make unique contributions to our understanding of galaxies in the early universe. IRAC surveys of the earliest clusters of galaxies probe the distribution of matter on very large scales and constrain the expansion history and energy content of the Universe. Even with only its two shortest wavelength channels Spitzer provides powerful probes of star formation and exoplanet physics. It seems clear then that the warm mission fits in well with our national priorities in astronomy and astrophysics and one could argue that it provides a timely and cost effective bridge between the current generation of facilities and future observatories such as the Wide-field Infrared Survey Explorer (WISE) and the James Webb Space Telescope.

The forefront questions in astrophysics increasingly require observational data from a wide range of the electromagnetic spectrum. The days of pure optical, radio, or infrared astronomers are passing and the youngest generation of observers is multi-wavelength oriented in its approach. In this context the warm mission has a great deal of scientific synergy with facilities operating at other wavelengths, and even those that will be soon observing overlapping regions of the spectrum. The synergy between HST and Spitzer is well documented; a number of the key legacy programs, such as GOODS, use both HST and Spitzer to sample the optical and IR regions of the spectrum. Similar synergies are likely to come to light, although in a time staggered manner, when ALMA and JWST begin operations.

4. SCIENTIFIC OPPORTUNITIES IN THE WARM MISSION

Spitzer has had a large impact over a wide range of science to date. The potential impact of the warm mission is curtailed somewhat by its restricted wavelength range. The potential for very large programs, however, opens up new possibilities that have been impractical during the heavily over subscribed early cycles. Before discussing detailed science programs it is instructive to ask where a 3-4 micron imager sits in relation to various astronomical phenomena and astrophysical processes.

4.1 IRAC Channels 1 & 2 in the Global Context

In Figure 2 we consider the location of the Spitzer IRAC channels 1 & 2 bands in the context of the extragalactic background light (EBL). The EBL has two broad peaks, one due primarily to direct radiation from the photospheres of stars and accretion disks. The energy density of this component peaks around ~ 1μm, suggesting that much of it comes from redshifts > 0.5 or so. The secondary peak at ~ 100μm is reprocessed radiation that has been thermalized by interstellar and circumstellar dust. This secondary peak has been the focus of much activity in infrared astronomy and IRAS, Spitzer and other missions have had an enormous impact here. IRAC channels 1 & 2 sample the long wavelength tail of the direct radiation peak. At 3.5 and 4.5μm IRAC samples the peak of the spectral energy distributions of evolved stars at intermediate redshifts, $1 < z < 4$, and the peak of the energy distributions of actively star forming galaxies at $z \sim 5$ and higher. Not surprisingly, these two areas – the study of red galaxies at $z > 1$ and young galaxies at $z > 5$ are fields in which IRAC has made major, and in some cases quite unanticipated, discoveries.

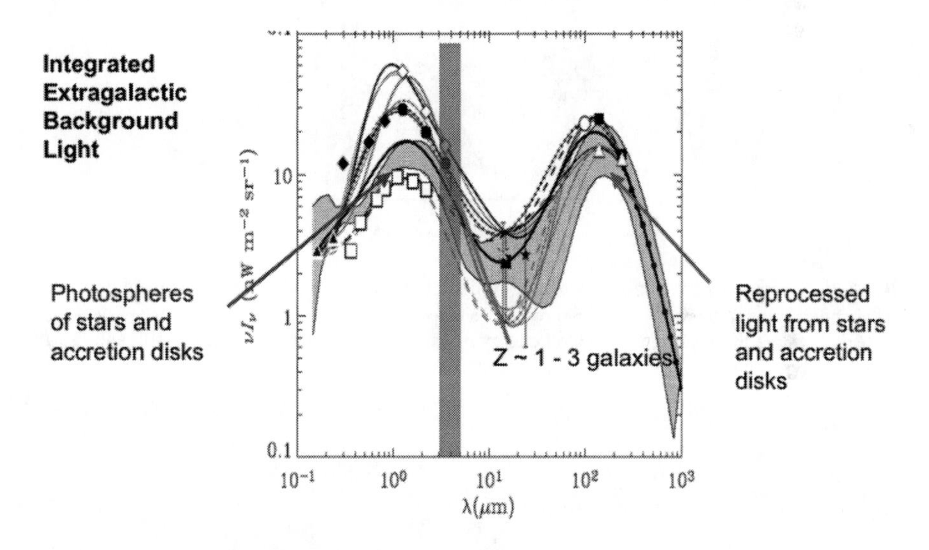

FIGURE 2. The integrated extragalactic background light in the UV to mid-IR range. The spectral range sampled by IRAC channels 1 & 2 is shown as the shaded region. The UV/visible and mid-IR peaks in the background arise from direct and reprocessed emission, respectively.

4.2 Studies of Old Galaxies at Intermediate Redshifts

The sensitivity of the IRAC bands to evolved stellar populations is illustrated in Figures 3, 4, and 5. For redshifts between about 1 and 3 the IRAC bands sample the broad H- induced peak in the spectral energy distributions of galaxies with ages greater than roughly 1 Gyr. This allows one to discriminate red galaxies at z ~ 1.5 – 3 from the foreground population in moderately deep IRAC images with ease. While K-band imaging has been used to identify red galaxies to z ~ 3 extremely long exposure times on 8m telescopes are required. IRAC can detect such sources with greater contrast in far shorter exposures.

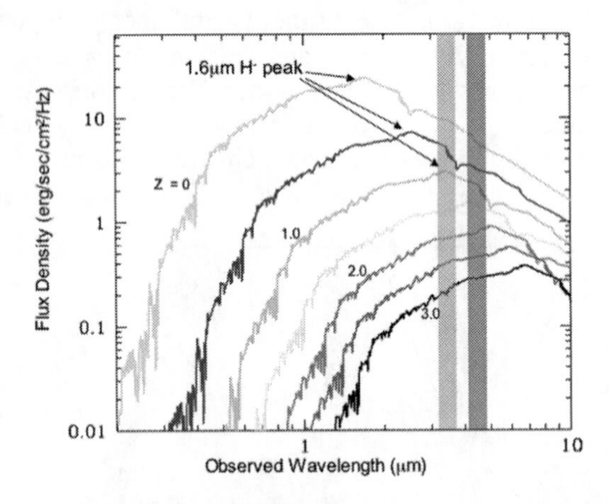

FIGURE 3. Spectral energy distributions, in F_ν units, of evolved galaxies at redshifts from 0 to 3. The IRAC channel 1 and 2 band-passes are shown as shaded regions At redshifts between ~ 1 and 3 the IRAC bands sample the peak of the spectral energy distribution for stellar populations with ages greater than about 1 Gyr.

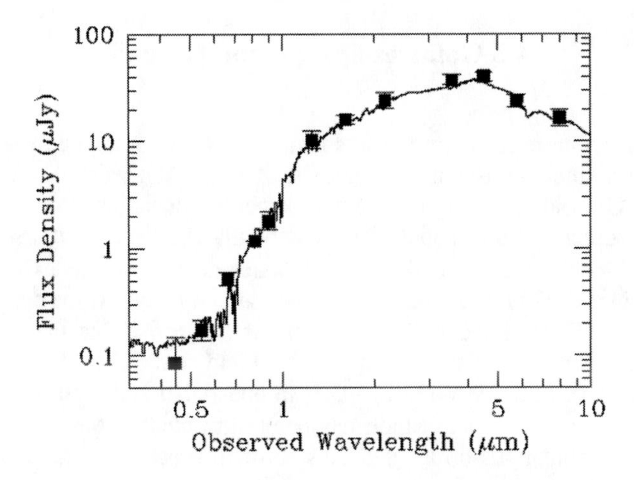

FIGURE 4. An example spectral energy distribution for a massive red galaxy at z = 1.5. The twelve bands cover B (0.4μm) through IRAC channel 4 (8μm). The broad peak of the spectral energy distribution at a rest-frame wavelength of 1.6μm falls in the IRAC 4.5μm band at this redshift.

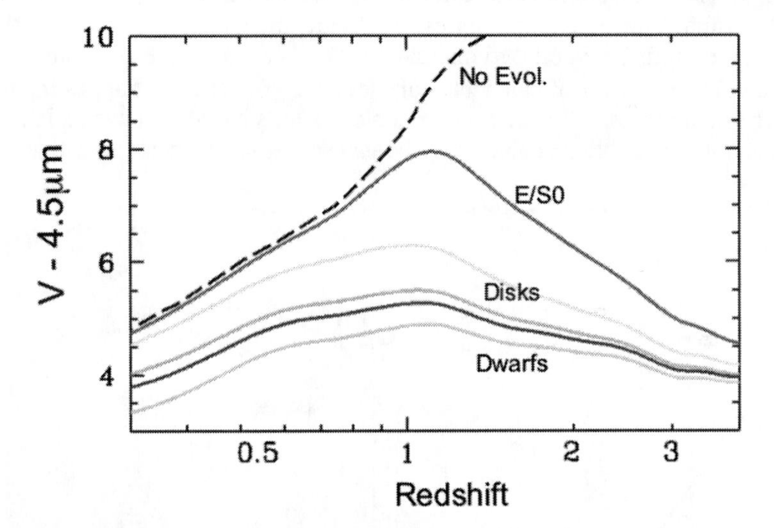

FIGURE 5. Optical V-band minus IRAC Channel 2 colors versus redshift for a variety of evolving galaxy models. The dashed black line shows the color trajectory for a non-evolving 13Gyr stellar population for reference. The other tracks show the color evolution for galaxies with exponentially decaying star formation rates with e-folding times of 1, 2, 3 and 5 Gyrs along with a constant star formation rate model. These are intended to illustrate the color evolution of elliptical/S0 galaxies, disk galaxies and dwarf galaxies in order of increasing e-folding times. The V-4.5μm color is a particularly powerful discriminator for early type galaxies at z ~ 1-2.

4.3 Galaxies at z ~ 6 and Beyond

One of the more surprising results from deep IRAC imaging programs was the detection of continuum emission from galaxies at z > 5. At a redshift of 6 the Universe is just under 1Gyr old. One might naturally expect that galaxies at this redshift would be extremely young and thus should have essentially flat (in F_ν units) spectral energy distributions. Most of the currently know galaxies at z ~ 5-6 have flux densities of ~200-300nJy in the visible and near-IR and thus should be below the detection limit of even the deepest IRAC images. Thus it came as a surprise that many of the z ~ 6 galaxies in GOODS, the Hubble Ultra-Deep Field and other deep surveys were detected with IRAC. The very red K – 3.5μm and rather flat 3.5μm – 4.5μm colors reveal a strong spectral break, which has been attributed to the Balmer continuum break in stellar populations dominated by stars with spectral classes from early B to mid-F. The best fitting ages for these galaxies derived from spectral synthesis models have been reported to be ~500Myr and the inferred stellar masses are quite large, reaching to a few x $10^{10}M_{sun}$ and perhaps higher. In Figure 6 I show an example from Eyles et al. [1], but there are also examples shown in Yan et al. [2] and others. The examples from Yan et al. and Eyles et al. are drawn from samples with high confidence redshifts based on Lyα emission. Mobasher et al. [3] and Wiklind et al. [4] have identified a number of candidate galaxies at z ~ 6 and higher appearing to have very large stellar masses. Secure spectroscopic redshifts are not yet available for these objects and it is possible that some may be galaxies at lower redshifts with unusually red colors.

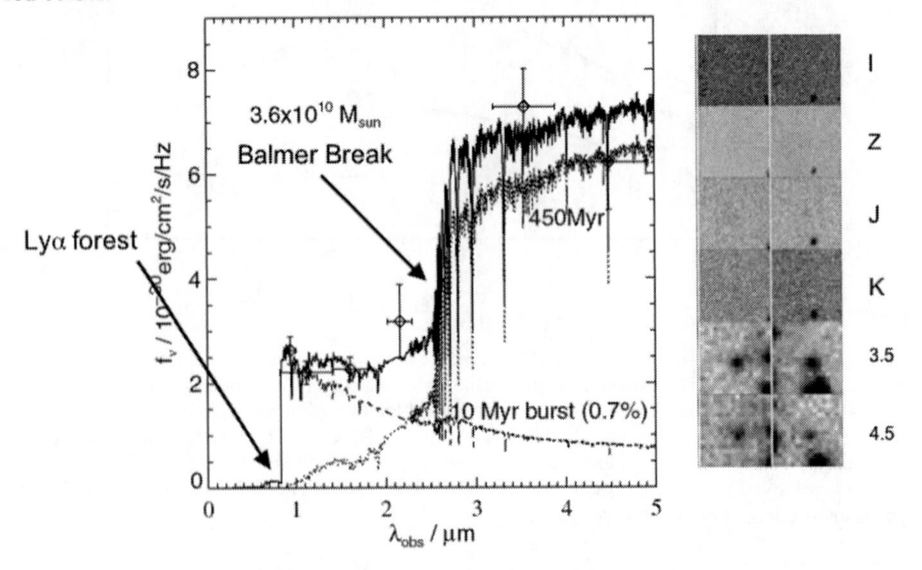

FIGURE 6. The spectral energy distribution of a z = 5.7 I-band drop out from Elyes et al. [1]. The grey-scale images on the right show detections in IRAC bands 1&2 and non-detections in all of the shorter wave-bands. The best-fit model SEDs is combination of a massive intermediate age population and a recent star burst. The total stellar mass is estimated to be more than 3 x 10^{10} solar masses.

The precise implications of the large breaks in the z ~ 6 galaxies detected by Spitzer are unclear. Recent spectral synthesis models by Maraston et al. [5] show that AGB stars can make fairly significant contributions, even at rather short wavelengths for ages of ~ 0.3 – 1Gyr. These models may lead to somewhat younger ages and significantly lower stellar masses for the z ~ 6 galaxies than first thought. Time will be needed before these issues are settled, but there is little doubt that Spitzer has provided a new and important view of galaxies in the early universe. It is particularly interesting to note that these discoveries arise entirely from channels 1 and 2 on IRAC, the capability that remain unchanged in the warm mission. Even in the cold state channels 3 and 4 lack the sensitivity required for these very faint objects.

4.4 Cool Stars and Hot Planets

The intrinsic L′ - M colors of main sequence and giant stars are of little diagnostic value as they are all near zero, being essentially Rayleigh-Jeans, for stars warmer than about L5. In the latest L dwarfs, T dwarfs and giant planets CH_4 absorption strongly impact the L-band while the M-band remains mostly clean. Thus L′ – M color provides a sensitive temperature indicator on its own. This can be seen in model atmospheres calculations (e.g. Burrows et al. [6]) as illustrated in Figure 7 below, and in the 3.6μm – 4.5μm vs. 3.6μm color magnitude diagram presented in Knapp et al., this volume.

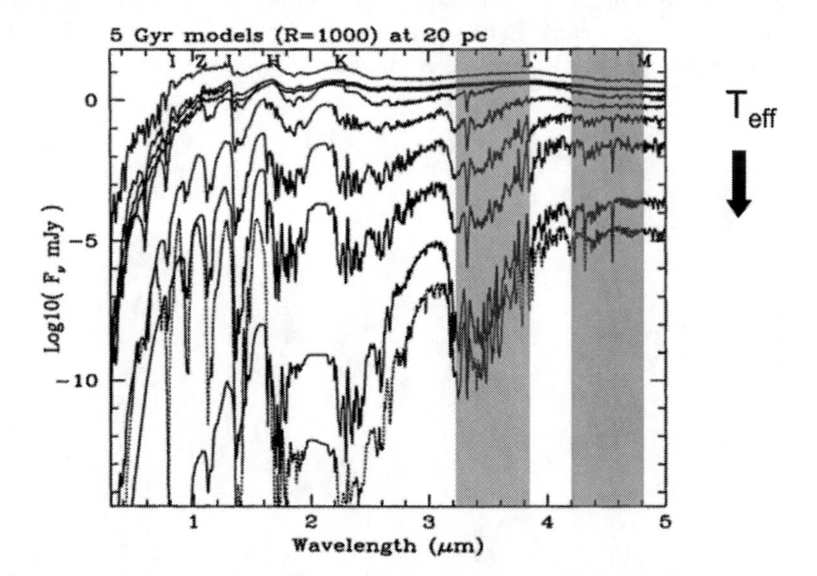

FIGURE 7. Infrared spectra from model atmospheres for old brown dwarfs and giant planets from Burrows et al. [6]. The IRAC channel 1 and 2 band-passes are shown by the grey shaded areas. The strong impact of methane absorption in the cooler objects and monotonic evolution of 3.6μm – 4.5μm color with temperature is evident.

The strong temperature sensitivity of the 3.6μm – 4.5μm color for cool objects suggests that an efficient survey for brown dwarfs and free-floating planets could be carried out with Spitzer in its warm phase. Knapp has suggested that this could be coupled to other imaging surveys at intermediate galactic latitude to leverage their scientific value. The WISE satellite has brown dwarf surveys as part of its core science mission. The faintest brown dwarfs in the WISE all sky survey will be single band detections. These will be confused with some number of artifacts, spurious detections and moving objects. There have been suggestions that Spitzer/IRAC channels 1 and 2 could provide efficient discrimination between genuine cool dwarfs and artifacts. This is an example of the scientific synergy discussed in the previous section.

4.5 Exoplanets

One might not think of an 85cm telescope as an ideal instrument for ultra-high precision photometry. Transit photometry is often a photon starved exercise and larger apertures on the ground benefit from both higher photon rates and reduced impact from scintillation. Spitzer, however, has become the instrument of choice for high precision exoplanet transit work. Knutson et al. [7] used IRAC at 8μm to measure the transit and secondary eclipse of HD 189733b with a precision of a few hundred micro-magnitudes with a time sampling of only 0.4seconds. This allowed them to determine the planet radius with a precision of 0.5% and map the brightness temperature distribution for the day and night side of the planet with uncertainties of only 10-30K. The great thermal stability of Spitzer is the key to its precision in repeated measurements. The heliocentric Earth trailing orbit of Spitzer frees it from the variable heating from Earthshine that impacts other spacecraft.

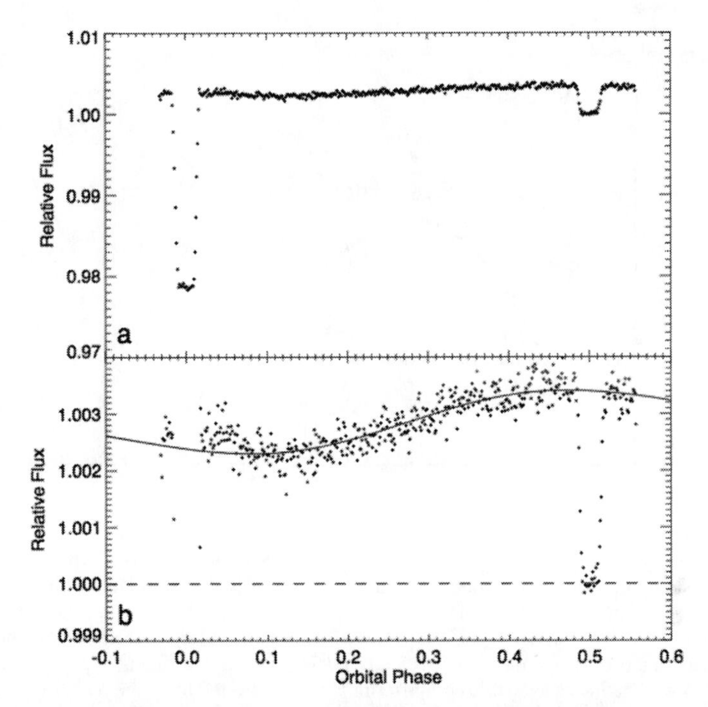

FIGURE 8. IRAC 8μm transit photometry of the exoplanet HD 189733b. The remarkable precision of the measurements (~ 500μ-magnitudes) allows not only detection of the primary transit and secondary eclipse, but leads to inferences regarding the distribution of brightness temperature on the face of the planet. Knutson et al. [7] used the slow variation in the flux between eclipses to derive a brightness temperature variation of ~ 250K due to heating of the planet from the central star. From Knutson et al. [7].

4.6 Active Galactic Nuclei

AGN are among the most broad-band, in terms of spectral energy distributions, of any discrete astronomical sources. The most extreme objects are detected from high-energy gamma rays to meter-long wavelengths. The wide range of physical scales and emission mechanisms in AGN are an important part of this broad spectral coverage. In Figure 9 I show schematic spectral energy distributions for AGN based on the composite energy distributions compiled by Elvis et al. [8]. The dominant contributors in the visible to mid-IR in broad-lined objects (type 1) are thermal accretion disk emission (the "blue bump"), non-thermal synchrotron emission and reprocessed thermalized UV continuum. In the narrow-lined or "type-II" objects the blue bump is either not present or, more likely, is highly obscured. If the type I and II objects are intrinsically similar but differ as the result of heavy obscuration, they should show the hot dust emission in comparable strengths. Thus the mid-IR is a prime testing ground for theories that unify broad and narrow-lined AGN.

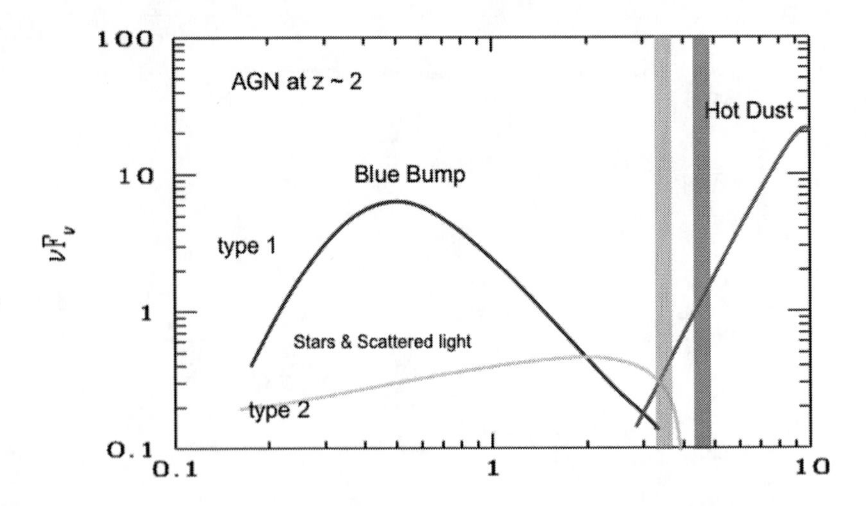

FIGURE 9. Schematic spectra of type I & II age in the visible and mid-IR. Type I, or broad-lined, objects show strong optical/UV emission that peaks near 1000A in the rest-frame, while Type II objects have weaker continuum that is often dominated by starlight. At mid-IR wavelengths both types of AGN show emission from dust with a range of temperatures. This emission arises from material heated by the central engine and provides a means of testing models that unify different types of AGN through orientation biases. The bands available on Spitzer in the warm era fall in on the short wavelength end of the dust spectrum for AGNs at z ~ 2, the peak of the quasar epoch. This figure is based on the composite spectral energy distribution for quasars compiled by Elvis et al. [8].

The spectral energy distributions shown in Figure 9 have been redshifted to z = 2, near the peak of the quasar epoch. At this redshift the IRAC channel 1 and 2 band-passes fall near the minimum between the direct and reprocessed emission and sample only the tail of the hot dust emission. IRAC channels 3 and 4 are better placed to sample the hot dust emission and a number of groups have used this fact to great advantage in identifying samples of highly obscured AGN. An example of the power of the four IRAC bands in separating obscured AGN from stars and galaxies in shown in Figure 10 from Stern et al. [9]. The two-color diagram in Figure 10 cleanly separates the broad-lined type-I AGN from other objects. The narrow-line type-II AGN are also separated, although not as cleanly as some fall within the region of color-color space occupied primarily by galaxies at modest redshift. Channels 1 and 2 alone would provide some discrimination, particularly between AGN and stars, but the galaxies and AGNs would be significantly more confused than in the case of the two-color approach using all four bands. It is likely that one could recover some of this sensitivity to obscured AGN during the warm mission by combining IRAC channels 1 & 2 with colors at shorter wavelengths where galaxies have spectral shapes that depart strongly from those of AGN. The drawback is that at shorter wavelengths obscured AGN are increasingly dominated by emission from the host galaxies and, particularly from the ground, are difficult to separate from normal galaxies. As one can see from Figures 4 and 9, ordinary galaxies have fairly flat 3.5 – 4.5μm colors over a wide range of redshifts, making redshift and spectral type discrimination difficult on the basis of these two bands alone.

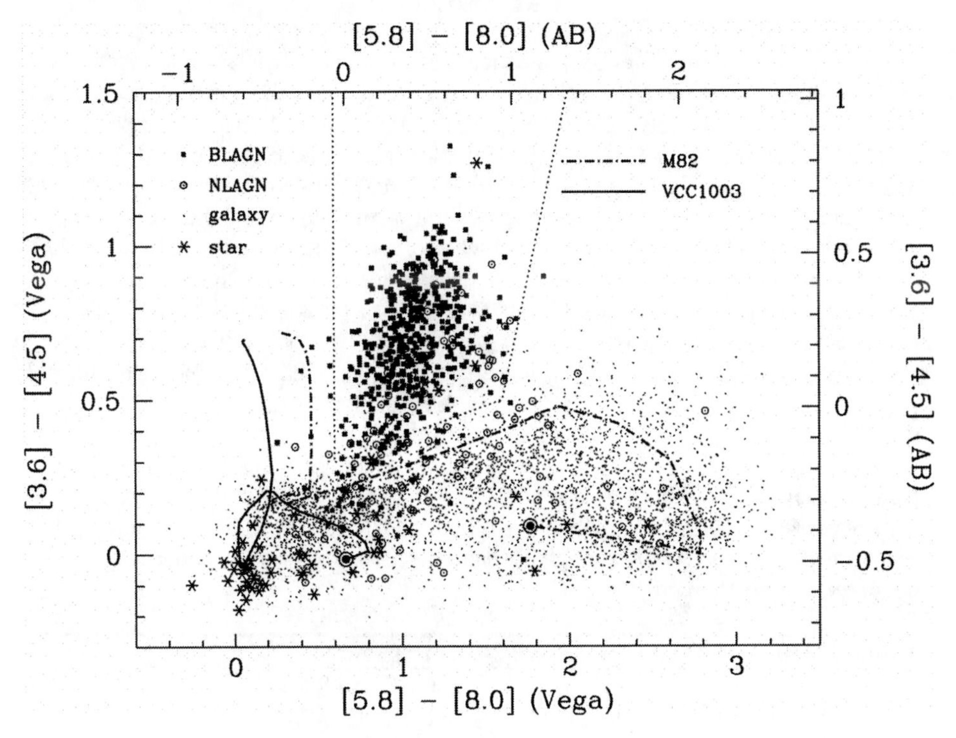

FIGURE 10. A two-color diagram comprised of IRAC channels 1 through 4 from Stern et al. [9]. The two-color approach cleanly separates galaxies at a range of redshifts from stars and AGN. The broad-lined type-I AGN very clearly separate from the galaxies while the narrow-line type-II objects are more widely spread in color, some falling within the galaxy locus.

The schematic spectral energy distributions shown in Figure 9 suggest that IRAC in its warm state is not all that well suited to studies of AGN physics. IRAS, ISO, and Spitzer have made important contributions to testing unification models and improving our understanding of the distribution of the obscuring material. Most of these advances have been made from observations at wavelengths beyond 5μm and from spectroscopy in the mid-IR. A good overview of the contributions of Spitzer to this field is provided by papers in the ApJ Spitzer special issue and the ASP conference proceedings from Spitzer workshops (Armus and Reach [10]). Similar arguments can be made regarding ultra-luminous galaxies where much of the interest is in using spectral diagnostics, such as PAH features, as probes of the energetics of the obscured sources.

5. WHAT SHOULD WE OBSERVE DURING THE WARM MISSION?

The previous section gives us some ideas regarding which areas of science are well suited to observations with Spitzer in its warm state. While there is little doubt that in its warm state Spitzer can still make important contributions to many, and perhaps all, areas of contemporary astrophysics, its fair to say that there are some areas that are more fruitful than others. Below I attempt to make a tabulation of astrophysical phenomenon that would make good targets for a warm Spitzer and those that are likely to be less compelling. There are undoubtedly other classes of objects that should be considered and exceptions to the broad classes listed below. Not surprisingly, the objects in the left column of table 1 tend to have their peak spectral output close to the IRAC 3.5 and 4.5μm bands, or have particularly favorable contrast compared to nearby objects (e.g. central star) at these wavelengths, while the objects in the right column have their peak output at shorter (e.g. hot stars) or at considerably longer wavelengths (e.g. ULIRGS). This harkens back to Figure 1 where we saw that the Spitzer warm bands fall between the direct photospheric emission and reprocessed peaks in the integrated background emission. The sources that will make good targets for Spitzer in the warm state either have been redshifted into this dark region of the spectrum or are just warm enough to naturally benefit from the reduced impact of hotter and cooler sources.

Table 1. What should we observe with warm Spitzer?

Good Targets	**Less Good Targets**
Exoplanets, Planets and small bodies	Hot stars
Cool stars and brown dwarfs	Star forming galaxies
Galaxy Clusters at $z > 1$	ULIRGS
Galaxies at $1 < z < 6$	Quasars & Seyferts (for AGN physics)

6. THE BIG QUESTIONS

Understanding which class of objects make the best targets is a good first step, but if we are going to make wise choices regarding the scope and structure of the science program in the warm phase, we need to identify key scientific question that can be addressed with precision imaging in the 3.5 and 4.5μm bands. Below I list ten of the big picture science questions that came out of the discussions at the workshop.

1. How are galaxies assembled from their constituent components of dark matter, stars and gas?
2. How does the halo mass distribution evolve as a function of time and environment?
3. When did the red sequence of galaxies form and how is it related to the collapse of groups and clusters?
4. What is the balance between star formation and accretion onto black holes in the global radiative luminosity density?
5. How are galactic disks structured, how are they built and what truncates them?

6. Do we understand the components and structure of the Milky Way?
7. What are the total current and past star formation rates in the Milky Way?
8. What is the stellar and sub-stellar mass distribution as a function of metalicity and environment?
9. How do disks around protostars evolve?
10. What sets the equilibrium radii of giant planets and how are they inflated?

The warm mission can make key contributions to addressing each of these questions by using one or more of its unique capabilities. I briefly consider each in turn.

How are galaxies assembled from their constituent components of dark matter, stars and gas? Spitzer is uniquely capable of weighing the stellar mass of galaxies. As shown in Figures 1 and 2, at 3.5 and 4.5μm IRAC samples the peak and long wavelength tail of the spectral energy distribution of red giants. The mass-to-light ratio in these wave-bands is close to unity and, more importantly, varies quite slowly with age. While there remains some uncertainty regarding the impact of AGB stars, this should ultimately be resolved leading to secure stellar masses and mass-to-light ratios for large stellar-mass-selected samples over a wide range of redshifts.

How does the halo mass distribution evolve as a function of time and environment? The great survey speed of IRAC and the potential for large observing programs during the warm mission offer the opportunity to determine the stellar ages, masses and spatial clustering amplitudes over a range of environments and epochs. The clustering lengths can be connected to the halo mass in a fairly robust manner (e.g. Giavalisco et al. [11]). The coupling of large redshift surveys with 3.5 and 4.5μm imaging surveys is a unique opportunity provided only by Spitzer via the legacy programs and future warm mission surveys.

When did the red sequence of galaxies form and how is it related to the collapse of groups and clusters? Understanding the origin of the massive and passive galaxies that form the red sequence today is a major goal of empirical and theoretical studies of galaxy evolution. The sensitivity to stellar mass and ability to survey large areas, as described above, provide an approach to overcoming the luminosity and large-scale structure biases that have impacted large surveys to date.

What is the balance between star formation and accretion onto black holes in the global radiative luminosity density? The classic Madau diagram provides a measure of the evolving global UV luminosity density. There are a number of uncertainties involved in converting this to a global star formation rate density. These include the impact of reddening and the unknown mix of contributions from hot stars versus AGN. Long-wavelength observations can help address both of these problems as they can recover the radiation that has been reprocessed by dust and, from the dust temperature distribution, distinguish between heating from stellar sources as opposed to AGN and disks with spectral energy distributions that extend to higher energies. The short wavelength channels on IRAC are particularly sensitive to the latter effect as they probe hot dust in the inner regions of accretion disks. IRAC studies coupled with deep and large area X-ray surveys can help determine the contribution of accretion to the global energy budget.

How are galactic disks structured, how are they built and what truncates them? Recent deep observations of the outer regions of galactic disks, particularly in HI and in the vacuum UV, have revealed that many disks extend much further than previously thought (e.g. Gil de Paz [12]). This has potentially profound impact on our understanding of how disk galaxies formed, how they evolve and how they are shaped by interaction with close neighbors. The importance of and interplay between tidal effects, disk flaring and ionization at the edges of galactic disks are not well understood. IRAC has outstanding sensitivity to low surface brightness features. The rather small aperture of Spitzer does not impact these surface-brightness limited studies. With its relatively wide field and fast survey speed the warm Spitzer will be a particularly powerful tool for this problem.

Do we understand the components and structure of the Milky Way? The big picture questions considered above treat galaxies as either test particles in large statistical ensembles or as integral systems to be probed from outside. Just as the theory of stellar evolution must be consistent with observations of the sun, our understanding of the structure and evolution of galaxies should be grounded in a firm understanding of the structure of the Milky Way. Our ability to produce a complete inventory of the contents of the galaxy, along with ages, compositions and dynamics, is hampered by our location in the plane. Long wavelength studies that penetrate the extinction (e.g. COBE) often lack the resolution required to study individual stars and clusters. The GLIMPSE program provided a powerful demonstration of the power of Spitzer to map in the inner region of the galaxy and the dense regions galactic plane. The outer regions of the galaxy and the vertical extent of the disk need further mapping, as both of these are critical to our understanding of the MW in the context of other galaxies.

What are the total current and past star formation rates in the Milky Way? It is now possible to reconstruct accurate and moderately precise star formation histories for the local group dwarf galaxies from color-magnitude diagrams. A detailed star formation history of the MW, while more difficult to produce, would be of great value. Star clusters provide a set of well-understood clocks and Spitzer could survey hundreds of these in the galactic plan, providing input for color-magnitude diagrams, age, reddening and distance determinations.

What is the stellar and sub-stellar mass distribution as a function of metalicity and environment? The slope of the bottom end of the stellar IMF and the transition to sub-stellar and planetary mass objects hold clues to the formation process and are vital to properly understanding the mass evolution of galaxies. Spitzer has unique capabilities to discover large numbers of cool low mass objects. The large gain in sensitivity compared to 2MASS or WISE opens an important niche for Spitzer. While lacking the full sky coverage of survey missions, Spitzer could nonetheless survey enough area to yield samples of hundreds of T-dwarfs and a few Y-dwarfs, as described by Knapp in this volume. Targeted observations of binary systems and clusters can also probe the coolest stars. Selecting clusters with a range of metalicities and densities should allow one to examine the role of environment in shaping the bottom of the IMF.

How do disks around protostars evolve? Disks play an important role in star and planet formation. Surveys of disks around young stars to date have been directed primarily at the youngest star forming regions. With the larger programs envisioned during the warm mission it should be practical to survey older star forming associations to assemble a more complete picture of disk frequency and evolution over a wider range of ages and derive a better understanding of disk lifetimes as a function of mass and environment.

What sets the equilibrium radii of giant planets and how are they inflated? Accurate determinations of the equilibrium sizes and gas giants is important not only for improving models of planet structure and formation, they also inform strategies for imaging exoplanets with the next generation of facilities. As we have seen above, Spitzer has an extraordinary power as a precise photometer. The number of known and potential transiting exoplanet systems should increase greatly in the next few years and the warm mission provides an opportunity for long stretches of uninterrupted observing campaigns to derive the structural parameters of both exoplanetary systems and giant exoplanets themselves.

6.1 Some Possible Key Science Programs

Throughout the course of the workshop we heard suggestions for a number of large programs aimed at key scientific questions. These were developed in varying level of detail in the pre-meeting white papers and most are discussed in these proceedings. I list several of these here in an attempt to provide an overview of some of the large programs under consideration and to set the stage for some of the discussion in the following sections.

- Complete survey(s) of the galactic plane
- Surveys of galactic open clusters
- Surveys of the structure and morphology of disk galaxies
- Studies of Exoplanet transits and eclipses
- Surveys of small bodies in the solar system
- Searches for T and Y dwarf stars
- Studies of IR excesses in white dwarf stars
- Ultra-deep survey of the end of the dark ages
- A Spitzer deep survey for galaxy and structure building
- Ultra-wide survey for galaxy clusters at $z > 1$

Some of these programs can be done in parallel, which is to say that multiple programs can be accommodated with a single data set. The T and Y dwarf survey, for example, could be combined with the galaxy cluster survey if an acceptable range of galactic latitudes could be agreed upon. Similarly, the galactic structure and open star cluster programs might be coordinated in a somewhat looser fashion that would still yield improved efficiencies.

7. OPPORTUNITIES FOR SURVEY PROGRAMS

One of the most attractive aspects of the warm mission is that its combination of restricted observing modes and high efficiency naturally lend themselves to large surveys. The largest Spitzer Legacy programs, while ambitious compared to typical GO programs, still fall short of enabling much of the science outlined at the workshop. The warm mission allows us to think in terms of thousands, rather than hundreds, of hours per program.

A number of ideas for survey programs have been suggested at this meeting. Most are aimed at moderately narrow science goals and nearly all are purely galactic or extragalactic in focus. It is potentially instructive to look at these as a whole and see what common threads can be found to link them into a few larger coherent programs. It is also instructive to consider where the proposed surveys lie compared to the Legacy and large GO and GTO programs. In Figure 11 I consider some of extant and proposed extragalactic surveys in a depth versus area projection. Existing programs are shown as dark rectangles, and suggested surveys are shown as light rectangles. The primary science thrust in various combinations of depth and area are shown as shaded ellipses. The range of surveys that are profitable is bounded on one side by efficiency considerations, as the spacecraft overheads become intolerably large for observations shallower than 5-10μJy. WISE will explore this region of parameter space efficiently. The all sky survey planned for WISE is expected to reach a 5σ point source sensitivity of ~ 100μJy at 3.5μm. At the other extreme one reaches the confusion limit somewhere below 100nJy where further integration fails to yield additional depth, or does so at a rate slower than $\sqrt{t}$. The sub-100nJy region of parameter space is a key part of the niche for JWST. A modest investment to push into the confusion limit with Spitzer might be useful both for immediate science return as well to aid in planning for observations with NIRCAM on JWST.

In Figure 11 I highlight regions of parameter space that address particular science areas. The boundaries are quite arbitrary and imprecise and there is significant overlap between adjacent areas. The ultra-deep survey discussed by van Dokkum in this volume would probe the stellar content of galaxies at $z > 5$ to a deeper level, and in larger numbers, than GOODS and other deep surveys. This program also directly probes the mass evolution of galaxies in the growth era from $3 < z < 5$ as well. The prime galaxy assembly epoch, from $1 < z < 3$, when the Hubble sequence appears and galaxies acquired many of the properties that differentiate them today is best probed with larger areas. Surveys of ~0.5 – 2 sq. degrees probe enough massive objects and a wide enough range of environments to allow one to address many of the critical issues related to feed-back and merging. The highest density regions can only be sampled with surveys covering several square degrees and large samples of massive clusters at intermediate and high redshift will require samplings on the order of 100 square degrees or more. Thus the three basic areas identified in Figure 11 map to the rough characteristics of the programs outlined in the contribution by the faint galaxy group in this workshop.

Comparing Figure 11 with Figure 1 shows that most of the surveys under discussion at the workshop require between 1000 – 2000 hours of spacecraft time. Survey more ambitious than these but potentially still well motivated, for example a 10 square degree survey to 350nJy to bridge galaxy assemble and large scale structure studies more effectively than the suggested 2 square degree survey, still require only ~ 10% of the warm mission duration.

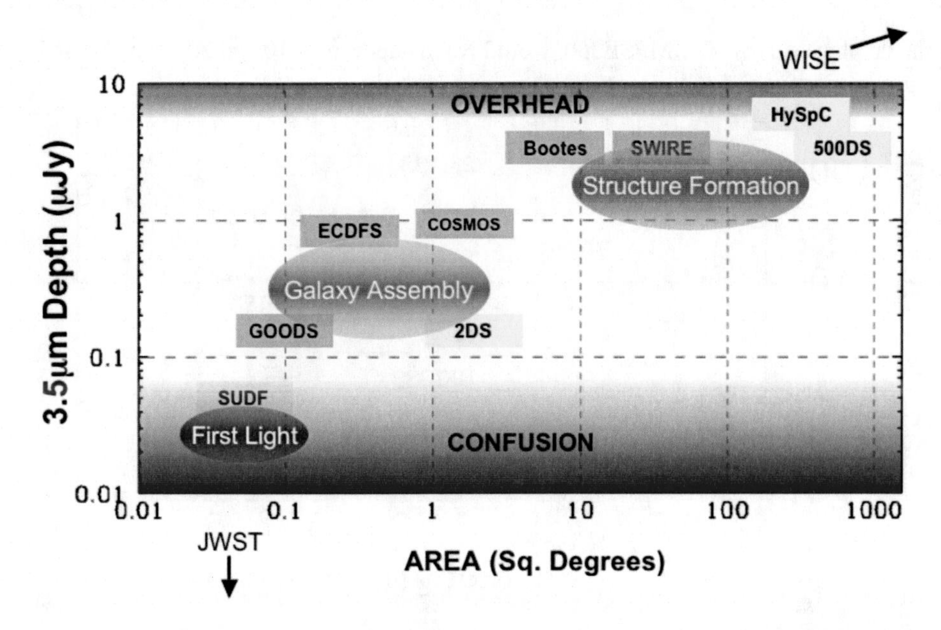

FIGURE 11. Depth versus area projection for current and proposed surveys with Spitzer/IRAC. The vertical axis is the 5σ point source limiting depth at 3.5µm in micro-Janskys. The rough limits imposed by spacecraft overheads and source confusion are shown at the top and bottom of the figure, respectively. Some extant surveys are shown as dark rectangles. These include GOODS, extended CDFS, COSMOS, NDWFS/Bootes and SWIRE. Straw-person surveys suggested at the workshop are shown as light rectangles. These include an ultra-deep field (SUDF), a deep two-square degree survey (2DS), a 500 square degree cluster survey (500DS) and an IRAC survey of the Hyper-Suprime-Cam Subaru/Princeton survey. Three basic science themes – first light, galaxy assembly and structure formation are shown as ellipses in the appropriate regions of the diagram. The suggested surveys typically require 500-2000 hours of spacecraft time, as can be seen by comparing this figure with the contours in Figure 1.

Comparing surveys in the galactic plane is somewhat more complex as all directions are not of equal scientific interest and the confusion limit can be a strong function of latitude and longitude. In Figure 12 I show the depths from a number of all sky or galactic plane surveys for wavelengths shorter than 10µm. This is adapted from a number of sources, including Bob Benjamin's contribution and Ned Wright's WISE web page. The GLIMPSE legacy survey has a depth well matched to 2MASS in the near-IR.

The UKIDSS survey of the galactic plane will cover 1800 square degrees to a K-band depth of 19.0 (Vega) and comparable depths in J and H. All of the plane easily visible from Hawaii will be covered over a period of ~ 7 years. This is quite a bit deeper than 2MASS. As discussed in the contribution by Benjamin in this volume, the UKIDSS survey is one of the motivations for a deeper 3.5 and 4.5µm survey of the galactic plane. WISE will have a raw sensitivity that is a factor of ~ 2 better than GLIMPSE, but the large pixels on WISE will make confusion a serious problem in high-density regions. The smaller pixels on IRAC, while still larger than one might like, offer a distinct advantage at the lowest latitudes. One suggested warm Spitzer galactic plane survey, GLIMPSE360, would reach depths of ~ 10 and 20μJy at 3.5 and 4.5μm, respectively.

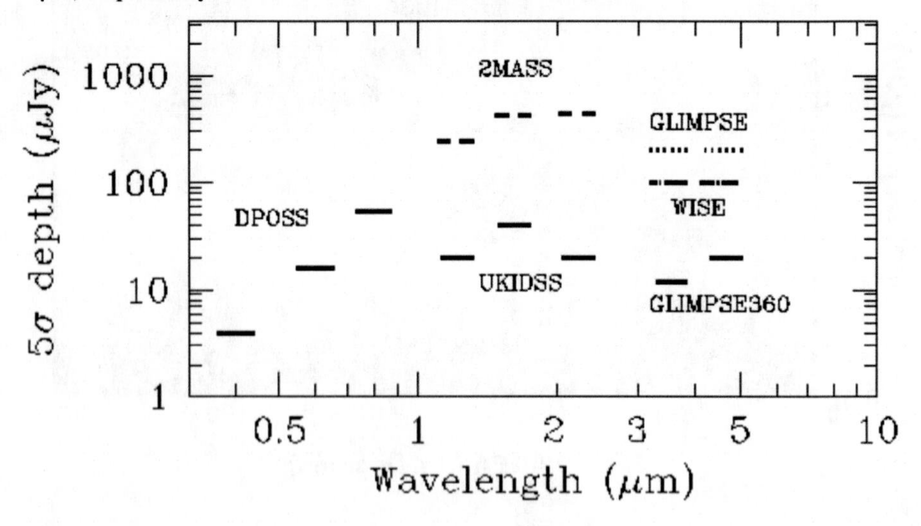

FIGURE 12. Depths of various all sky and galactic plane surveys at visible and infrared wavelengths below 10μm. The UKIDSS survey will miss key regions of the plane, as they are not within reach of observatories in Hawaii. WISE will provide full sky coverage, but crowding is likely to be an issue in the heart of the plane and near the galactic center. GLIMPSE360 offers a combination of depth, area, and sampling that is well matched to the next generation of optical and near-IR surveys. This figure is adapted from figures on the WISE home page and the contribution from Benjamin in this volume.

8. HOW SHOULD WE OBSERVE?

To this point we have considered what targets and combinations of area and depth might be of interest for large programs in the warm era. One of the goals of the workshop was to take a fresh perspective on how the community might plan to use the observing and data resources enabled by a warm mission in a more efficient manner. To date Spitzer has been used by the community at large in two rather different modes. In the Legacy Programs large teams carry out surveys with well-defined deliverables in the form of high-level data products and ancillary data. The teams produce scientific papers from these data, as does the community at large. These survey programs spawn follow-up programs and truly provide a legacy for the observatory. The Legacy programs have varied in size, the largest have been ~ 600 hours, more typical programs are ~100 – 200 hours in duration. One the order of ~ 1/3 of the telescope time have been devoted to these large programs. The remainder of the public time goes towards smaller programs that typically have more limited scope. The GO programs vary widely in scope, but many are 10-20 hours or less in duration. These lead directly to science papers and often to follow-up programs and observations. While data from the small programs are in the Spitzer archive, the PIs of small GO programs are not required to produce high-level data products for the community.

As we think about the warm mission we may wish to reconsider the balance between large and small programs. By the time the cryogen is exhausted Spitzer will have been in operation for roughly five years and many high priority small programs will have been carried out. Many of the important projects that will not been completed at that time will be those whose scope was outside even that of the legacy programs. Hence there is a strong argument for considering a scale of program that make use of the unique aspects of the warm mission while also reducing the level of support needed per hour of spacecraft time.

Many, if not all, of the most successful large survey projects in astronomy, and those involving observation from orbit, have arisen organically from single principal investigators or from a small group of scientists. The alternative, planning science from the top down by committee is fraught with risks. Meshing top-level science goals and grass roots science priorities is a fine balancing act that requires regular updating. This approach has been quite successful in setting national priorities for large missions and it may be helpful, at some level, in the case of the warm mission where a unique resource is available over a limited time interval.

8.1 Striking The Right Balance

One of the questions raised at the start of the workshop related to the balance between large and small programs. While I have attempted to make the case above for large ambitious survey programs, it is clear that there is no single prescription that works for all areas of science. It seems fairly clear that studies of distant extragalactic targets are best served by large and very large survey programs. Many of the outstanding questions require large databases and in recent years the power of large programs (e.g. SDSS, DEEP2) in breaking long-standing impasses in our understanding of galaxy evolution has been made abundantly clear. This remains true whether one defines large programs in terms of areal coverage (e.g. SDSS) or by shear number of hours thrown at a single deep field (e.g. Hubble Ultra-Deep Field). Both wide and deep surveys have important roles to play in the far extragalactic science.

The near-field extragalactic programs are also transitioning from studies of single objects to large statistical programs. The legacy programs aimed at nearby galaxies have been highly successful in large part because they allow one to identify underlying properties of galaxies that become clear with samples large enough to beat down the "noise" associated with the individual character of each object. Further advances in this area will require large, but not enormous, programs and may well work best in a staged approach where ~500 hours programs are planned and carried out in series and build on results and lessons learned as they progress.

Galactic programs have some of the characteristics of both the distant and near extragalactic programs cited above. Many of the interesting targets are single stars or star clusters that are widely distributed on the sky. While large samples are needed, it is not clear that a new paradigm for observing is called for. Galactic structure programs, on the other hand, typically require areal coverage that is large by the standards of extragalactic programs. A blend of survey programs aimed at mapping large contiguous areas, preferably in concert with similar programs at other wavelengths, and statistical samplings of individual objects is likely to be the way forward here.

Lastly, the planetary and exoplanet science programs appear to be working very well in the current mode. While one can argue that in the galactic and extragalactic areas we know fairly well how to best use the cameras in a technical sense and we have a fairly clear understanding of the appropriate targets, this may not be true for the exoplanet science. Techniques and target lists are evolving on a time scale that is short compared to the lifetime of the warm mission. One might argue that the timescale for the current proposal and scheduling process is a bit too slow for this field.

8.2 A Different Approach to Time Allocation?

Observatory scheduling is a slow and inefficient process. The time lapse between when a proposal is written and when observations are executed is typically 6–18 months. Each year some ~350 person-years are expended globally in writing and reviewing astronomical observing proposals that are rejected. This is appropriate because observing time is a scarce and expensive resource and science is a highly competitive marketplace of ideas. The expectation of declining support resources suggests that we might reexamine this process. If funds supporting the proposal process and those supporting technical activities are indeed fungible, we might think carefully about how to balance our priorities. It has been suggested that the proposal review process at Spitzer costs the equivalent of 2-4 FTEs and that this could be reduced by ~ 1-2 FTE without harm.

Most observing proposals are roughly even combinations of scientific justification and experimental design. Investigators describe and motivate the science they wish to address and layout a set of observations in terms of targets, wavelengths, filters, spectral resolution, and depths that they will use to advance their science. Proposal writing for Spitzer in the warm era will be a somewhat different process. There will be no choice of instrument, no options on filters, and only one spectral mode – R ~ 5 imaging at 3.5 and 4.5μm. The only experimental design choices left to the observers are: where to point the telescope and how long to expose. Reviewers will consider whether the proposing teams made a strong scientific case for their project and if they are pointing in a sensible place and exposing long enough. Technical considerations are likely to be of relatively minor importance in evaluating proposals. Rather, proposing teams are likely to stress their ability to extract science from the data efficiently and to bring ancillary data to bear on the problem in question. There will likely be multiple teams proposing to carry out the same basic science, much of which is discussed in some detail in this volume. It would not be difficult to image that reviewers will be faced with numerous proposals for which the only substantive difference is the composition of the proposing teams.

One could consider alternative approaches to scheduling Spitzer in the warm era. A number of key science areas could be identified as high priorities for the observatories and some guidelines issued for programs that address these. An open call for proposals would invite interested parties to propose 1) where to point, 2) how deep to image, and 3) which ancillary data are useful to addressing the particular problem. The teams could be invited to deliver high-level data products and tools to the Spitzer archive in exchange for financial and technical support from the SSC. This process could enable coordination with large programs and data sets at other wavelengths, both on the ground and in orbit. Naturally individuals or teams would be allowed, and encouraged, to propose for programs outside of the identified key science areas.

Much of the effort described above and the person-years of effort devoted to it involve relatively small requests for telescope time. The typical observing proposal for an 8m class telescope on the ground, or one of the great observatories in orbit, requests ~ 10 - 20 hours of actual on-source integration time. Most of the effort involved in processing and reviewing observing proposals is devoted to these small programs. In the case of HST, for example, roughly 90% of the effort is spent reviewing the small and medium programs. This is sensible for an observatory with several functioning instruments each of which has multiple modes. It may not make sense for Spitzer in the warm era.

One alternative approach is to decouple the review process for small and legacy class proposals. Programs below a predefined threshold, say 50 hours, could be reviewed by a standing committee that need not meet in person. This proposal track could operate on a different cadence than the large proposals and could be more responsive to rapid developments in fast moving fields. Staggering the large and small programs could also enhance pilot programs and follow-up programs in a way that improves the success rate and impact of the legacy class programs. The large proposals would continue to be reviewed by the external panels that meet face to face on a roughly annual basis.

9. SUMMARY

The scientific potential of Spitzer post-cryogen is not only clear, it is exciting and has the potential to energize the community by enabling a scale of projects that is usually only possible with dedicated missions. The science enabled by large programs using IRAC channels 1 & 2 is well matched to our community's priorities and offers important synergy with current and future facilities. The primary challenges may lie in choosing which of the many interesting programs to carry forward, maintaining the balance between large and small programs, and finding the most efficient way to operate the facility without compromising the vital competitiveness that keeps our field alive and vital.

ACKNOWLEDGMENTS

I'd like to thank Lisa Storrie-Lombardi and her team for organizing and supporting such an interesting workshop. I also thank all of the members of the steering committee and contributing authors for their dedicated effort in producing the pre-meeting white papers and the contributions to these proceedings.

REFERENCES

1. L. Eyles, A. Bunker, E. Stanway, M. Lacy, R. Ellis, M. Doherty, *MNRAS*, **364**, 443-454, (2005).
2. H. Yan, M. Dickinson, M. Giavalisco, D. Stern, P. Eisenhardt, H. Ferguson, *ApJ*, **651**, 24-40, (2005).
3. B. Mobasher, et al. *ApJ,* **635,** 843-844, (2005).
4. T. Wiklind et al., "Massive Galaxies at z > 5", in *Galaxy Evolution Across the Hubble Time,* Proceedings of IAUS 253, Cambridge University Press, 2007, pp. 368-372.
5. C. Maraston, *MNRAS*, **362**, 799-825 (2005).
6. A. Burrows, D. Sudarsky, J. Lunine, ApJ, 596, 587-596 (2003).
7. H. Knutson, et al. *Nature,* **447**, 183-186 (2007).
8. M. Elvis, B. Wilkes, J. McDowell, R. Green, J. Bechtold, S. Willner, M. Oey, E. Polomski, and R. Cutri, ApJS, 95, 1-68 (1994)
9. D. Stern, et al. *ApJ*, **631**, 163-168 (2005).
10. L. Armus and W. Reach, *The Spitzer Space Telescope: New Views of the Cosmos",* ASP Conference Proc. 357, 2006, Astron. Soc. Pacific.
11. M. Giavalisco, C. Steidel, K. Adelberger, M. Dickinson, M. Petteni, and M. Kellog, ApJ, 503, 543-552. (1998).
12. A. Gil de Paz, et al., *ApJ*, **627**, L29-L32, (2005).

2. – OVERVIEW MATERIAL PRESENTED AT THE WORKSHOP

Spitzer Warm Mission Workshop Introduction

Lisa J. Storrie-Lombardi and Sean Carey

Spitzer Science Center, California Institute of Technology, MS314-6, Pasadena, CA 91125 USA

Abstract.
The Spitzer Warm Mission Workshop was held June 4-5, 2007, to explore the science drivers for the warm Spitzer mission and help the Spitzer Science Center develop a new science operations philosophy. We must continue to maximize the science return with the reduced resources available, both using (a) the shortest two IRAC channels, and (b) archival research with the rich Spitzer archive. This paper summarizes the overview slides presented to the workshop participants.

Keywords: Spitzer Space Telescope
PACS: 98.85.Hp

1. INTRODUCTION

1.1. Workshop Format

Monday

- Introduction to the Warm Mission
- Overview
- Mission plans and questions
- Solicited white paper reports
- Contributed white paper summaries
- Splitzer group discussion
  - Solar Systems
  - Our Galaxy
  - Nearby Galaxies
  - Distant Galaxies

Tuesday

- Archive Presentation
- Splinter group summaries
- Discussion

CP943, *The Science Opportunities for the Warm Spitzer Mission Workshop,*
edited by L. J. Storrie-Lombardi and N. A. Silbermann

1.2. Workshop Steering Committee

Chair	Pat McCarthy (OCIW)
Nearby Universe	Daniela Calzetti (U. Mass)
Extrasolar Planets	Drake Deming (GSFC)
Stars, Brown dwarfs	Jill Knapp (Princeton)
Solar System	Carey Lisse (JHU-APL)
Galactic Structure and ISM	Mike Skrutskie (U. Virginia)
Star Formation	Steve Strom (NOAO)
Distant Universe	Pieter van Dokkum (Yale)

2. OVERVIEW

2.1. Life After Helium

- Observatory has ample reserves on consumables, power, etc.
- Cryo-telescope assembly expected to equilibrate at $\sim$25-29K
- IRAC will have essentially unchanged sensitivity at 3.6 and 4.5 μm
- All other detectors non-operational
- Spitzer archive will still be brimming with data
- Community will be in the first round of extracting science from the archive

2.2. Our Vision: To Fully Exploit NASA's and the Community's Investment in the Spitzer Mission

We will do this by:

- Capturing the full legacy of Spitzer into a robust, permanent archive.
- Expanding the science from Spitzer beyond the Liquid Helium lifetime through a vigorous archival research program.
- Utilizing the continuing observatory capabilities for unique, vital science possible only with Spitzer.

I. The Data Archive

- At the end of the Spitzer cryo-mission, we must reprocess the full data set to uniform calibration and minimal artifacts.
  - Will leave a legacy for science utilization that will remain vital for decades.
  - Will apply the full knowledge and understanding of Spitzer.
  - Usefulness of Spitzer and return on investment will be enhanced by new generations of users.
  - Exact contents will depend on resources (therefore NASA environment), and community needs and inputs.

II. Exploiting the Spitzer Data Archive: Community Support

- Quality and uniformity are critical for new science leveraging the entire archive.
  - Optimized calibration, minimized artifacts in the final processing.
- Full realization of the science potential of the permanent archive requires:
  - Adequate funding to the science community.
  - Support by active scientists at SSC, providing expertise and adapting software.
- Without a dedicated support plan, archival research funding would be available only through ADP, and technical support at the SSC would be minimal.
  - ADP funding was $2 million in 2004 for $\sim$30 mission data sets
  - As currently established the ADP funding is inadequate to support a meaningful Spitzer archival program

III. Warm Spitzer: A Unique Asset

- At end of cryogenic phase, Spitzer will still be a unique space observatory
  - Telescope should equilibrate at < 30K in solar orbit.
  - IRAC $5' \times 5'$ FOVs at 3.6 and 4.5 μm will operate in parallel.
  - **3 - 5 μm sensitivity essentially unchanged from cryogenic phase, unmatched until JWST flies**
  - No measurable degradation in the IRAC arrays to this point.
  - Observatory represents over a billion dollars cumulative investment.
- Powerful capabilities
  - Finely tuned, calibrated science instrument.
  - Wide-field, superb mapping engine.
  - Time-domain access on all scales from milli-seconds to years.
- Well-honed operations
  - $\sim$6 years of experience and optimal efficiency.
  - Stable, efficient ground support and data analysis system.

IV. Warm Spitzer Sensitivity – Figure 1

- IRAC 3.6 and 4.5 μm bands match WISE bands 1 and 2 and lie in the JWST sweet spot.
- $\sim$3 orders of magnitude between WISE and JWST sensitivity will be the domain of warm Spitzer/IRAC as the tool of choice.
- Shallow integrations can follow-up on WISE discoveries.
- Deepest integrations will provide path-finding science for JWST.

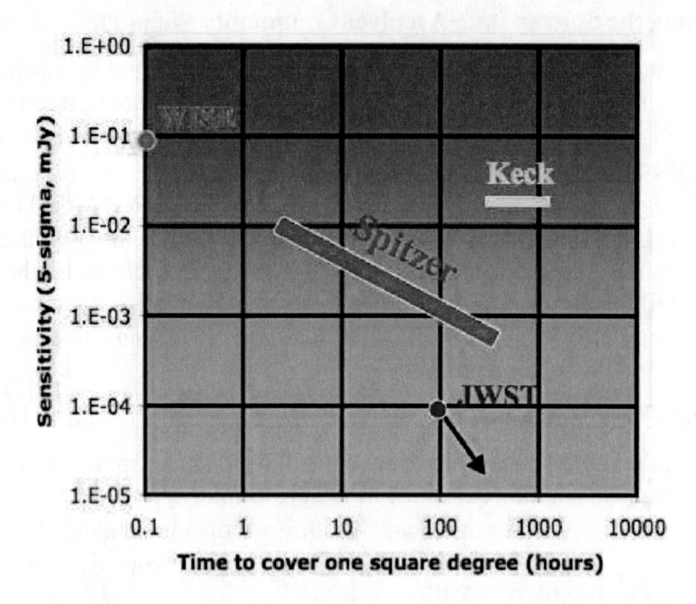

FIGURE 1. Spitzer sensitivity during the warm mission phase compared to predicted WISE and JWST performance and measured mid-IR performance with Keck. This figure is adapted from the Spitzer-WISE memo at http://ssc.spitzer.caltech.edu/documents/wisememo.doc.pdf.

3. IRAC PERFORMANCE AND OPERATIONS IN THE WARM SPITZER MISSION

Operating Environment Assumptions

- Spitzer will be passively cooled after cryogen runs out.
  - Telescope $\sim 24-25$K
  - Multiple Instrument chamber (IRAC) $\sim 25-29$K
- Warm up above MIPS, IRS and 5.8 and 8.0 μm operating temperatures occurs within 12 hours of cryogen running out.
- Telescope temperature equilibrium occurs within 4 weeks.
- OPZ (operational pointing zone) remains the same.
- Same effective downlink rate as cryogenic operations.
  - IRAC data rate is halved.
- Pointing system exhibits same stability and accuracy.

Predicted IRAC Performance

- Observations with 3.6 and 4.5 μm (InSb) arrays only.
- Temperature of arrays actively controlled.
  - Arrays heated to operating temperature of 30K.
- Testing of similar arrays at 30K at University of Rochester

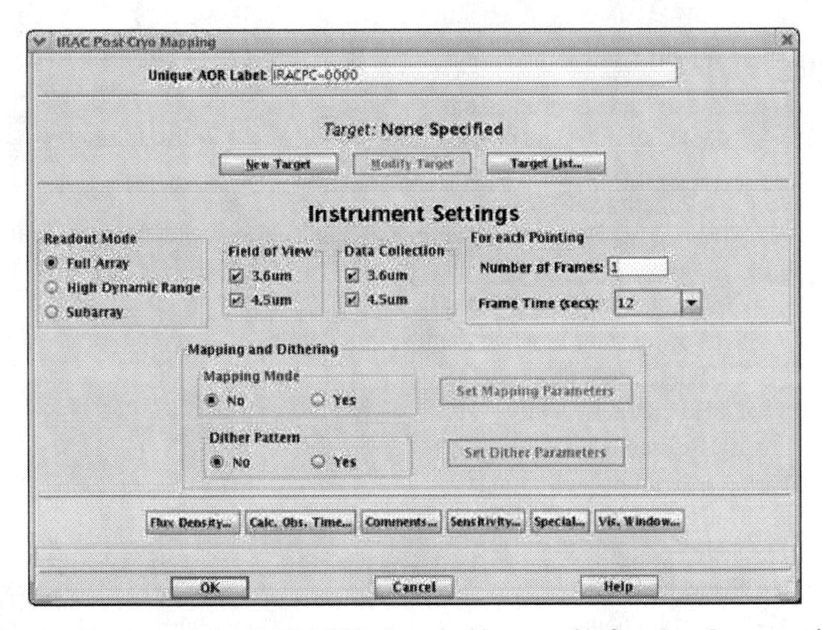

FIGURE 2. The warm mission IRAC AOT is shown in this screen shot from Spot. It operates similarly to the IRAC AOT in Spot today.

- No significant increase in dark current
- Slight increase in read noise
- Most observations should still be background/photon dominated
  - For frame times > 30 seconds, arrays should be background limited for all backgrounds
- Latents could be more significant but possibly decay faster
  - Redundancy will be important

Warm IRAC Astronomical Observing Template (AOT) – Figure 2

- Data taking at 3.6 and 4.5 μm only.
  - Can choose to take data in only one channel to reduce data volume.
- Dither and mapping options remain the same.
- Full frame mode
  - 0.4*, 2, 6*, 12, 30, 100, 200, 400* second frame times (*potential new frame times)
  - Use of 200 and 400 second frame times contingent on improved noise properties for deep images
- High Dynamic Range mode
  - 12, 30, 100, 200, 400* second frame times
- Subarray mode
  - 0.02, 0.1, 0.4 second frame times

Warm Instrument Characterization

- First month of the warm mission
  - Sample, simple science programs during weeks 2-4 to fill gaps during functional observations
- Week 1 – Functional checkout
  - Aliveness test, Determine temperature set points, optimize array biases
- Week 2 – Array properties
  - First month of the warm mission
  - Set Fowler sampling and finalize frame times
  - Calculate noise properties
  - Latent characterization
  - AOT checkout
- Week 3 – Baseline calibrations
  - Dark and Flat calibrations
  - Stellar calibrations
  - Focus check
  - Distortion map
  - PRF measurement
- Week 4 – Science Verifications
  - Deep image
  - Photometric monitoring
  - Galactic shallow survey

4. MISSION PLANS

- Pre-launch mission plan
  - 5 to 5 – 1/2 year cryogenic mission
  - 1 – 1/2 to 2 year warm mission
  - 1 year close-out
- Current Proposal
  - 5 – 1/2 year cryogenic mission
  - 5 year warm mission
  - 1 year close-out
- Bottom Line
  - **$110 million for 3 additional years of warm observing operations**
  - **$50 million for operations + $60 million for user community**

5. OPERATIONAL ASPECTS

- Expect high observing efficiency to continue
  - Execute 6500 – 7000 hours of science per year
- **The challenge is to conduct this mission cost-effectively: maximize the science to cost ratio.**
- Planning is based on model of **half current staff** at SSC, JPL and LMA when final reprocessing of cryo-mission data is complete
  - To operate mission with this work-force requires substantial simplifications of operations
    - Substantially reduced number of supported programs
    - Simplification of planning and scheduling
    - Fewer scheduling interrupts
    - Reduced engineering staff for performance analysis and anomaly response
- Look for economies of scale without sacrificing the science
  - Maintain peer-review process
    - Make it less expensive
    - Annual review costs $250k not counting the FTEs supporting it
- Shift emphasis to large and huge programs, since well have 7000 hours per year to allocate
- Fund data analysis and archival research at an appropriate level
- Streamline science and mission operations to the max
- Engage the community in the planning process – **You are Here!**

6. COMMUNITY SUPPORT

Continue providing substantial support for the community.

- **Currently send $30-35 million year to the User Community**

GO-Legacy	$18-22 million
Archive-Theory	$2 million ($2.7 in Cycle-4)
GTO science funding	∼$7 million
Fellowship program	∼$1.6 million
Overhead	$3 million (∼ 10%)

Spitzer constant ($ per hour) ∼$3k

- **Warm Mission Plan – $20 million per year to User Community**

Warm Observing	$13 million
Archive/Theory	$4.8 million
Fellowship program	$1 million
Overhead	$1.2 million (∼ 6%)

Spitzer constant ($ per hour) ∼$1.8k

7. STRAWMAN PLANS FOR DISCUSSION AT THE WORKSHOP

7.1. Transition to the Warm Mission

- Best estimate for cryogen depletion is March-April 2009.
- We must have $\sim$ 1000 hours ready to execute by February 2009
- Program selected and advertised in advance based on input from Workshop or separate selection process
- Execute in HDF style
- Observations designed by science committee (External+SSC)
- No proprietary period
- No direct funding
- Archival funding available via regular review process

7.2. Observing Proposals

Proposal Categories - Observing

- Small $<$ 100 hours (Should this be 50 hours?)
  - AORs required
  - 1-year proprietary period
  - 1000 hours maximum per cycle (2175 hours awarded in Cycle-4)
  - Directors time (5-10%)
    - Could this be used for small category? ($<$ 10 hrs?)
- Medium 100 – 500 hours
- Large 500 – 2000 hours
- Huge $>$ 2000 hours
- Big programs
  - Template AORs with proposals
  - No proprietary period
  - Really big programs could be executed over 2 years
- No direct funding for proposals $<$ 10 hours
- Page charges for successful Spitzer proposers paid directly by the SSC?

7.3. Archive-Theory Proposals

- Continue to offer 1-year Archive-Theory proposals ($\sim$ 50 – 100k)
- Legacy Archive
  - Multi-year archive programs
  - Return enhanced data product deliverables to SSC-IRSA

- Up to $500k

- Large Archive
  - Multi-year archive programs
  - No enhanced data product deliverables
  - Up to $300k
- Multi-year Theory – should we support these too?
- What fraction of the total community funding should go to Archive/Theory?
- Should the amount for each category be preordained?

7.4. Review Process

- Hold annual proposal calls and review meeting
  - Does the process need to be annual (would 18 month centers do?)
- Do the review process in two phases
  - **Save $200k per year on review costs = one FTE**
  - Phase 1: remote review of all proposals and submission of grades
    * Top 10% of small proposals awarded time?
    * Additional small allocation determined by lottery from proposals ranked 10-XX%
    * Top 20 – 25% (or highest ranked 1000 hours) of small proposals awarded time?
    * Big programs (medium, large and huge) forwarded to TAC to provide oversubscription factor of 2
  - Phase 2: face-to-face meeting of TAC to select big programs
  - Archive-Theory-Observing all reviewed together
- Variant:
  - Review Archive-Theory six months out of phase with observing
  - Same review panels and TAC
  - TAC meets remotely to select Legacy archive programs

7.5. Program Support

- Program reviews
  - Cursory technical checks
  - No duplication checks after selection
- Scheduling
  - Continue to schedule in weekly blocks
  - 24-36 hr PAOs (periods of autonomous operations)
  - Low impact ToOs – no restrictions
  - Select one-high impact ToO/year (currently we select up to 10)

- Archive/Data Rights
  - No embargo checking for large, public surveys
  - Advertise this in Cycle-5 as it may impact those programs

8. SUMMARY OF QUESTIONS FOR THE WORKSHOP

- What are the most important science drivers for a warm Spitzer mission?
- What should be the duration of the warm mission?
- What public HDF-style program should be prepared for the cryo/warm transition period?
- What is the appropriate balance between smaller and larger programs?
- Are ToOs an important component of the warm mission? If yes, at what level?
- Should any science programs be specifically solicited for the warm mission?
- Are there any huge (> 5000 hours) projects that should be done? If yes, how should they be selected and organized?
- How does the community participate in science of big projects if not part of the executing teams?
- Can most of the review process be done remotely instead of bringing 100 people to Pasadena annually for a week?
- Should the review of observing proposals and archival-theory proposals be held at the same time or 6 months out of phase?
- Warm Transition Program
  - Should specific enhanced data products be produced by SSC?
  - Should we carve out a specific dollar amount to support archival research with these data?
- Observing Proposals
  - What hour range should be defined as small? .. < 50 ? <100?
  - Use DDT for small category?
  - No direct funding for very small programs?
  - Do we have the right breakdown in categories?
  - Should there be a preordained distribution of time between categories?
- Archive/Theory Proposals
  - What fraction of the total community funding should go to Archive-Theory?
  - Should the amount for each category be preordained?
- Review Process
  - How often do we need to select programs?
  - Should we use a lottery element for any of it?

ACKNOWLEDGMENTS

The Spitzer Science Center would like to thank everyone who participated in the workshop and the preparation of the white papers. We greatly value the input we have received from the community.

3. – SPITZER SCIENCE CENTER WHITE PAPERS

The Spitzer Warm Mission Science Prospects

John R. Stauffer*, Vincent Mannings*, Deborah Levine*, Ranga Ram Chary*, Gillian Wilson*, Mark Lacy*, Carl Grillmair*, Sean Carey*, Susan Stolovy*, David Ciardi† and Joe Hora**

*Spitzer Science Center, Caltech 314-6, Pasadena, CA 91125, USA
†Infrared Processing and Analysis Center, Caltech 220-6, Pasadena, CA 91125, USA
**Harvard-Smithsonian Center for Astrophysics, 60 Garden St., Cambridge, MA 01238, USA

Abstract.
After exhaustion of its cryogen, the Spitzer Space telescope will still have a fully functioning two-channel mid-IR camera that will have sensitivities better than any other ground or space-based telescopes until the launch of JWST. This document provides a description of the expected capabilities of Spitzer during its warm mission phase, and provides brief descriptions of several possible very large science programs that could be conducted. This information is intended to serve as input to a wide ranging discussion of the warm mission science, leading up to the Warm Mission Workshop in June 2007.

Keywords: Spitzer Space Telescope, infrared astronomical observations, exoplanets, AGN, galaxy clusters, asteroids, star-formation regions, protostars
PACS: 95.55.Fw, 95.85.Hp, 98.62.Py, 98.35.Ln, 98.35.Ac, 97.82.Fs

1. INTRODUCTION

The Spitzer Space Telescope passed a milestone of three years in orbit in August 2006, with enough cryogen for another approximately 2.5 years in its nominal mission profile. We are now starting the detailed process of considering how Spitzer will be operated in the post-cryogenic period, when the long-wavelength instruments will no longer provide useful data but the 3.6 and 4.5 μm channels of IRAC will still perform well. This document provides a brief review of Spitzer's current status and its expected performance capabilities during the post-cryo period. We then describe plausible changes in the way Spitzer will be operated in order to reflect the post-cryo capabilities and the expected decrease in ground-operations support.

An important issue for the warm mission is the mix of observing projects undertaken. We believe it is untenable to continue both the time allocation and scheduling exactly in the manner of the cryogenic mission because this would require a staff size at the Spitzer Science Center (SSC) that is only modestly reduced from the current level. We believe that the only way to significantly reduce the SSC staffing level is for most of the observing time to be allocated to a few very large projects. We briefly outline how we believe this would work below. Given this model, it then becomes particularly important to have an early and wide-ranging discussion of the types of large projects best suited, and the mechanism by which those projects will be selected and managed. We address the first of those issues by providing outlines of four large projects as potential models for the type of science that can be done during the warm mission. We stress that

CP943, *The Science Opportunities for the Warm Spitzer Mission Workshop,*
edited by L. J. Storrie-Lombardi and N. A. Silbermann

these projects are for purposes of illustration, and that they have not been selected. The details of how the large projects will be selected are still evolving, but the process will involve a mostly-external steering committee and an open workshop which they will organize, community-led white papers to develop detailed plans for specific potential large programs, and calls for proposals with peer-reviewed selection.

The range of science that could be done during the warm mission is extremely broad. The total amount of observing time for science could be of order 30,000 to 40,000 hours. If all of that time were devoted to a moderate depth survey (10 μJy, 5σ) - i.e. about ten times deeper than WISE (*Wide-Area Infrared Survey Explorer*), of order 20 to 30% of the sky could be surveyed. Alternatively, a deep survey (< 1 μJy, 5σ) intended to serve as a precursor to JWST observations could be obtained for of order 100 square degrees if that were the only project conducted during the warm mission. Obviously, anything in between those two extremes could also be conducted, including programs aimed at temporal variability or proper motions.

2. EXPECTED SPITZER SPACECRAFT CAPABILITIES IN THE WARM MISSION

The Spitzer Space Telescope has achieved exceptional science return and efficiency due in large part to the excellent performance of the spacecraft and cryo-telescope assembly (CTA). The spacecraft and cryostat have met or exceeded requirements, notably in the areas of pointing control and cryogenic lifetime. In the first 7 months of 2006, Spitzer spent 85.4% of the total elapsed time doing science or science instrument calibration. Based on our in-orbit experience to date, we believe that the Spitzer spacecraft hardware should perform well for many more years. There is no evidence that any of the redundant units on board have failed. The only system where there is some uncertainty is the spacecraft computer. In August 2006, Spitzer underwent an autonomous side-swap of the spacecraft computer (from Side-A to Side-B). No cause for the side-swap was identified, and the most probable explanation is a single-event upset, with the strong expectation that Side-A would operate normally if used. However, since the side-swap, we have continued operation on Side-B because that was deemed the conservative course of action.

During the nominal science mission, Spitzer operates one science instrument at a time, with instrument transitions occurring every 1-2 weeks on average. A sequence of preplanned science is uploaded weekly, and the spacecraft orients its high-gain antenna towards earth once or twice each day to downlink science data. At 2.2 Mb/s these downlink periods take 20-60 minutes depending upon the data volume. Spitzer cannot tolerate over-filling its mass memory, and sustained high data volumes necessitate downlinks twice per day for all MIPS campaigns and some fraction of IRAC operations. Spitzer is in an earth-trailing solar orbit and it recedes from earth at a rate of about 0.1 AU/year. This affects telecommunications, particularly toward the end of the nominal mission and during post-cryo operations.

Spitzer's limiting consumable for the nominal mission is the superfluid helium which cools both the instruments and the telescope (via venting from the cryostat). Current predictions for the end of the cryogenic mission place the date at which the cryogen will be expended at March 2009 $\pm$ a month. This date is estimated from a measurement of the remaining helium mass based on the thermal response of the cryogen to heat dissipated into the cryostat. The only other consumable on Spitzer is cold N_2 gas for angular momentum management. Given current and expected use rates, there is enough N_2 on board for at least another ten years of operation.

Once the cryogen is expended the temperature of the telescope and the cryostat will rise. The telescope is expected to reach a steady-state temperature of 24-25K after 10-15 days, and is projected to remain in focus at that temperature. Neither MIPS nor IRS will be able take useful science data because of the elevated dark currents. The two shortest wavelength channels for IRAC at 3.6 and 4.5 μm, however, will be only slightly affected by the warmer telescope and cryostat, and it is predicted that they will operate with essentially the same sensitivity as they have for the cryogenic portion of the mission. Those two channels are the most sensitive IRAC channels now, and will remain the most sensitive cameras at those wavelengths until JWST is launched.

The spacecraft itself is basically unaffected by the depletion of the cryogen. Operations will continue with only minimal changes from the nominal operations scenario. Most of the small number of changes that are planned will help facilitate operating the mission with the reduced ground resources expected to be available during the post-cryogen phase. We expect to operate typically with just one downlink per day, and possibly with two-week-long sequences rather than one-week long sequences. For a constant data volume, the single downlink per day would become longer as the distance from earth increases and the downlink rate decreases. However, compared to now, the data volume will be effectively halved because only two of IRAC's four channels will be operating. The two factors approximately cancel each other. If both transmitters aboard the spacecraft are used, it can support a downlink rate of half the nominal rate, or 1.1 Mb/s out to the end of post-cryogenic operations. With a single transmitter as in nominal operations, the final data rate would be about half that. The maximum length of a downlink is 2.5 hours. While not a certainty, it is entirely possible that downlink data volume will not limit the observing cadence with Spitzer during the post-cryogen period any differently than has been true during the cryogenic mission.

Spitzer's operational pointing zone (OPZ) is limited by the sun-Spitzer geometry, and that does not change over the course of the mission. However, in order to downlink data to Earth, the spacecraft must be able to orient its fixed high gain antenna to point towards the Earth. The high-gain antenna is located on the bottom of the spacecraft and points approximately opposite to the telescope boresight direction. As Spitzer drifts away from the Earth, the orientation it has to assume in order to downlink data gradually causes the telescope pointing direction to approach the edge of the OPZ. Toward the end of the warm mission, the nominal geometry begins to place the downlink orientation outside the OPZ. This can be tolerated initially by pointing the high-gain antenna slightly away from the nominal direction, reducing bandwidth in exchange for lengthening the duration of the mission. As another telecommunications constraint, should Spitzer enter safe mode, the signal from its low-gain antennae when it is rotating about the sun-line must be receivable at earth, and the spacecraft must be able to receive uplink signal from earth in order to recover from safing. Given our current best estimates, it is this latter limitation that we will run up against first and that will signal the end of the post-cryo mission.

Based on current estimates, the end of the post-cryo phase will occur in January 2014. To operate past early November 2013, the OPZ will need to be extended by about 5 degrees to permit downlink orientation, but analysis demonstrates this will be acceptable. At or near the end of 2013, the angle between the low-gain antennas and earth in safe mode exceeds 40 degrees, and the uplink bit rate would drop below the minimum allowed - and this is not mitigable.

More details about the design and operation of Spitzer and the characteristics of the IRAC camera can be found at the Spitzer website[1]. A detailed description of the IRAC camera design and performance can be found in Fazio *et al.* [1].

3. EXPECTED CAPABILITIES OF IRAC IN THE WARM MISSION ERA

3.1. Summary

During the warm mission the IRAC 3.6 and 4.5 μm channels will operate near a temperature of 30K, and will experience little or no degradation compared to their nominal mission performance. Based on thermal and optical models of the telescope and IRAC, a focus adjustment using the telescope secondary mirror mechanism is not expected to be required to maintain the nominal mission optical performance.

3.2. Telescope and IRAC Optics

According to a thermal analysis of the observatory, after the cryogens are exhausted the telescope and multiple instrument chamber (MIC) are expected to warm up over a period of about 10 days, and reach an equilibrium temperature of 25 – 29K. This range is for a nominal IRAC focal plane power dissipation.

The Spitzer telescope does have the ability to adjust its focus by moving the secondary mirror assembly. The IRAC camera itself does not include a focus mechanism. During the initial in-orbit checkout period in 2003, the secondary mirror assembly was moved twice in order to achieve a location which produced acceptably focused images for all three instruments. The telescope focus was monitored during the cooldown after launch, and no significant change in focus position was detected below 55K. The difference between the case of cooldown after launch and the post-cryo mission is that the MIC and inner cryostat were at 1.2K during the initial cooldown, whereas after the cryogens are expended the MIC and cryostat will warm up with the telescope and stabilize at $\sim$25K. However, this is not expected to cause a significant focus shift. In the IRAC instrument, the lens materials will experience a small change in the index of refraction which will cause a small shift in the instrument focus and point response function (PRF), but analysis performed by the IRAC team indicates that this should not significantly affect

[1] see http://ssc.spitzer.caltech.edu

the instrument performance. Therefore, the intention is to obtain image characterization measurements at the new temperature equilibrium point (i.e. 10+ days after cryogen exhaustion), but a refocus of the telescope should not be required. As a precaution, however, the project recently completed a planning exercise to make sure all of the tools and expertise would be available to refocus Spitzer during the post-cryo check-out phase if that were necessary. The conclusion of that exercise was that the team was ready for this eventuality, and that it should not consume a significant amount of telescope time.

3.3. Detector Performance

It was known from pre-launch testing of the InSb detectors used for the 3.6 and 4.5 μm channels that the dark current in the arrays was sufficiently low to allow for their use in a post-cryo mission. In order to better characterize their performance near 30K, a series of tests were performed recently at the University of Rochester by McMurtry, Pipher, and Forrest [2]. The characteristics of these devices are considered to closely match the in-flight devices.

The results showed that operation at 30K did not significantly degrade the detector performance. The dark current is not significantly higher; it is expected to remain below 1 electron/sec until about 37K, so there is ample margin. The quantum efficiency of the devices should be unchanged. The read noise will likely be the same or slightly higher. However, in long exposures where the zodiacal background will dominate the noise, it will not affect the instrument sensitivity. The level of residual images caused by saturating sources is seen to increase, but they also decay faster so on balance are not significantly worse. The current IRAC pipeline contains flags for marking residual image-affected pixels and excluding them from mosaics made with IRAC images. No significant new types of image artifacts or instabilities were observed during the 30K tests over what is currently seen in the IRAC nominal mission data.

We have also monitored the number of noisy, hot, dead, and otherwise scientifically unusable pixels in the IRAC detectors during the nominal mission up through the end of 2005. This period has included two extremely intense solar flares in November 2003 and January 2005. In the 3.6 and 4.5 μm channels, no increase in the number of noisy, hot or dead pixels was noted. Therefore we would expect no significant increase in their number during the warm mission - even considering the fact that the next solar maximum should occur around 2011 (in the middle of the warm mission).

4. SPITZER/IRAC COMPARED TO OTHER FACILITIES

The Spitzer warm mission will be conducted in an environment where it is expected that COROT will have been launched a few years earlier, Kepler will have just been launched, WISE will be launched during the same year, and where JWST would be expecting launch within about four years. The Spitzer warm mission should be planned to complement what can be done with WISE and to help prepare scientists to utilize JWST as efficiently as possible.

TABLE 1. Sensitivity And Areal Coverage For Different AOR Types

AOR Type	Number Dithers	Frame Time(sec)	Ch1 (5σ,μJy)	Ch2 (5σ,μJy)	Area in 1000 hr Sq.Deg.	Example
Shallow	3	2	100.	110.	400	GLIMPSE,WISE
Wide	4	12	8.5	12.5	225	360GLIMPSE
Moderate	4	30	3.5	6.	125	z>1 clusters
Deep	6	100	1.3	2.5	25	
Very Deep	900	100	0.1	0.2	0.25	High-z

WISE is designed to provide a map of the complete sky in four passbands (3.3, 4.7, 12 and 24 microns). Therefore, its two short wavelength bands approximately duplicate the two IRAC bands available for use in the warm mission. However, because WISE has a smaller primary mirror and is designed to obtain a shallow survey of the entire sky, it has a comparatively low sensitivity - the required 5σ sensitivity levels are 120 and 160 μJy, respectively, at 3.3 and 4.7 microns. The lower angular resolution for WISE (PSF about 2.5 times broader than for IRAC) means that WISE will have more difficulty in crowded fields than IRAC (e.g. the galactic plane or star clusters).

By contrast, JWST's NIRCAM is **very** sensitive and has a very good angular resolution, but is not well-suited to covering large areas of the sky. NIRCAM will have a large number of broad-band filters covering the 1-5 micron wavelength region and a PSF size of order 0.1 arcseconds in the 3 to 5 micron wavelength range. The five sigma sensitivities for a 1000 second exposure (the nominal exposure time expected for NIRCAM) are of order 10 and 20 nJy for the broad-band filters near 3.5 and 4.6 microns. JWST's observing capabilities are still being determined, but it is expected that a combination of data volume issues, commanding issues and slew/acquire times will significantly limit the ability to image large areal regions. A current estimate is that an area of approximately 0.5 square degree could be imaged in a 24 hour period down to a 5σ point-source sensitivity of about 100 nJy at 4.6 microns using an integration time of around 50 seconds per field (shorter integration times would not appreciably increase the survey area because most of the clock time is consumed in overheads and slew/settle). The depth would be comparable to the Spitzer GOODS program. One of the most valuable contributions for the Spitzer warm mission may be to provide deep enough surveys of relatively wide regions to provide the first targets for JWST once JWST is in orbit. That is, obviating a need for JWST to use its precious observing time for wide-shallow surveys of its own in order to provide target lists for followup deeper imaging or spectroscopy.

Figure 1 provides a graphical comparison of the combination of sensitivity and areal coverage for WISE, JWST/NIRCAM and Spitzer/IRAC. The figure illustrates that the niche for Spitzer/IRAC is for surveying to depths not reachable with WISE and for areal coverage which would be prohibitively expensive in time for JWST. Table 1 provides sensitivity levels for several different integration times for IRAC Ch1 (3.6μm) and Ch2 (4.5μm), with links to the possible large projects which we describe later.

As one example, consider the case of a program to survey for clusters of galaxies with $1 < z < 2$ (see section 6.1). In order to identify enough new clusters to significantly

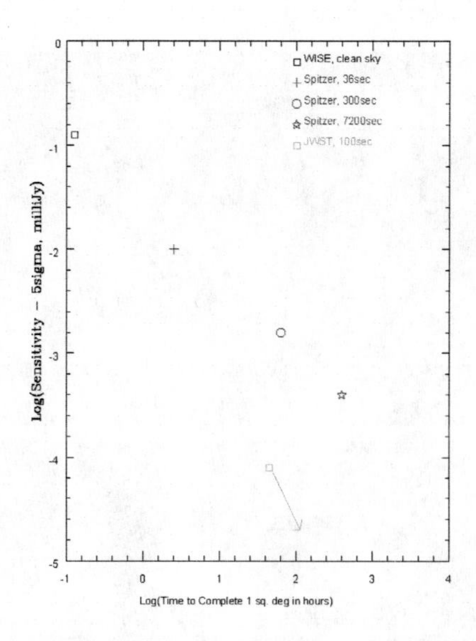

FIGURE 1. Comparison of the time needed to conduct imaging surveys of a given area of the sky for WISE, Spitzer/IRAC and JWST/NIRCAM. For Spitzer, the three symbols correspond to surveys with total integration times per point on the sky of 36, 300 and 7200 seconds.

advance this field, the area surveyed needs to be of order 200 square degrees, or more. In order to identify these clusters requires being able to detect cluster members down to two and a half times fainter than an L$*$ galaxy, corresponding to about 4 μJy, five sigma at 3.6 μm. These clusters will therefore be completely undetected in the WISE all-sky survey. To complete the required sky coverage with JWST, however, would take of order a year of dedicated observing. Spitzer/IRAC could complete the survey in three to four months, and do so in time to complete full analysis of the data and to propose follow-up detailed NIRCam imaging once JWST is launched.

As another example, consider a project to confirm candidate Y dwarfs identified with WISE. If WISE only completes one epoch of observations for the full sky before it exhausts its cryogen, only a few of the nearest Y dwarfs will be detected in two or more bands - most will be one-channel detections (the 4.5 μm band). It is expected there will be tens of thousands of such 4.5 μm-only sources, the great majority of which will be spurious. Obtaining good S/N IRAC 3.6 μm and 4.5 μm photometry for these sources would be sufficient to confirm or reject these sources as Y dwarfs, and would take only of order 1 minute of observing time each. A couple dozen such sources could be observed each day with IRAC in a snapshot mode with relatively little impact on the primary program(s) conducted that day, and hence thousands could be covered each year. Attempting to do the same thing with JWST/NIRCAM would require much more telescope time due to the much slower slew/settle times for JWST, and so would not be feasible. A similar program with ground-based facilities would also be prohibitive in terms of number of nights on big telescopes.

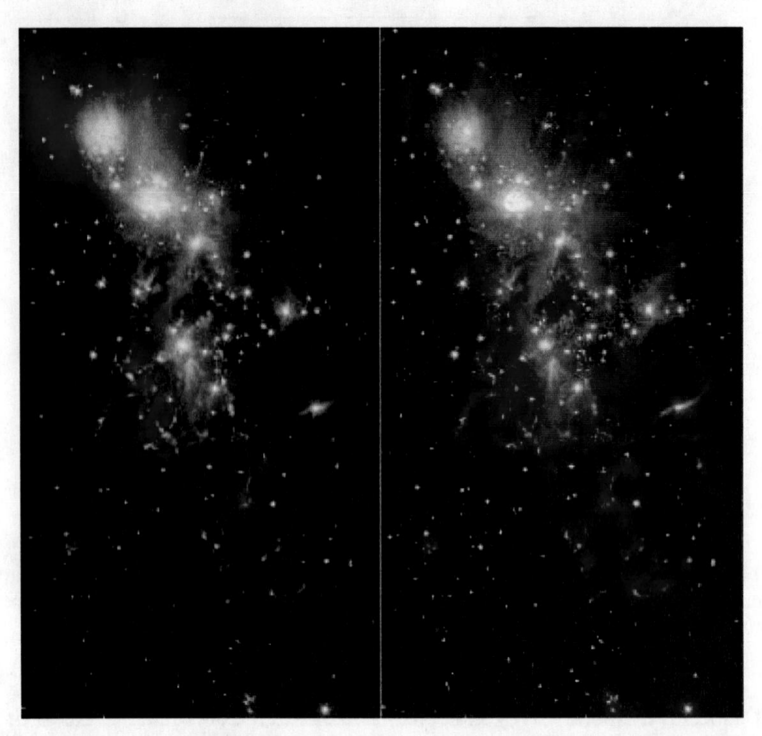

FIGURE 2. (a) Three-color (IRAC 3.6, 4.5 and 5.8 μm) image of the NGC1333 star-forming region (NASA/JPL-Caltech/R.A. Gutermuth). (b) The same region, but using 2MASS K-band as the blue filter, and IRAC 3.6 and 4.5 μm as the green and red filters.

For some types of science, it is useful to have data in more than two bandpasses. A pertinent example is the search for young stars with primordial, circumstellar disks in star-forming regions. Such searches are well-suited to the cryogenic IRAC because a two-color diagram (Ch1 minus Ch2 vs. Ch3 minus Ch4) clearly separates pure photospheres, from classical T Tauri stars with disks and from even younger stars with infalling envelopes. In the warm mission, an IRAC-only survey with just channel 1 and 2 would not be able to discriminate stars that are simply heavily reddened from those with disks or envelopes. Much of this science could be recovered, however, by combining the IRAC imaging with a ground-based survey in the near-IR (J, H or K). From a purely aesthetic viewpoint, Figure 2 illustrates that the latter type of survey can produce 3-color images with as much visual impact as the current IRAC-only images.

5. OPERATIONAL PLANS DURING THE WARM MISSION

The current staffing level for the SSC is sufficient to support the nominal mission – but just barely. Once the cryogen runs out, support of new Spitzer observations will no longer require staffing for the two retired instruments (MIPS and IRS). However, this represents a small fraction of the anticipated steady-state staffing level required for the SSC during the warm mission period. A significant fraction of the work for the Science User Support team (SUST) is directly proportional to the number of approved programs, with a premium for programs led by inexperienced Spitzer users. By mandating that all but a small fraction of the observing time be devoted to very large projects, the workload, and hence the staffing level, for the SUST should be significantly reduced. The Operations Planning and Scheduling Team (OPST) workload would also decrease with fewer projects (fewer special cases, less interaction with observers), and the workload of the Community Affairs team could decrease significantly if the yearly TAC meeting could be simplified and if the number of proposals submitted, reviewed and approved could be reduced appreciably.

It is expected that the IRAC AOT will not undergo significant changes for the warm mission (other than to reflect the fact that only two arrays will be used). Because there will no longer be transitions between instruments, it should be possible to reduce the number and frequency of calibration observations, thus providing even more observing time for science. It is probable that we will try to minimize the number of observations scheduled for targets-of-opportunity . In order to make the most of the large programs, we will likely require the teams proposing those programs to produce enhanced data products (in particular, source lists and photometry) or we will solicit archival proposals to perform that function.

It is possible that one or more large projects may be identified which are most efficiently implemented by the SSC as a service to the community. In that case, the data from the program would be made public without any proprietary period and special effort would be made to fund archival proposals to mine the data from that program; enhanced or higher-level data products could also be generated by the SSC or by proposing teams.

6. EXAMPLES OF LARGE "LEGACY" PROJECTS FOR THE WARM MISSION

As a means to begin the conversation about the science programs that should be pursued during the Spitzer warm mission, we provide below a few illustrative, large programs. These are *not* meant to pre-empt other ideas or to establish precedence for any particular teams. They are simply explorations of ideas for large and scientifically exciting projects that would have significant impact. The programs that will in fact be conducted will be chosen via an open selection process whose nature will be defined with community input during the upcoming year.

6.1. The Case for a Wide, Moderate Depth Extragalactic Survey

The tremendous power of deep multipassband datasets has become increasingly apparent over the last few years. Multipurpose surveys (e.g., GOODS, COSMOS, AEGIS) have very successfully synergised datasets to study galaxy formation and evolution to high redshift. However, the fields studied have been very small, ~ 1 deg^2 and are therefore subject to cosmic variance. Moreover, the fields are of limited use for studying the most clustered and/or rare sources e.g., extremely red objects (EROs), ULIRGs, and superclusters. Larger area surveys are clearly necessary.

The SWIRE Legacy Survey[2], was the largest of the original six Spitzer Legacy Programs. SWIRE's six wide-area, high galactic latitude fields were imaged by Spitzer at 3.6, 4.5, 5.7, 8.0, 24, 70 and 160μm over a total of $\sim$ 50 deg^2. The SWIRE dataset has been used to trace the evolution of dusty, star-forming galaxies, evolved stellar populations, and active galactic nuclei as a function of environment (Lonsdale *et al.* [3, 4], Oliver *et al.* [5], Babbedge *et al.* [6], Rowan-Robinson *et al.* [7], Farrah *et al.* [8]). However, even SWIRE is not sufficiently large to discover more than a handful of the most massive clusters at $z > 1.5$.

Astro-F and WISE will perform shallow all-sky IR surveys and JWST will perform very deep pencil beam surveys. This leaves an obvious unfilled niche - deep multipass-band surveys on scales of hundreds of square degrees. Note that the IR imaging component of such a deep wide-field survey can only be performed from space. An IRAC SWIRE-depth survey of 200-300 square degrees would take 2000 – 3000 hours (i.e., three to four months) to perform. CFHT's WIRCAM, the largest existing ground-based IR camera, would require $\sim$ 2000 **nights** to reach the same depth at K for an equivalent survey, and would produce more variable photometry across the fields than can be achieved with the ultrastable IRAC camera.

MegaCam on the CFHT is in the process of filling this large-area niche at optical passbands, observing what will be the foremost multi-passband optical dataset for many years to come, the $u^{\star}g'r'i'z'$ CFHT 170 deg^2 "Wide" Legacy Survey[3]. This area is much too large for redshifts to be determined spectroscopically, one must rely on photometric redshift estimation techniques. However, accurate estimation of photometric redshifts at $z > 1$ is impossible with an optical dataset alone (Ilbert [9]). And yet, frustratingly, the redshift range $1 < z < 2$ is exactly the redshift regime one would like to study. It is of particular importance for three reasons, a) It is the redshift at which dark energy begins to dominate over dark matter. b) It corresponds to the "redshift desert" where prominent spectral features are notoriously difficult to measure in the optical: more is known about galaxy evolution at $z > 2$ (from e.g. Lyman-break studies) than at $1 < z < 2$ c) It is the epoch of mass assembly in galaxies.

The CFHTLS fields (and indeed, those proposed by several future large ground-based optical surveys e.g., PanSTARRS, LSST), would be well-matched to an IRAC imaging survey. The CFHTLS optical survey should be complete by 2008, a year or more prior to the beginning of the Spitzer warm mission. The addition of IRAC passbands would facilitate reliable photometric redshifts to $z \sim 2$ [the redshift regime where the 4.5 and/or 3.6μm channel unequivocally sample longward of the 1.6 μm bump feature, found in all "normal" galaxies (Sawicki [10])]. The resulting combined optical-infrared survey, would be a one-of-a-kind dataset, opening up the $1 < z < 2$ universe, and enabling a variety of galaxy evolution and cosmological applications. This dataset would be impossible to surpass for at least another decade.

---

[2] http://swire.ipac.caltech.edu/swire/swire.html

[3] http://www.cfht.hawaii.edu/Science/CFHLS/

Clusters Of Galaxies at $1 < z < 2$ As Dark Energy Constraints: Galaxy clusters form at the highest peaks in the primordial density field, and as a result, their abundance and spatial distribution are very sensitive to the underlying cosmology. Clusters at $z < 1$ are now proving themselves useful tools for constraining cosmological parameters, especially with respect to the matter density, Ω_m, and the amplitude of the matter power spectrum, σ_8 (Gladders et al 2006). A sample of clusters in the redshift range $z < 1 < 2$, however, would provide strongest constraints on the nature of the dark energy, arguably the most important problem in cosmology today.

Combining Cosmological Constraints: The improved accuracy which IRAC would bring to the estimation of photometric redshifts would translate into improvements in the constraints achievable from any weak lensing cosmic shear analysis (Refregier [12], Schneider [13], Hoekstra [14]). One would ultimately wish to combine the constraints from the cluster and lensing surveys with those available from Baryon Acoustic Oscillations[4] and Type Ia Supernovae, and of course from the Cosmic Microwave Background.

Stellar Mass Density Evolution: Mid-IR observations are ideally suited to tracing the accumulation of stellar mass in galaxies (Labbé *et al.* [15], Shapley *et al.* [16], Rigopoulou [17]). Note that at $z > 0.4$, even a K-band selected dataset can only sample shortward of restframe 1.6μm, and so stellar mass estimates for galaxies above $z = 0.4$ are subject to increasing uncertainties, especially w.r.t. dust content. A wide-field seven-passband dataset would allow an unprecedented measurement of the build-up of stellar mass in galaxies from $z = 2$. Galaxies could be classified by luminosity, morphological type (early or late) and environment (field or cluster), and the epoch of mass assembly determined for each type.

The Relationship Between Stellar Mass And Total Mass: A main science driver behind the CFHTLS Wide Survey (and future surveys such as PanSTARRS and LSST) is the application of of various weak gravitational lensing techniques e.g., the CFHTLS Wide Survey will measure galaxy dark matter halo evolution to $z \sim 1$, but will still be limited to measuring galaxy luminosities in the restframe optical. The IRAC observations will complement the weak lensing analysis in two important ways. Firstly, IRAC imaging will greatly improve photometric redshifts estimates of the source (and lens) galaxies (most of the source galaxies will lie at $1 < z < 2$). This is vital to break the degeneracy which exists between the source redshift distribution $N(z)$ and the mass normalization. Secondly, since the IRAC observations will allow stellar mass estimates of the lens galaxy population, and these can be subdivided by redshift, luminosity and spectral type, this dataset will provide a measurement of the evolution of baryons relative to the dark matter out to at least $z = 1$. These data will therefore allow a direct test of assumptions relating the physics of feedback e.g., cooling flow suppression by AGN in massive galaxies (Croton *et al.* [18]) to the evolution of the dark matter halos.

On larger scales, the dark matter distribution can be mapped using standard mass reconstruction techniques. The ability to reliably group source galaxies by redshift, allows one to perform "mass tomography". By careful selection of the source galaxy

[4] We might also be able to make the case for the dataset discussed here for the warm mission as being of use as a BAO probe or pathfinder survey.

redshift distribution, one can tune the redshift range of the (intervening) dark matter of interest. By selecting several source galaxy redshift distributions in turn, one can create 3D maps of the mass distribution. These mass maps could then be compared to the distribution of galaxies as a function of redshift, luminosity and morphological type, revealing the evolution of baryons versus dark matter on large scales.

Galaxy Clustering Evolution: It will also be possible to measure galaxy clustering as a function of redshift, luminosity, scale, and morphological type. These measurements would provide constraints on theories of galaxy formation and evolution.

Obscured AGN: Although the most highly-obscured AGN require longer wavelengths to be identified, there is a significant population of moderately-reddened quasars which can be found even at near-infrared wavelengths (Cutri *et al.* [19], Glikman *et al.* [20]). An IRAC survey based either on photometric techniques for identification of quasars in the infrared (Warren *et al.* [21]) or on matching to radio surveys such as FIRST or upcoming surveys with LOFAR would enable us to find several thousand moderately-obscured quasars. With a sample of this size to compare with the existing SDSS sample of normal quasars (whose flux limit is well-matched to the IRAC surveys), we can search for evidence of an evolutionary link between dusty and normal quasars through studies of their host galaxies (Sanders *et al.* [22]) and test models for the dependence of the fraction of dusty quasars on the luminosity of the AGN (Lawrence [23]).

6.2. A Deep Survey for High-z Galaxies and Supernovae

6.2.1. Galaxies From $0.5 < z < 7$*: Clustering and Evolution*

Spitzer IRAC observations in the 3.6μm and 4.5μm passband, by virtue of tracing redshifted optical/near-infrared light, provide an excellent measurement of the stellar mass in galaxies out to $z \sim 6$. Spitzer observations conducted as part of the Great Observatories Origins Deep Survey (Dickinson *et al.* [24]) have revealed that many galaxies build up stellar masses comparable to that of the Milky Way within 1 Gyr of the Big Bang (Chary *et al.* [25], Yan *et al.* [26], Eyles *et al.* [27]). Although massive by themselves, the stellar mass in these galaxies is only a fraction ($\ll$1%) of the total co-moving matter density which is dominated by dark matter. This dark matter primarily resides in the halos of galaxies and within 10^{13-14} $M_\odot$ structures and superstructures. Detecting the former requires observing the spatially resolved dynamics of stars within galaxies which is nearly impossible in the distant Universe. The formation and evolution of clusters and superclusters on the other hand can be traced by simply measuring the redshift evolution in the clustering properties of galaxies on angular size scales of 3-10 Mpc. Comparison of the redshift evolution of galaxy clustering with those estimated from simulations of the large scale structure such as the Millennium simulation (Springel *et al.* [28]) can then be used to test the "hierarchical structure formation in standard cold dark matter" paradigm (Figure 3). This paradigm is under threat due to the tentative evidence from the dynamics of nearby dwarf spheroidal galaxies which indicates the presence of "warm" dark matter in the cores of these galaxies (Wilkinson *et al.* [29]). The effect of such "warm" dark matter would be to weaken the clustering signal on short

TABLE 2. Deep Survey Fields

Field	Area	Coordinates	No. of Redshifts	Current IRAC coverage
GOODS-N	$30' \times 30'$	12h36m, +62d14m	~3000	25 hr Central 165 arcmin^2
GOODS-S	$30' \times 30'$	03h32m, -27d48m	~2200	3 hr depth entire field
				25 hr Central 165 arcmin^2
EGS	$10' \times 90'$	14h19m, +52d43m	~12000	3 hr depth entire field

co-moving distances in the distant Universe and enhance the clustering signal on large angular scales

Accurately measuring the evolution of the clustering amplitude of galaxies as a function of redshift requires the detection of ~1000s of typical $\gtrsim 10^9$ $M_\odot$ galaxies out to $z \sim 6$. This can be achieved by observing fields which have the best spectroscopic redshift surveys and the best multiwavelength imaging data which also enable accurate photometric redshifts to be determined. There are three fields (Table 1) which match these requirements - the GOODS-N field, the GOODS-S field and the Extended Groth Strip (EGS). These fields encompass an area of 900 arcmin^2 each resulting in a total areal coverage of 2700 arcmin^2 which is $1.5\times$ the solid angle subtended by the full moon.

The 400 arcmin^2 of deep Hubble/ACS observations in the GOODS fields have yielded ~500 candidate objects at $z \sim 6$ (Bouwens *et al.* [30]). Of these, only about 12% have been spectroscopically confirmed. The 2700 arcmin^2 area would therefore yield more than 3000 candidate objects at $z \sim 6$ which could be spectroscopically followed up by JWST.

6.2.2. *Supernovae From the First Stars at* $z > 6$

Wilkinson Microwave Anisotropy Probe results indicate that the process of reionization started early, maybe around $z \sim 10$. The star-formation histories of $z > 5$ field galaxies are insufficient to account for the reionizing flux unless the duration of the starburst is extremely short, lasting less than 6% of the time (Chary *et al.* [25]). An alternative to the short starburst timescale is that the stellar mass function is biased towards high masses which would result in a greater flux of ionizing photons for each unit of baryon that goes into stars. Since star-formation at high redshift likely takes place in low metallicity environments, the optical depth of the star to its own flux is low. Unlike at low redshift, where massive stars blow themselves apart from their radiation pressure, the absence of metals at high redshift results in the star being stable till the fuel is exhausted (Heger *et al.* [31]). Such massive stars, referred to as Population III stars, are thought to end up as pair-creation supernovae whose light curve and spectrum are uncertain (Scannapieco *et al.* [32]). At redshifts of 10 and beyond, the light from these supernovae is time-dilated and can only be detected as near-infrared sources that are variable over timescales of months. One of the biggest problems in searching for such distant supernovae in the GOODS field has been the limited area covered in two epochs separated by a few months. Only 1/3 of the GOODS area, i.e. ~100 arcmin^2 are sensitive to objects that are variable over long timescales. Even in this limited area, due to the rotation of the asymmetric Spitzer point spread function with time, the identification of variable sources is limited to isolated sources which are at relatively bright flux densities. The model light curves of pair-creation supernovae from $>$100 $M_{\odot}$ stars at $z \sim 10-20$, although uncertain, are thought to be at the detection limit of 25 hrs of Spitzer observations at 3.6μm and 4.5μm (Figure 4).

By observing the 2700 arcmin^2 area of the three fields in Table 1 for 25 hrs - separated as two epochs of 12.5 hrs depth, it would be possible to observe the same patch of sky with the same position angle. This will minimize residuals due to the PSF rotation and allow better pairwise subtraction of data taken at different epochs enabling the sensitivity to variable sources to be pushed down the background limited values rather than the confusion limited value. Even non-detections at these faint flux limits will place strong constraints on the spectrum and energetics of these exotic objects and assess if they are sufficient to account for reionization. Such a uniform, "wide" area survey will also enable an accurate measurement of the fluctuations in the extragalactic background light. The measurement of these fluctuations in a deep Spitzer IRAC observation have placed the first constraints on the existence of high redshift Population III stars (Kashlinksy *et al.* [33]).

To reiterate, the deep survey would be comprised of imaging for the total 2700 arcmin^2 region of the three fields in Table 1 to the GOODS depth of 25 hours in the 3.6μm and 4.5μm passbands. Including overheads, we estimate this would take about $\sim$3600 hrs of observing time. The areal extent of these fields is determined by the quality of the Hubble/ACS optical imaging data, the Chandra X-ray data, the existing Spitzer MIPS far-infrared imaging data and the VLA/radio, Keck and ESO/VLT spectroscopic data. By ensuring a perfect areal overlap between the deepest surveys performed by NASA's 3 Great Observatories and the wealth of multiwavelength data from the largest ground-based telescopes, this would provide a lasting legacy in the study of galaxy evolution, structure formation and first generation of stars in time for follow-up with future facilities like the James Webb Space Telescope.

6.3. A 360 Degree GLIMPSE of the Galactic Plane

Mid-infrared, large scale, shallow surveys of the Galactic plane and other regions of our Galaxy have already provided a wealth of data on Galactic structure, the physics of star formation and the evolution of the Milky Way. The GLIMPSE project (Benjamin *et al.* [34]) used IRAC to survey the inner Galactic plane from $-65 < l < 65$ and $|b| < 1$ (in combination with the Galactic center observations of Stolovy [35]). GLIMPSE has revealed new star clusters hidden by the extinction of the Galactic plane (Mercer *et al.* [36]), numerous previously uncataloged star forming regions (Mercer *et al.* [37]) and furthered our understanding of Galactic structure (Benjamin *et al.* [38]).

The proposed project is to finish the Spitzer mid-IR census of the Galactic plane by mapping the extent of the thick disk around the midplane of the Galaxy over the entire longitude range. The observations will trace the warp of the Galactic disk in the outer Galaxy. The survey would address three main science topics: the extent of star formation

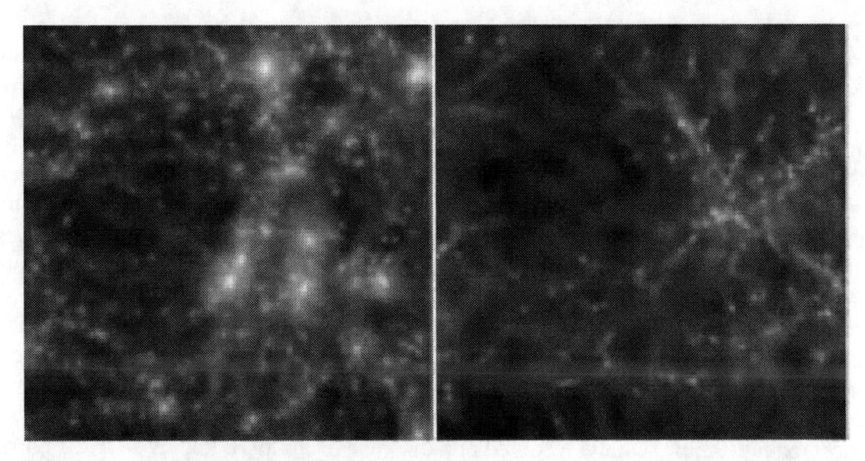

FIGURE 3. Evolution in the dark matter distribution in a 30 arcmin x 30 arcmin between $z \sim 1$ (left) and $z \sim 5$ (right) from the Millennium simulations (reprinted from Springel *et al.* [28]). Clearly, the dark matter and therefore, the galaxies are much more strongly clustered at $z \sim 1$ than at $z \sim 5$. By tracing the evolution of clustering of galaxies in redshift bins and comparing them to simulations of structure formation in dark matter dominated models, we can constrain the nature of dark matter. In "warm" dark matter scenarios, the clustering of galaxies will have more power on larger angular scales and less power on small scales. A survey of the type proposed would be able to measure the clustering amplitude of galaxies on scales up to $10h^{-1}$ Mpc while the current GOODS observations are limited to correlation lengths of $\sim 3h^{-1}$ Mpc.

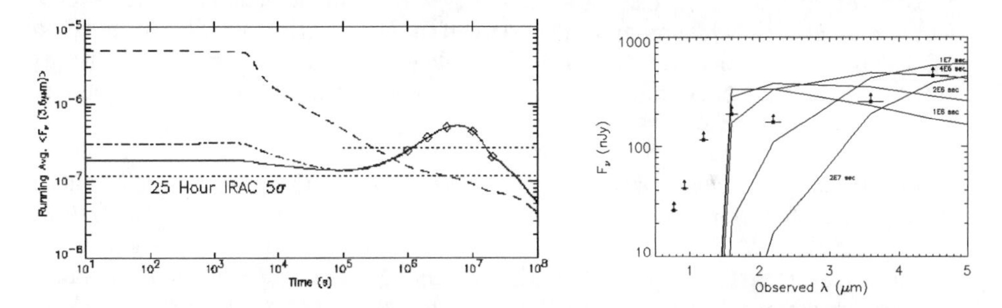

FIGURE 4. Plot showing the characteristics of a pair creation supernova from a 200 $M_\odot$ star exploding at redshift of 10. The left hand panel shows the light curve in the 3.6μm channel (solid line). The right hand panel shows the time evolution of the spectral energy distribution of such a supernova. Also shown in the right hand panel are the sensitivity limits of the GOODS Hubble and Spitzer observations as well as the ESO/VLT near-infrared imaging data. The ability to constrain the presence of these objects is currently hindered by the 100 arcmin^2 area in GOODS which has two epoch coverage and the rotation of the Spitzer point spread function over these two epochs. If present, these variable sources might be remnants of the first generation of stars forming in the Universe.

in the outer Galactic disk, the number of evolved stars (particularly massive evolved stars) in the Galactic disk, and the structure of the Galaxy along heavily extincted lines of sight.

Far-outer Galaxy star formation. Ongoing star formation at Galactocentric distances up to 19 kpc has recently been identified (Kobayashi *et al.* [39]). The lower metallicity, low volume density of molecular clouds and longer time between spiral arm passages in the outer Galaxy provides a very different environment for star formation than inside the Solar circle. The role of environmental factors and triggering on star formation efficiency can be examined by comparing star formation as a function of Galactocentric radius. A uniform census of star formation at the periphery of our Galaxy can be used to understand the distribution of star formation at the edges of other spiral galaxies.

In addition to being sensitive to the point source emission from protostars, the 3.6 μm band is sensitive to the diffuse dust emission from PAHs excited by young stars. Molecular outflows associated with class 0 and class I protostars can be detected by an enhancement in the 4.5 μm band which contains the H_2 v=0-0 S(9) line. Molecular outflows of modest extent (0.1 pc) can be resolved at a distance of up to 10 kpc by IRAC. Figure 5 displays a young star forming cluster towards l $\sim$ 180 degrees (NGC 1893; distance of $\sim$ 6 kpc) at K band, 3.6 and 4.5μm. The image is representative of the quality of the proposed IRAC observations and demonstrates how the mid-IR bands reveal embedded star formation and that interesting star-forming regions exist well outside the longitude range covered by the existing GLIMPSE surveys.

Evolved stellar populations. The number of identified, evolved massive stars in the Galaxy is an order of magnitude smaller than predicted from observations of other spiral galaxies (Shara *et al.* [40]). The proposed survey will be able to identify all Wolf-Rayet stars in the surveyed field throughout the extent of the Galactic disk. Previously unknown AGB stars will be detected in abundance. The role of these stars on replenishing metals in the ISM and their effect on the Galactocentric metallicity gradient can be examined with this unbiased survey.

Galactic structure along extincted lines of sight. Near-infrared investigations have recently been able to map well the distribution of stars in the thick disk of the Galaxy (Cabera-Lavers *et al.* [41]). However, there is substantial extinction in the Galactic plane even at K-band (Marshall *et al.* [42]) particularly along lines of sight containing massive star forming regions such as Cygnus X and spiral arm tangent points. Recent work by Frieswijk *et al.* [43] has shown that near-infrared extinction is also important along lines of sight in the 2nd and 3rd quadrants. The IRAC bands are much less affected by extinction ($A_K \sim 5 \times A_{4.5}$; Weingartner and Draine [44]) and can probe the distribution of stars more effectively along high extinction lines of sight.

Context of proposed survey. A mid-infrared survey of the entire Galactic plane is complementary to existing and proposed surveys at shorter and longer wavelengths. A uniform, systematic survey at 3.6 and 4.5 μm will catalog protostars, evolved stars and highly obscured stars. Shorter wavelength observations (2MASS, UKIDSS) are not able to unambiguously detect and/or identify intrinsically red sources such as protostars and evolved stars. Figure 6, a color-color diagram from Whitney *et al.* [45], displays the location of main sequence stars, evolved stars and protostars on near and mid-IR color spaces. With just the addition of the 3.6 μm band, evolved stars are more clearly separated from reddened early-type main sequence stars and protostars are distinguished from reddened late type giants. The 4.5 μm band provides even more leverage in color space (Gutermuth *et al.* [46]) distinguishing between class II and class III protostars.

Far-infrared and submillimeter observations are more sensitive to young protostars but current and future instruments are unable to match the resolution of IRAC. IRAC provides resolution 4$\times$ that of Herschel's 70 μm band and 7$\times$ that of the shortest passband of the JCMT Galactic plane survey. High resolution imaging is important for determining the multiplicity of star-forming regions (Allen *et al.* [47]). The multiplicity of star formation, particularly in massive star forming regions, is very sensitive to the formation mechanism and can be used to determine the relative roles of competitive accretion, turbulent fragmentation, etc.

Implementation Plan. To complete the IRAC survey of the galactic disk, a program could be conducted during the Spitzer warm mission which would result in a 1250 square degree survey of the Galactic plane in two epochs. As the inner Galactic plane has already been observed by the GLIMPSE team, only one epoch for this region would be required. This survey would follow the warp and flare in the outer Galaxy as traced by molecular line surveys (Wouterloot *et al.* [48]). The survey region would include most of the molecular disk and the majority of star formation in the Galaxy. Figure 7 displays the proposed survey coverage in Galactic coordinates. The region of the plane within $|l| < 5^{\circ}$ has been well mapped by Spitzer and does not require additional coverage.

The original GLIMPSE fields (light grey regions in Fig 7) require a single pass of one two second frame to improve radhit and solar-system object rejection. As the remainder of the survey region has a lower object density, we can integrate deeper before becoming confusion limited. With four 12 second integrations, the limiting magnitudes at 3.6 and 4.5 μm are 18.4 and 17.5, respectively. The sensitivity of the $|l| > 65^{\circ}$ portion is well-matched to the ongoing UKIDSS survey and is considerably deeper (10$\times$) than the all-sky survey of WISE. The portion of the survey region in dark grey in Fig. 7 will be done with two epochs of two 12s high-dynamic range (HDR) integrations. HDR mode is necessary to recover the flux of bright sources (up to $\sim$ 1 Jy). Four dithers are required for robust radhit rejection.

The GLIMPSE survey covered 220 square degrees in 400 hours with IRAC. The proposed observations would use a similar mapping strategy but would include an additional epoch and use 12 second HDR mode observations (which require 50% more time than the 2 second GLIMPSE observations). The estimated time for this very complete survey of the Galactic disk is $\sim$ 5500 hours.

6.4. Thermal Imaging of Extrasolar Giant Planet Transits

Despite exquisite ground-based and spacebourne measurements, current models of giant planets cannot even constrain the mass of the putative solid core at the center of Jupiter to a range smaller than 0 to 12 Earth masses. Yet core masses of giant planets are the most important parameter distinguishing the two currently competing theories of planet formation; core accretion (Mizuno [49], Hubickyj *et al.* [50]) and gravitational

FIGURE 5. Three color composite image of the young open cluster NGC 1893. 2MASS K_s band data is displayed in blue, IRAC 3.6 μm is in green and IRAC 4.5 μm in red. Numerous protostars (red objects) are visible. The IRAC data is representative of the quality of the proposed survey.

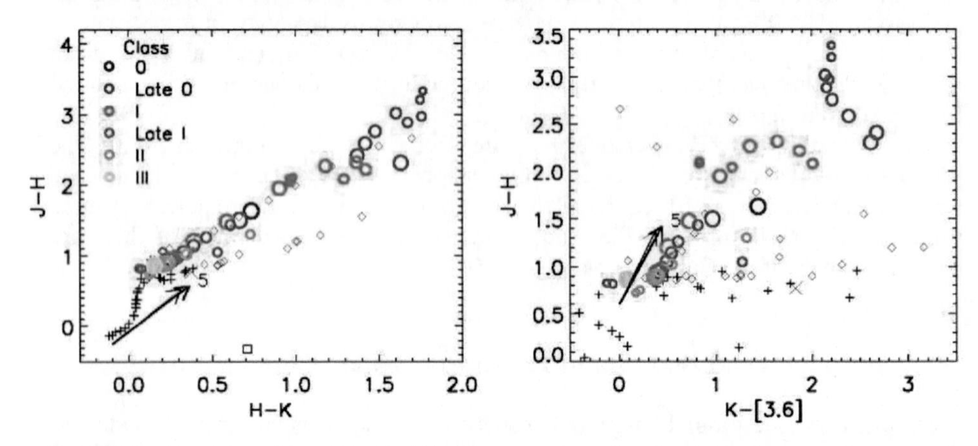

FIGURE 6. Near-IR and Near-IR with IRAC 3.6 μm color-color plots indicating the color space of main sequence stars (crosses), red giant and supergiant stars (diamonds), AGB stars (squares), planetary nebulae (asterisks), reflection nebulae (small crosses), T-tauri stars (large cross) and protostars (colored circles). Reprinted from Whitney *et al.* [45].

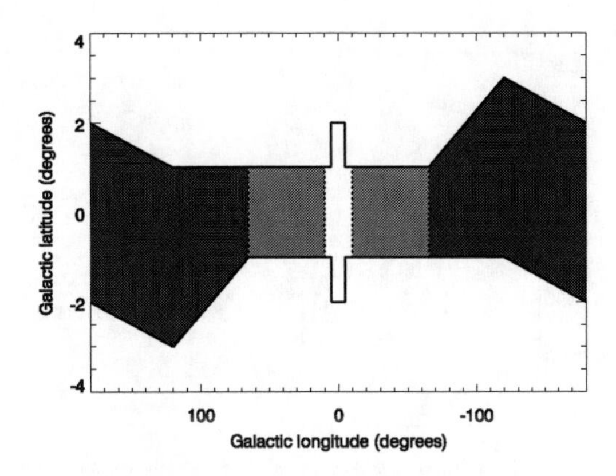

FIGURE 7. Coverage of the proposed Galactic plane survey. The field to be mapped with IRAC is outlined. The light gray regions need one additional epoch of 2 second integrations. The dark gray regions will be mapped by two epochs of two 12 second HDR exposures.

instability (Boss [51, 52]). The detection of the photometric signature of "hot Jupiters" transiting across the face of their parent stars has opened up a new and exciting means to study extrasolar giant planets (EGPs). By combining information from the radial velocity orbital solutions and the depth of the primary transit, it is possible to directly estimate the radii of the transiting EGPs. While the masses and radii determined for the 9 known EGPs are generally in accord with theoretical expectations, there are exceptions that continue to challenge our current understanding. By virtue of their size distribution, their extreme temperature ranges, and their growing number, transiting EGPs may be the most promising avenue of attacking the problem of planet formation.

For close-orbit, extrasolar giant planets (EGPs), the flux density contrast ratios between planet and star are approximately 2 orders of magnitude greater at 3 and 5 μm than they are at optical wavelengths. This relatively favorable flux contrast makes it possible to detect the secondary transit where the hot Jupiter is eclipsed by the parent star, if accurate mid-infrared photometry is available. IRAC has already demonstrated the required photometric stability by directly detecting the photons emitted by an EGP. Figure 8 shows the 4.5 μm time series photometry of the secondary eclipse of TrES-1 (Charbonneau *et al.* [53]), where the parent star is a K0 dwarf with V = 11.8. Together with similar 8 μm observations, these data have provided the first observational constraints on models of thermal emission and atmospheric constituents in hot Jupiters (Burrows [54]). There are now several Spitzer programs in progress to carry out photometry of transiting EGPs using IRAC, MIPS, and IRS.

By providing insights into the atmospheric constituents and thermal properties of EGPs, including heat transport from the day to night sides by winds and jet streams (which may measurably shift the light curves with respect to the orbital ephemeris), IRAC photometry can enable more stringent constraints to be placed on important model inputs, including bulk composition, the equation of state, and the degree of inhomogeneity (Guillot [55]). "Hot stratospheres", which are common to gas giants in

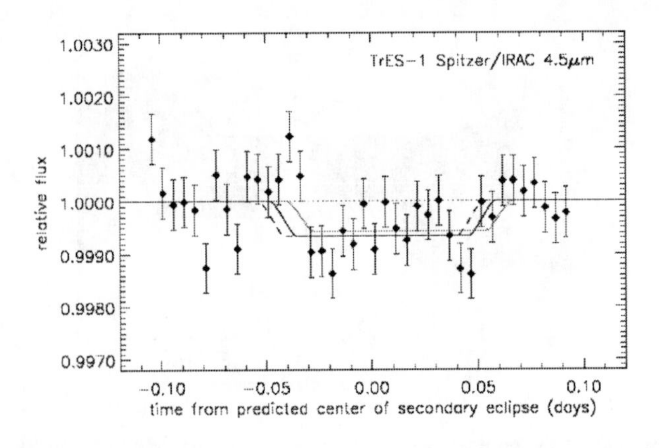

FIGURE 8. Binned IRAC 4.5 μm time series photometry of TrES-1. Reprinted from Charbonneau *et al.* [53].

our Solar System, and which have the effect of reducing planetary radii by absorbing heat high in the atmosphere rather than in the lower regions where stellar insolation can slow the normal contraction of the planet, will be most clearly seen in IRAC photometry (Fortney *et al.* [57]). Figure 9 compares measured IRAC fluxes for TrES-1 (Charbonneau *et al.* [53]) with model atmosphere predictions (Sudarsky *et al.* [60]) and clearly demonstrates weaknesses in the model. Agreement between theory and the observations can be significantly improved by increasing the metallicity of the model atmosphere by a factor of between 3 and 5 over solar (Fortney *et al.* [56]). 3.6 μm observations will enable much stronger constraints to be placed on the amount of atmospheric methane, and by extension on the efficiency of heat transport to the night side of the planet (Fortney *et al.* [57]).

For brighter systems such as HD 189733b (V = 7.7, sp. type K0V) and HD 209458 (V = 7.7, sp. type G0V), time series IRAC photometry will enable a search for longitudinal variations indicative of planetary weather. By gathering high-cadence photometric observations during secondary ingress and egress, it should be possible to detect differences in the shape of the light curve from the predictions of a uniformly illuminated disk. Model predictions (Williams [58]) have shown that the 3.6 and 4.5 μm channels should be particularly sensitive to realistic variations in the distribution of surface flux for systems like HD 209458b. Other models (Burrows *et al.* [59]) have shown that the relative flux in the 3.6 and 4.5 μm channels is sensitive to the details of the chemical composition in the stratosphere. Both the surface inhomogeneity detection and the chemistry model constraints will require high quality IRAC observations of many secondary eclipses per target.

Another application with a tremendous potential payoff is the measurement of temperature gradients across the surfaces of hot EGPs. Understanding how the intense stellar insolation is absorbed and reradiated is fundamental to an understanding of their atmospheric physics. Using time series, 24 micron MIPS photometry, Harrington *et al.* [61]

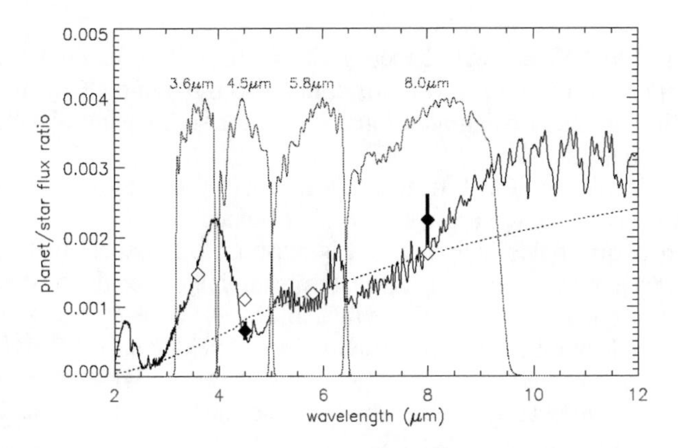

FIGURE 9. Predicted planet/star flux ratios from a model hot jupiter spectrum (Sudarsky *et al.* [60]), compared with IRAC measurements (Charbonneau *et al.* [53]) for TrES-1 (filled diamonds). The dotted line shows a blackbody spectrum corresponding to a temperature of 1060 K. The open diamonds show the expected flux in each IRAC waveband. Note the substantial overprediction of flux at 4.5 μm , suggesting a significant additional source of atmospheric opacity near 4μm. Reprinted from Charbonneau *et al.* [53].

recently demonstrated the existence of a large asymmetry in the temperatures of the day and night sides of υ Andromeda b. The conclusion they draw is that most of the incident stellar radiation is reemitted at the subsolar point, with very little heat transport to the night side of the planet. The technique relies solely on the brightness modulation of the planet as it orbits its parent star and does not require a transiting geometry. The method is thus potentially applicable to all known hot EGPs. While the planet/star flux ratio is more favorable at 24 microns for EGPs in the $\sim 1000-2000$ K range, this ratio is no more than 1.7 to 5 times smaller at 4.5 microns (depending on relative temperatures of star and planet), with typical values of around ~ 2.5. There are currently 53 EGPs known with semi-major axes of less than 0.1 AU (that of υ Andromeda is 0.06 AU). With a sufficient investment of observing time, IRAC should therefore enable us to better characterize the atmospheres of a significant number of hot, non-transiting EGPs.

How many transiting EGPs will be available during Spitzer's warm mission? The COROT mission (scheduled for launch in December, 2006) is expected to yield on the order of 45 transiting EGPs with $11 < m_v < 16.5$ over the course of its 2.5 year mission (Gillon *et al.* [62]). The first public data release is currently expected to occur during the summer of 2008, with subsequent data releases every 6 to 12 months. The time frame is thus well matched to post-cryogen, IRAC followup observations.

The launch date for Kepler currently remains November, 2008. Whereas the detection of terrestrial-mass planets is expected to require most of the 4 year mission, hot Jupiters will be readily detectable within the first year. The Kepler project has offered to communicate such detections in advance of publication (Borucki, 2007, private comm.). The time frame for the first such detections would therefore be late 2009, near the time when Spitzer exhausts its cryogen supply. Though estimates are necessarily model dependent, the expectation is that Kepler will detect on the order of 100 giant inner planets in the first year for stars with $9 < m_v < 14$ (for comparison, the parent star for TrES-1 has $m_v = 11.8$). The discovery of an additional ~ 30 outer-orbit giant planets is expected over the life of the mission.

In addition to Kepler and COROT, there are at least a dozen ground-based, photometric transit surveys active at the present time (Charbonneau *et al.* [63]). With a sustained and distributed effort of this magnitude, it is reasonable to expect that the number of known transiting giant planets will increase substantially over the next several years, even without the Kepler detections. Roughly half of these transit surveys utilize a small-aperture, wide-field approach, and are primarily sensitive to relatively bright parent stars (V< 10). These stars will be particularly good targets for precision photometry with the IRAC 3.6 and 4.5 μm channels, which will be the most sensitive available probes of EGP atmospheres until the launch of JWST.

Each monitoring observation of an EGP (one filter, one eclipse) takes of order 8 hours to cover the ingress, eclipse and egress and to define well the out-of-eclipse continuum level. If for a given IRAC channel five secondary-transits are monitored (to improve the S/N over a single measurement), and if data are obtained in both IRAC channels, the average exposure time per target would be of order 80 hours. Obtaining complete secondary transit data for 20 target stars would therefore require of order 1500 hours of Spitzer observing time. We note that at the current rate of detection of transiting planets by ground-based campaigns, we would expect more than enough targets to be available even without data from COROT or Kepler.

ACKNOWLEDGMENTS

This work is based (in part) on observations made with the Spitzer Space Telescope, which is operated by the Jet Propulsion Laboratory, California Institute of Technology, under contract with NASA. This publication makes use of data products from the Two Micron All Sky Survey, which is a joint project of the University of Massachusetts and the Infrared Processing and Analysis Center/California Institute of Technology, funded by the National Aeronautics and Space Administration and the National Science Foundation.

REFERENCES

1. Fazio, G., *et al.*, 2004, ApJS, 154, 10
2. McMurtry, C., Pipher, J., and Forrest, W., 2006, SPIE, 6265E, 6
3. Lonsdale, C. J., *et al.*, 2003, PASP, 115, 897
4. Lonsdale, C., *et al.*, 2004, ApJS, 154, 54
5. Oliver, S., *et al.*, 2004, ApJS, 154, 30
6. Babbedge, T. S. R., *et al.*, 2006, MNRAS, 676
7. Rowan-Robinson, M., *et al.*, 2005, AJ, 129, 1183
8. Farrah, D., *et al.*, 2006, ApJL, 641, L17
9. Ilbert, O. 2006, A&A, 457, 841
10. Sawicki, M. 2002, AJ, 124, 3050
11. Gladders, M. D., and Yee, H. K. C. 2000, AJ, 120, 2148

12. Refregier, A. 2003, ARA&A, 41, 645
13. Schneider, P. in *Gravitational Lensing : Strong, Weak & Micro. Saas-Fee Advanced Course 33*, edited by G. Meylan, P. Jetzer and P. North, Springer-Verlag, Berlin, 2006, 273 (astro-ph/0509252)
14. Hoekstra, H. 2006, ApJ, 647, 116
15. Labbé, I., *et al.*, 2005, ApJL, 624, L81
16. Shapley, A. E., *et al.*, 2005, ApJ, 626, 698
17. Rigopoulou, D. 2006, ApJ, 648, 81
18. Croton, D. J., *et al.*, 2006, MNRAS, 365, 11
19. Cutri, R. M., Nelson, B. O., Francis, P. J., and Smith, P. S., in *AGN Surveys*, IAU Colloq. 184, edited by R. F. Green, E. Y. Khachikian, and D. B. Sanders, ASP Conference Series 284, ASP, San Francisco, 2002, 127
20. Glikman, E., Gregg, M. D., Lacy, M., Helfand, D. J., Becker, R. H., and White, R. L. 2004, ApJ, 607, 60
21. Warren, S. J., Hewett, P. C., and Foltz, C. B. 2000, MNRAS, 312, 827
22. Sanders, D. B., Soifer, B. T., Scoville, N. Z., and Sargent, A. I. 1988, ApJL, 324, L55
23. Lawrence, A. 1991, MNRAS, 252, 586
24. Dickinson, M. Giavalisco, M., and The GOODS Team, "The Great Observatories Origins Deep Survey" in *The Mass of Galaxies at Low and High Redshift*, edited by R. Bender and A. Renzini, Springer-Verlag, Berlin, 2003, pp. 324-331.
25. Chary, R. *et al.*, 2007, ApJ, submitted
26. Yan, H., Dickinson, M., Giavalisco, M., Stern, D., Eisenhardt, P. R. M., Ferguson, H. C., ApJ, 651, 24
27. Eyles, L., Bunker, A., Stanway, E., Lacy, M., Ellis, R., and Doherty, M. 2005, MNRAS, 364, 443
28. Springel, V. *et al.*, 2005, Nature, 435, 629
29. Wilkinson, M. *et al.*, 2006, EAS, 20, 105
30. Bouwens, R. J. *et al.*, 2006, ApJ, 653, 53
31. Heger, A., *et al.*, 2003, ApJ, 591, 288
32. Scannapieco, E., *et al.*, 2005, ApJ, 633, 1031
33. Kashlinsky, A., *et al.*, 2005, Nature, 438, 45
34. Benjamin, R. A. *et al.*, 2003, PASP, 115, 953
35. Stolovy, S., *et al.*, 2006, J. Physics, Conf. Vol 54, pp. 176-182
36. Mercer, E. P., *et al.*, 2005, ApJ, 635, 560
37. Mercer, E. P., *et al.*, 2004, ApJS, 154, 328
38. Benjamin, R. A. *et al.*, 2005, ApJL, 630, 149
39. Kobayashi, N., Yasui, C., Tokunaga, A., Saito, M., 2005, Protostars and Planets V LPI Contribution 1286, 8639, http://www.lpi.usra.edu/meetings/ppv2005/pdf/8639.pdf
40. Shara, M. M., Moffat, A. F. J., Smith, L. F., Niemela, V. S., Potter, M., and Lamontagne, R. 1999, AJ, 118, 390
41. Caber-Lavers, A., Garzón, and Hammersley, P. L. 2005, A&A, 433, 173
42. Marshall, D. J., Robin, A. C., Reylé, C., Schultheis, M., and Picaud, S. 2006, A&A, 453, 635
43. Frieswijk, W. F. W., Teyssier, D., Shipman, R. F., Hily-Blant, P., 2005, Protostars and Planets V LPI Contribution 1286, 8582, http://www.lpi.usra.edu/meetings/ppv2005/pdf/8582.pdf
44. Weingartner, J. C., and Draine, B. T. 2001, ApJ, 548, 296
45. Whitney, B. A., Wood, K., Bjorkman, J. E., and Cohen, M. 2003, ApJ, 598, 1079
46. Gutermuth, R. A., Megeath, S., Muzerolle, J., Allen, L., Pipher, J., Myers, P., and Fazio, G., 2004, ApJ, 154, 374
47. Allen, L. E., *et al.*, 2005, in *Protostars and Planets V*, edited by B. Reipurth, D. Jewitt and K. Keil, Univ. Arizona Press, Tucson, 2007, pp. 361-376
48. Woulterloot, J. G. A., Brand, J., Burton, W. B., and Kwee, K. K. 1990, A&A, 230, 21
49. Mizuno, H. 1980, Prog. Theor. Phys., 64, 544
50. Hubickyhj, O., Bodenheimer, P., and Lissauer, J. 2004, in *Gravitational Collapse: From Massive Stars to Planets* edited by G. Garcia-Segura et al., Rev. Mex. AA Ser. Conf., pp. 83-86
51. Boss, A. P. 1997, Science, 276, 1836
52. Boss, A. P. 2004, ApJ, 610, 456
53. Charbonneau, D., *et al.*, 2005, ApJ, 626, 523
54. Burrows, A. 2005, Nature, 433, 261

55. Guillot, T. 2005, Ann. Rev. of Earth and Plan. Sci., 33, 493
56. Fortney, J. J., Marley, M. S., Lodders, K., Saumon, D., and Freedman, R. S. 2005, ApJ, 627, L69
57. Fortney, J. J., Saumon, D., Marley, M. S., Lodders, K., and Freedman, R. S. 2006, ApJ, 642, 495
58. Williams, P. K. G., Charbonneau, D., Copper, C. S., Showman, A. P., and Fortney, J. J. 2006, ApJ, 649, 1020
59. Burrows, A., Sudarsky, D., and Hubeny, I. 2006, ApJ, 650, 1140
60. Sudarsky, D., Burrows, A., and Hubeny, I. 2003, ApJ, 588, 1121
61. Harrington, J., *et al.*, 2006, Science, 314, 623
62. Gillon, M., Courbin, F., Magain, P., and Borguet, B. 2005, A&A, 442, 731
63. Charbonneau, D., Brown, T. M., Burrows, A., and Laughlin, G., in *Protostars and Planets V*, edited by B. Reipurth, D. Jewitt and K. Keil, Univ. Arizona Press, Tucson, 2007, pp. 701-716

Spitzer Warm Mission Archive Science Opportunities

Lisa J. Storrie-Lombardi*, John R. Stauffer*, Bidushi Bhattacharya*, Sean Carey*, Dave Frayer*, Mark Lacy*, Victoria Meadows*, Alberto Noriega-Crespo*, Luisa Rebull*, Erin Ryan†, Inseok Song*, Susan Stolovy*, Harry Teplitz*, David Trilling** and Schuyler van Dyk*

*Spitzer Science Center, Caltech 314-6, Pasadena, CA 91125, USA
†Dept. of Astronomy, U. Minnesota, 116 Church St. SE, Minneapolis, MN 55455, USA
**Steward Observatory, U. Arizona, 933 N. Cherry Ave., Tucson AZ 85721, USA

Abstract. The rich data archive from the Spitzer cryogenic mission will be comprised of approximately 25 TB of data. A five-year warm mission would add an additional 15-20 TB. All of these data will be processed and archived to form homogeneous, reliable database to support research for decades after the end of the Spitzer mission. The SSC proposes a robust archival research program during the warm mission phase. A sampling of possible archival programs are described.

Keywords: Spitzer Space Telescope, archives, infrared astronomical observations
PACS: 95.55.Fw, 95.80.+p, 95.85.Hp

1. INTRODUCTION

The Spitzer Space Telescope is approaching the milestone of four years in orbit (at the time of the workshop), with enough cryogen for a 5.5 year nominal mission profile. After the cryogen is exhausted, the 3.6 and 4.5 μm channels of IRAC will still perform well, but the other Spitzer detectors will no longer produce useful data. Because the sensitivity in the two shorter IRAC channels is better than any other existing or planned facility until JWST is launched, we expect that Spitzer will continue to operate for an additional five year period as an IRAC 3.6 and 4.5 μm imaging mission. We refer to this as the Spitzer Warm Mission.

One of the Spitzer Science Center's top priorities after the cryogen is depleted is to do a final reprocessing of all of the cryogenic mission data as quickly possible in order to provide the most homogenous, most reliable database possible for archival usage by the community. The cryogenic archive is expected to comprise of order 25 TB of data. Most of the observing time during the warm mission will be devoted to large, legacy-type programs, which will generate very large volumes of data particularly amenable to archival research.

We believe that the Spitzer combined cryogenic and warm mission archive will be heavily used during (and after) the warm mission time period (2009-2014), and we are planning to support this usage by building a robust, flexible archive interface and by providing a significant amount of funding to support the analysis of Spitzer archival data. Spitzer is able to observe $\sim$7000 hours per year which is more than twice as much time as any space or ground-based optical/infrared facility.

CP943, *The Science Opportunities for the Warm Spitzer Mission Workshop*, edited by L. J. Storrie-Lombardi and N. A. Silbermann

In this white paper, we outline the expected order-of-magnitude data volume in the Spitzer final archive; the types of data that will be provided; the planned capabilities of the archive interface; and provide some illustrative examples of the types of science programs we expect would be motivated by the Spitzer data.

2. THE FINAL SPITZER DATA ARCHIVE

The cryo archive contains data from all three instruments, IRAC, MIPS and IRS. Currently, the archive interface ("Leopard") allows users to obtain raw, basic calibrated data (BCD) and post-BCD data from individual Astronomical Observation Requests (AORs), with their associated calibration files on a per-AOR basis.

Data from Spitzer Legacy programs, processed beyond the level of the current SSC pipelines by the Legacy teams, are currently made available for direct download from the SSC website, and also ingested into the Infrared Science Archive (IRSA). These data include mosaics of multiple AORs, spectral data cubes and band-merged source lists.

The current rate of ingestion into the archive is about 4.5 TB per year, with 4 TB per year from data received and processed by the pipeline, and 0.5 TB per year from the Legacy teams. By the end of the 5.5 year cryo-mission we expect approximately 25 TB of data in the archive. Currently the actual volume stored by the archive is larger than these rates would suggest as 1-2 processing versions are retained on disk before moving them into our long-term tape archive. The warm mission data rate is not expected to be much lower – although IRAC will only be taking data in two channels, the data rates will still be much higher than currently from IRS. Our data rate estimate for the warm mission is approximately 3.5-4 TB per year, but will vary depending on the character of the executed science observations.

After the cryo mission ends, the cryo data will undergo a final processing. This is expected to be completed in 12-18 months after the end of the cryogenic mission. These data will be entered into a "final" Spitzer archive, whose design is underway. The main differences between the final archive and the current one derive from the fact that the structure of the current archive is strongly related to processing requirements, rather than long-term storage and ease of access to the data. The final archive will be designed as a long-term archive, curation of which will be handed over to IRSA at the end of the Spitzer mission.

The main difference between the final archive and the current archive from the user perspective will be that data will be available not only on a per-AOR basis, as currently, but also on a per-BCD basis. The archive will also be fully compatible with interoperability standards (e.g. it will support existing Virtual Observatory protocols). We will ensure that a full provenance of each file is maintained, so that the details of the processing of each product are made clear. Calibration files will also be available online so that users can re-reduce data using different calibrations from those used in the final processing should they feel it is necessary.

As part of the final archive design the SSC is formulating the list of data products expected to be delivered to the final archive. These will, of course, include the existing products generated by the pipelines, but will also include higher-level products which we hope to produce by the end of the cryo mission if time and resources allow. One example of this type of product is source lists for imaging data.

The final archive may also contain some of the tools currently available to Spitzer users (e.g. Mopex, Spice, GERT, CUBISM and IRSCLEAN). However, it is yet to be determined how maintenance of these tools will be funded past the end of the Spitzer mission, and, consequently, whether they will be served by IRSA.

3. SUPPORT FOR MINING THE SPITZER DATA ARCHIVE

An active archival research program has been supported through the annual calls for proposals since Cycle-1. The Legacy programs executed in the first year of the Spitzer mission provided large, coherent, publicly available data sets for the general community to mine very early in the mission. Additional Legacy programs selected in each subsequent cycle, in addition to the substantial amounts of data becoming public each year, ensure a very rich archive for supporting research programs beyond the science proposed by the original investigators (e.g. Deguchi *et al.* [1]). We have just scratched the surface of the archive so far and we are already seeing an increase in archival research.

The SSC normally recommends that 10% of the available data analysis funds for each cycle be awarded to archive and theory programs ($\sim$\$2 million annually). The typical award is \$50,000 to \$100,000. In Cycle-4 the time allocation committee specifically recommended awards of \$2.7 million due to the high quality of the archival programs. Previous time allocation committees and the Spitzer User's Committee have requested that the SSC solicit large archival proposals. The SSC has been unable to add these large programs during the cryo mission due to annual budget cuts which have impacted the User Community funds.

During the warm mission the SSC proposes substantially increasing the funding available for archival programs. The focus on the observing programs will be larger programs and with a smaller dollar/hour for data analysis funding. During the cryo-mission the SSC awards \$20-25 million annually for general observer, archival and theory funding with 90% going to observing data analysis support and 10% for archive/theory support. For the warm mission we propose to award \$20 million annually with 75% going to observing data analysis support and 25% for archive/theory support. In addition to one-year archive proposals we would like to add a new program supporting Legacy archival proposals. These would be multiyear archival programs to tackle problems that can't properly be addressed with a typical one-year archive program.

4. SPITZER WARM MISSION SCIENCE: CONSTRAINTS, AREAL COVERAGE AND LIKELY DATA GENERATION RATE

A companion white paper entitled "The Spitzer Warm-Mission Science Prospects" describes in detail the expected capabilities of Spitzer after the cryogen runs out, and our plans for operating the observatory during the warm mission. That white paper also describes several possible large legacy-type programs that could be carried out during the warm mission. Here, we provide a brief review of that material in order to indicate the scope and character of the data from the warm mission period that will be available for archival usage.

The Spitzer warm mission has a natural and well-defined beginning point, which is the point when the cryogen is exhausted. The best estimate for the end of cryogen is March-April 2009. The warm mission also has a natural and well-defined ending point based on geometry. Spitzer is slowly drifting away from the Earth. In 2014 the spacecraft reaches a distance such that we will be unable to recover from a safe mode using the low-gain antenna. This constraint provides a warm mission duration of order five years.

Based on our current plans and knowledge, the Spitzer observing efficiency should be unchanged in the warm mission, implying that there will be of order 6500-7000 hours of time available each year for science observations. The sensitivities and noise-characteristics of the two shorter-wavelength IRAC channels should also be unchanged from the cryo-mission. Because only two of the four channels will be operating, however, the data generation rate will be of order half that of the cryo-mission. From an extrapolation of the IRAC campaigns executed to date, we expect a data generation rate in the warm mission period of order 3.7 TB/year.

Both as a necessary means to reduce operating costs and as a means to encourage different types of science programs from those carried out during the cryogenic mission, there will be a strong preference for large and very-large programs during the warm mission. IRAC warm mission programs will occupy a niche in sensitivity-area space between that for WISE (all-sky but comparatively very shallow) and JWST-NIRCAM (very deep but pencil-beam). This suggests that most of the warm mission observing programs are likely to cover regions of the sky from tens to hundreds of square degrees, with sensitivities of order $1-10\mu$Jy, 5σ. If **all** of the observing time in the warm mission were devoted to a relatively shallow survey with 10 μJy, 5σ sensitivity, of order 1/4 of the sky could be covered. Other possible programs could image the same region of the sky two or more times for proper motion purposes or to detect variable sources. In some cases, in order to address the primary science goals, it will be necessary for the proposing team to obtain complementary optical or near-IR imaging of the surveyed region, and those data would also be expected to become part of the Spitzer archive.

A direct consequence of allocating the majority of the warm mission observing time to very large programs is that archival science will take on a significantly greater importance. By their nature, programs that cover large areas of the sky uniformly can address multiple science topics. In order to encourage full exploitation of the archive we intend to allocate a greater percentage of the user community funding funding for archival science during the warm mission time period, compared to the cryo-mission.

5. EXAMPLES OF LARGE ARCHIVAL PROJECTS - GALACTIC SCIENCE

5.1. MIPS 24μm Archival Search for Exo-Zodiacal Dust

A number of programs used Spitzer to detect debris disks around nearby dwarfs in order to help understand the processes involved with planet formation. At 24 μm, even relatively short integration observations can detect small amounts of warm dust to a hundred parsecs; therefore, any scan-map observation could detect such debris disks around field stars that happen to fall within the mapped area. An analysis of appropriate archival MIPS data could thus add significantly to the statistical study of the frequency and evolution of debris disks around main sequence stars. An automated spectral energy distribution (SED) fitting could be performed on all stars in MIPS 24 μm images with integration times per point on the sky $>$ 24 seconds; the MIPS photometry could be combined with optical/near-IR photometric data from Tycho-2, SuperCOSMOS, USNO-B1, NOMAD, and 2MASS catalogs. A SED fitting technique with synthetic stellar spectra can determine stellar photospheric flux at 24 μm with a 3σ uncertainty of $\lesssim$ 20 mJy. This high fidelity SED fitting allows identifications of stars with very small warm IR excess, analogous to our solar system's asteroidal belt or zodiacal dust grains at somewhat younger ages. A very short lifetime of warm grains around these stars requires either continuous, violent, multiple collisions among planetesimals (i.e., heavy bombardment) or a catastrophic huge collision between two planets (e.g., BD+20 307, Song *et al.* [2] The latter mimics the collision which created our own Moon. These stars may represent a population of adolescent stars currently undergoing a heavy bombardment in their inner planetary zone. Using about ~8,000 MIPS 24 μm images from GO Cycle-1, First Look Survey, and Legacy data, about 50 mid-IR excess stars have already been identified. From the entire 5 year Spitzer cryo mission data, we anticipate a couple of hundred warm excess stars would be detected allowing better estimates of the occurrence rate of heavy bombardment at other stars (or the rate of catastrophic planet-planet collisions).

5.2. Y-Dwarf Search from IRAC Archival Images

IRAC archival images provide a large data set to search for late-T and Y-dwarfs. Late T-dwarfs occupy a distinct region in the IRAC color-color diagram and are well-separated from essentially all other stars (even PMS stars with primordial disks). The data shown in Fig. 1 are from an archival image (AOR 3954944) of a known T8 dwarf (2MASSI 041519-0935). A GTO observation by G. Fazio targeted this specific T8 dwarf. Figure 1 demonstrates that late T-dwarfs like this one can be easily identified from an IRAC archival search. The T8 dwarf is designated with horizontal and vertical error bars while other background sources were plotted as dots without error bars. A criterion to be plotted in this figure was S/N> 10 detection in all 4 bands. Although no Y-dwarfs have been identified, contemporary theoretical spectral models of these very cool dwarfs predict that at least early Y dwarfs should occupy a similar region of the IRAC color-color space as late T-dwarfs. It is almost certainly true that Y dwarfs will occupy a very distinct region of color space because of their very cool temperatures and presumably exotic molecular features. High redshift quasars can have similar colors to late T dwarfs and so a survey with IRAC or with IRAC plus ground-based near-IR photometry could be contaminated by such quasars – but those are interesting objects also, and their detection is valuable in their own right. Any Y dwarf candidate that can be identified with a wide-area IRAC survey will be near enough to Earth to have relatively large, and

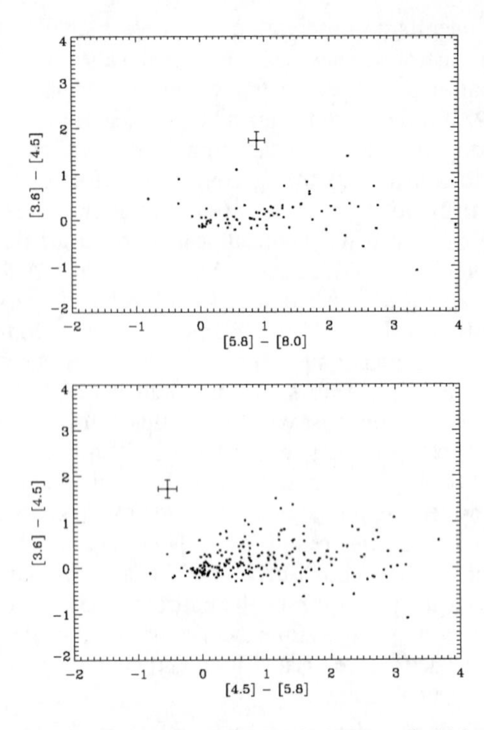

FIGURE 1. A known T8 dwarf (2MASSI 041519-0935) is designated with horizontal and vertical error bars while other background sources were plotted as dots without error bars. A criterion to be plotted in this figure was S/N> 10 detection in all 4 bands.

measurable proper motions – allowing the separation of the Y dwarf candidates from the distant quasars with suitable two-epoch data. A few test studies of several square degrees of IRAC survey data from First Look Survey and Legacy programs have been performed in order to find these very cool dwarfs, but have so far not found any. In the combined cryogenic and warm mission archive, we expect several hundred square degrees of imaging capable of detecting Y dwarfs to 50 pc or more - and in such a large area, we would expect to find a few Y dwarfs (combining the IRAC observations with moderate depth J band imaging for the warm mission survey data).

5.3. A Census of Star-Cloud Collisions Using IRAC

The disk of the Milky Way is composed of a mixture of stars and gas. The stars come in a wide range of mass from > 100 solar masses to below 0.01 solar mass, with the study of the stellar IMF being a flourishing subfield of its own. The clouds of gas in the galactic disk also come with a wide range of masses. Radio surveys provide data from which the mass function of these clouds can be derived. Cloud census surveys are good at finding the most massive clouds, and to some extent in measuring structures within large cloud complexes. However, it is much more difficult to make a census of small, relatively isolated clouds. Therefore their frequency and properties are much less well characterized. One of the best ways to identify such clouds is when they pass near luminous stars. The nebulae surrounding the Pleiades is an example of the collision of a small ($\sim$10 solar mass) molecular cloud with a star system. Examples of individual field A stars also encountering small clouds have been reported in the literature (e.g. Kalas *et al.* [3]; Gaustad and van Buren [4]). Observations with IRAC channel 4 (8μm) provide a particularly sensitive means to detect such encounters because of the strong PAH emission in that band. As shown by Stauffer *et al.* [5], some of the Pleiades stars without evidence of interaction with the Merope cloud at optical wavelengths do have easily detected 8 μm excesses (and spatially resolved 8 μm emission at Pleiades distance). Figure 2 shows one of these star-cloud interactions as observed by IRAC. A survey of the entire cryogenic IRAC database for other examples of cloud-star interactions via the signature of excess or spatially resolved 8 μm excesses would better constrain the volume filling factor of small molecular or atomic clouds, and in favorable cases allow one to estimate the properties of those clouds.

5.4. Cleaning the IRAS PSC and FSC

The IRAS point source catalog (PSC; Beichman *et al.* [6], IRAS) and faint source catalog (FSC; Moshir et al. [7]) provide extremely rich resources for all-sky searches for interesting objects such as stars with circumstellar disks or AGN. However, the structure of the background emission at any one location can provide a significant source of confusion to the large-aperture IRAS measurements. A Spitzer archive project to compare IRAS and Spitzer data over large scales could create a much higher-quality catalog of such interesting objects, weeding out knots of nebulosity or spurious sources (such as asteroids), and finding objects that break into multiples when viewed with MIPS at 24 μm as illustrated here in Fig. 3

In a 10.6 square degree MIPS survey of the Perseus molecular cloud Rebull *et al.* [8] assessed how successful the IRAS survey was in identifying point sources in a potentially confusing region - in this cloud, complex extended emission is present at all three MIPS bands (and the IRAC bands for that matter). Simply examining the PSC and FSC objects in this region finds objects that are listed as detections at 60 or 100 um, with only upper limits at 12 and 25μm; even without Spitzer data, one might suspect that such sources correspond to texture in the extended emission. Though the PSC and FSC do contain warning flags for cirrus the Spitzer data comparison clearly show that significantly fewer spurious sources appear in the FSC than in the PSC. Overall, 20-30% of the "point sources" at 12 and 25 μm resolve into knots of nebulosity. There are many more "point sources" reported at 60 and 100 μm (than at 12 and 25 μm) in the PSC. A much lower fraction of these objects are recovered; many more of these are completely missing or fall apart into nebulosity when viewed with MIPS. More than half of the 100 μm point sources are completely missing, and 25-40% of the 100 and 60 μm point sources are clearly confused by nebulosity. Overall, a much larger

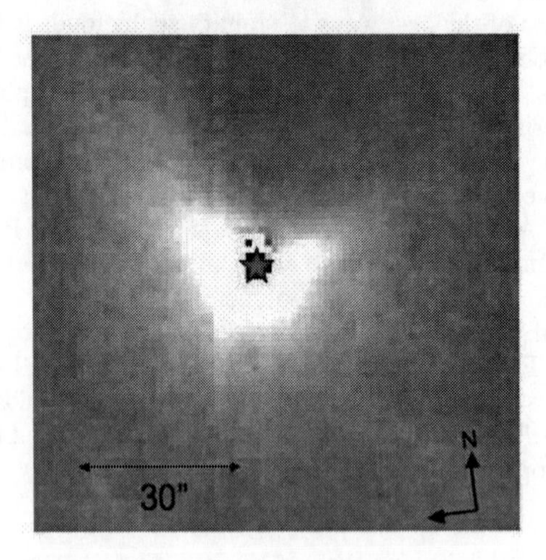

FIGURE 2. Example of an A star impacting a small interstellar cloud. This is an IRAC 8 μm image of HII1234, an A dwarf member of the Pleiades. We have subtracted an empirical PSF from the image to enhance the PAH emission from the star-ISM interaction. The bow-shock shaped nebular emission is due to a shell of material created by radiation pressure from the star sculpting the material in the cloud as the Pleiades moves through the cloud.

fraction of FSC objects are recovered by MIPS than PSC objects. Surprisingly few IRAS sources are resolved into multiple objects by Spitzer, but those that are resolved are often "interesting"(see discussion in Rebull *et al.* [8]).

A Spitzer archival project to do a similar comparison of IRAS and Spitzer maps over the 3750+ hours of MIPS scan observations (very roughly 1000+ square degrees distributed over the whole sky) in the archive could create a vastly improved IRAS PSC that weeds out the 20-40% of objects that are simply knots of nebulosity, and finds the interesting objects that break into multiples. For the Perseus analyses, the comparisons of the IRAS catalog and the MIPS images were done by hand, but an automated routine could be developed to look for true MIPS point sources within the IRAS error ellipse for every IRAS point source.

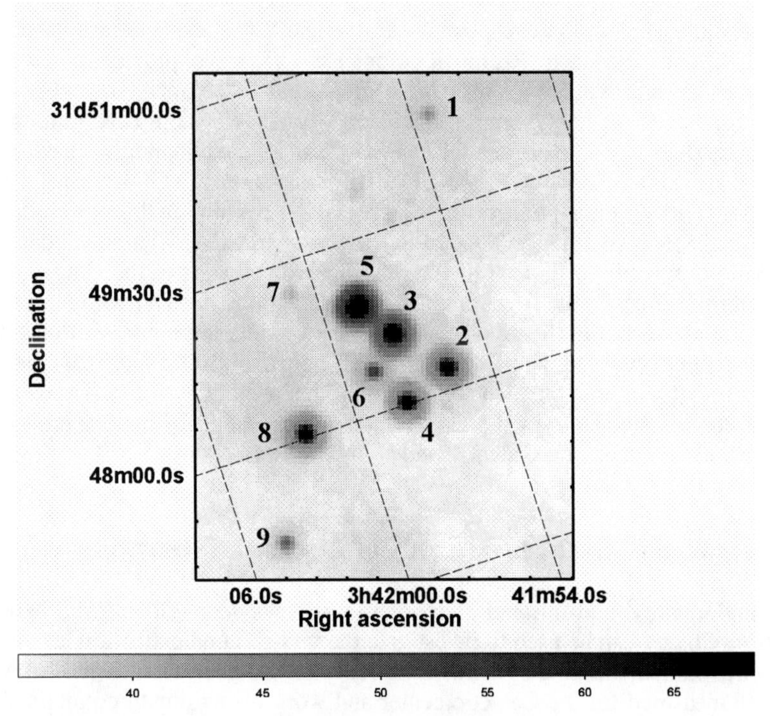

FIGURE 3. Example of an "object" identified as a single point source with IRAS data, which is resolved by MIPS (24μm) into a small cluster. Reprinted from Rebull *et al.* [8].

5.5. Identifying "Missing" Evolved Stars

Massive stars profoundly influence the evolution of galaxies, through energy input and chemical enrichment of the interstellar medium via stellar winds and supernovae. Stars with $M_{\rm init} > 20\,M_\odot$ are believed to evolve, depending on $M_{\rm init}$, as O-type ZAMS $\rightarrow$ Be/B[e] $\rightarrow$ (red supergiant ?? $\rightarrow$) luminous blue variable (Be/B[e]/Of) $\rightarrow$ Wolf- Rayet (WR) star $\rightarrow$ supernova/collapsar. The high luminosities ($L \sim 10^6\,L_\odot$) and short duration ($\sim$0.5 Myr) of the WR phase, and short lifetimes (<4 Myr) of WR precursors make WRs excellent markers and chronometers of very recent star formation in galaxies, as well as sensitive metallicity indicators through their observed relative line strengths. WRs also potentially serve as probes of the IMF in galaxies. The number of observationally identified WRs in the Galaxy so far (approximately 300) is much lower than model predictions by factors of 4–10. This discrepancy motivates the search in the infrared (where, e.g., $A_K/A_V \approx 10\%$ for WRs likely "missing," or "hidden" optically by the rapid increase of line-of-sight A_V with distance (0.5–2 mag kpc^{-1}) in the Galactic plane (almost all of the known WRs have $|b| < 5°$). All WR stars exhibit infrared excesses, due in large part to the free-free emission originating from electron scattering in the neighborhood of (mainly) He^{++} ions in the outer regions of their dense winds; circumstellar heated dust shells can also contribute, especially for the late WC stars. What we have found is that using the combined near-IR/mid-IR color space as a 'sieve' provides a very successful means of revealing hidden WRs—the near- to mid-IR colors well sample the large infrared excess all WR stars exhibit (with $f_\lambda \propto \lambda^{-2.7 \text{ to } -3.2}$), compared to normal photospheres (with $f_\lambda \propto \lambda^{-4}$). From a GLIMPSE+2MASS combination we have found 21 new WRs, as well as about 180 new evolved, high-mass stars in Of, Be, B[e], LBV, and Of/WN, i.e., *pre-WR* transitionary, phases. We will be continuing this search in Cycle-4 using GLIMPSE-II, GLIMPSE-3D, MIPSGAL, and MIPSGAL-II archival data.

We will not be able to find *all* of the missing evolved stars in the Galaxy, but over the long term, primarily the IRAC and MIPS observations from any program targeting the Galactic plane during the cryo mission will provide a large reservoir of data from which to identify new candidates. The data would help in both characterizing the massive star formation throughout the plane *and* testing more accurately the stellar evolutionary scenarios for massive stars in the Galaxy. This is essential for understanding galaxy evolution both in the nearby Universe and at high redshift.

5.6. Systematic Study of Infrared Extinction

The high-quality mid-infrared photometric surveys (e.g. GLIMPSE, MIPSGAL, C2D) from Spitzer can be used to investigate the mid-infrared extinction law as a function of environment and line of sight. Until recently, extinction in the mid-infrared had only been measured for the Galactic center and work is ongoing to compare observed extinction laws for particular lines of sight to current grain models (Román-Zúñiga *et al.* [9]). Initial results (Indebetouw *et al.* [12]) suggest an extinction curve in rough accord with current models (Weingartner and Draine [11]) but suggest that the mid-IR extinction has a flatter wavelength dependence (see Fig. 4). A key question to address is whether the mid-IR extinction curve is universal or has a strong dependence on line of sight. A comprehensive investigation of the extinction law along many lines of sight will permit modeling the law as a function of environment and will constrain the grain size distribution, grain growth and other physical processes.

One method of determining the extinction to a particular star would be to simultaneously fit the available 1.2-24 μm photometric points with an model spectral energy distribution and extinction law Robitaille *et al.* [13]. Point source catalogs for the entire inner Galactic plane ($|l| < 60°$, > 260 square degrees) exist from the 2MASS, GLIMPSE and MIPSGAL surveys. This project would also make use of existing high quality photometry from the c2D legacy program. A wide range of extinctions could be probed ($A_V \sim 1$ to > 100). The extinction determination can be augmented by archival IRS data of available targets.

With a well parameterized extinction curve, the total dust column can be estimated even for very large columns ($A_V > 100$). The derived column map then can be used in investigating Galactic structure by helping to constrain a deconvolution of far-IR through mm continuum emission into phases of the ISM (warm ionized, warm neutral, cold neutral). In addition, the three-dimensional distribution of the extinction can be modelled using appropriate stellar populations models and the existing photometry

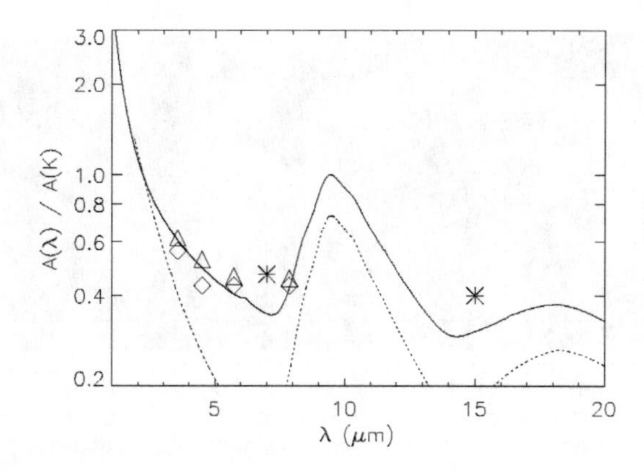

FIGURE 4. Measured mid-infrared extinction curves from Indebetouw *et al.* [12] (diamonds), Román-Zúñiga *et al.* [9] (triangles) and Jiang *et al.* [10] (asterisks). The R_V = 5.5 model extinction law from Weingartner and Draine [11] is plotted for comparison.

(Marshall *et al.* [14]). Extinction measurements provide a temperature independent way of estimating the column of big interstellar grains. This method is complementary to the future millimeter/submillimeter observations of interstellar dust where the emission is a convolution of the column density with the temperature distribution and emissivity.

5.7. Measurement of Galactic Star Formation and Cataloguing of Infrared-Dark Clouds

High quality source lists and images of the Galactic plane have been produced by the GLIMPSE and MIPSGAL surveys at wavelengths of 3.6 to 70μm. The $>$ 260 square degrees these surveys cover encompass most of the Galactic disk and encapsulate most of the star forming regions of the Galaxy. These data are complemented by existing spectral line surveys and will be augmented by upcoming surveys of the Galactic plane in the millimeter/submillimeter. Models of protostellar SEDs exist and are currently being refined (Robitaille *et al.* [13]). A very ambitious synthesis of the available multi-wavelength data of the Galactic plane and theory of star formation would be to catalog all detectable protostars in the Galactic plane and estimate the current star formation rate and initial mass function of the Galaxy. Distances of protostars can be estimated from millimeter spectral line data and association of the diffuse emission/absorption around a protostar with the morphology of the molecular gas as traced by spectral line mapping. One class of star forming objects for which this method has proven fruitful is infrared-dark clouds (IRDCs; Fig. 5). With the multi-wavelength infrared mapping of the Galactic plane, IRDCs should be easily identified using contrast-color filters as has been done previously using the MSX Galactic plane data (Simon *et al.* [15]). A catalog of these objects alone will provide a current snapshot of the earliest stage of star

FIGURE 5. Sample infrared-dark cloud at 3.6, 8.0 and 24.0 μm. The extinction lane is well correlated with millimeter wave line emission; therefore, a kinematic distance to the object can be well estimated (3.6 kpc in this case). The majority of red point sources in the image are class I and II protostars.

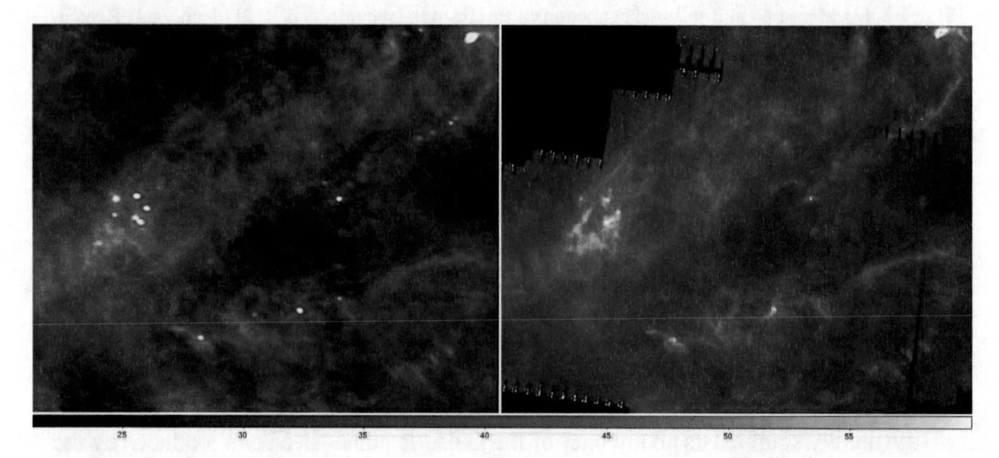

FIGURE 6. The Taurus Molecular Cloud (TMC) has been surveyed with IRAC and MIPS (Padgett *et al.*2007; in preparation). The image shows a nearly four degree field of the TMC, comparing an IRAS HiRes map at 100μm (left; $\sim 1.5'$ resolution) and the MIPS 160μm image (right;$\sim 40''$ resolution). The structure of the cold dust in this classical molecular cloud displays a fascinating and complex morphology.

formation in the Galaxy.

The protostellar identification and extinction studies are complementary and can be done jointly with existing source catalogs and using the existing spectral energy distribution models. Much of the analysis can be done in an automated fashion through least-squares fitting of SEDs and extinction models to the photometry.

5.8. Turbulence and The Structural statistics of the ISM

There is a wealth of research on the morphology and structure of molecular clouds, based on the analysis of two-dimensional images obtained through several tracers. These emission maps of the interstellar medium (ISM), either in ^{12}CO, or dust extinction or IRAS infrared emission, show a remarkable property: they are self-similar on a wide range of scales, from a few hundreds of parsecs down to ~ 0.02 pc (see e.g. Bensch *et al.* [16], Falgarone *et al.* [17]).

A way to describe this self-similarity is through the angular power spectrum, which is a function of inverse scale size (wavenumber) with an exponent ranging from $\beta \sim -3.6$ to $\beta \sim -2.5$. It is this self-similarity that has allowed us to make the connection between morphology and turbulence. The classical cascading energy dissipated from large to small scales is also a power-law with an index closely related to that of the density or velocity dispersion (through their emission and absorption tracers). The connection between structure or morphology of the gas and dust, either cirrus-like or molecular, with that of the feedback processes that sculpt their shape and the evolution of the ISM, is one the most fertile and exciting fields in Galactic astronomy.

In the infrared, in particular, the data from the Spitzer Galactic First Look Survey (GFLS) (Burgdorf *et al.* [18]) demonstrated how far one can push the limits. In a section of the Gum Molecular cloud, the IRAC and MIPS images have permitted study of the turbulent density field to scales of $\sim 5''$, and shown that HI and dust power spectra, traced by the 24μm & 70μm emission, are essentially identical down to spatial scales ~ 0.01 pc (Ingalls *et al.* [19]). In this regime, excitation, ionization and extinction are key players in interpreting the observations. The MIPS 24μm images offer more than an order of magnitude improvement in the spatial resolution of turbulent processes over other infrared surveys (eg. IRAS).

A similar analysis to that of the GFLS is being currently undertaken on some areas of the MIPSGAL survey and Taurus molecular cloud (Fig. 6). These data sets are but a small fraction of what is available in the Spitzer archive. Data from the C2D and GLIMPSE legacies are obvious candidates to work with, but cirrus-like emission or structures have also been detected in the SWIRE and Extra-GFLS (for example).

An archival project to study the structural statistics of the ISM (diffuse, clumpy, molecular) will allow a better understanding of turbulent density fields and to test our ideas of energy transfer and dissipation, not only at interesting angular scales, but also in the most diverse environments.

6. EXAMPLES OF LARGE ARCHIVAL PROJECTS - EXTRA-GALACTIC SCIENCE

6.1. Spitzer's Combined Extragalactic 16 μm Survey

The IRS provides science-quality 16 μm imaging with its Peak-Up (PU) camera. While the number of dedicated PU imaging programs is small ($\sim$ 5% of IRS time), short low (SL) spectra and PU data are taken together on a single detector. There are about 170,000 SL exposures in the Spitzer archive. These include almost a square degree of low-background PU imaging with exposure times of at least 100 sec; about 100 square arcminutes of these have exposure times more than 15 minutes. These data will include 5–10 thousand galaxies with $f_{16} > 120$ μJy. No pointed Spitzer survey could achieve this areal coverage due to the small PU field of view, so the archival data represent previously unexplored phase space for the mission.

These PU images can greatly improve the 16 μm source counts. The flux limit of the archival data is comparable to the deepest ISOCAM surveys, and has better spatial resolution. At the limit these counts will be dominated by LIRGs and ULIRGs at redshifts approaching unity. Proper measurement on the extragalactic source counts are required across all available passbands in order to constrain the shape of emission (PAH) and absorption (silicate) features in template SEDs. Current analysis of 24 μm source counts by different groups still favor disparate SEDs; additional wavelengths may resolve the issue.

Many, if not most, SL spectra are taken of objects in fields also observed by IRAC and MIPS. Thus, the 16 μm photometry of many sources can be used together with other Spitzer data to constrain the MIR slope of extragalactic sources. Kasliwal *et al.* [20] used archival 16 μm images to show that the 16/24 μm ratio can be used to identify "silicate dropouts" (Takagi and Pearson [21]), that is $z \sim 1.5$ objects with strong silicate absorption in the MIPS-24 passband, in order to identify the most strongly extincted starbursts and AGN. Teplitz *et al.* [22] found that 16 μm excess is also indicative of PAH emission at $z \sim 1$.

6.2. Spitzer's Combined Extragalactic 70 μm Survey

All of the MIPS-70μm extragalactic survey data could be reprocessed using optimized offline data reduction scripts. From the ultra-deep programs of GOODS-N and GOODS-S (PI: Frayer, PID3325+20147) and the S-COSMOS (PI: Sanders PID20070+30143, TC: Frayer) and FIDEL (PI: Dickinson, PID30948, TC: Frayer) Legacy programs, it has been demonstrated that offline processing can improve the rms sensitivity by 20-25% over the SSC online products (Frayer *et al.* [23]). More importantly, the techniques clean up the non-Gaussian wings of the noise distribution, enabling the extraction of sources which are about 1.5-2 times fainter at the same level of reliability (e.g., >99.5%). These techniques would be difficult to implement within the online SSC system since they involve iterative reduction steps that would require a complete re-design of the SSC downlink infrastructure. The data from all programs (including targeted GO observations) would be optimally combined for all of the popular survey fields [e.g., xFLS, SWIRE, Bootes, GOODS, EGS, IRAC-deep, Lockman-deep, GTO, and COSMOS] to produce the deepest possible data over each field.

An improvement of a factor of 1.5-2 in depth in the derived source catalogs would have a wide-range of science applications. For instance, one could produce the definitive 70μm source counts from Spitzer. The wide-area xFLS, SWIRE, and Bootes programs can be used to measure the bright source counts, while the deep S-COSMOS, FIDEL, GTO surveys, and the publicly available GO programs (e.g., IRAC-deep field and Lockman) can be used to measure the turn-over (~5-12 mJy) of the Euclidean-normalized differential source counts. Accurate source counts severely constrain the models of galaxy evolution and their spectral-energy-distributions (SEDs). Both the evolutionary models and the underlying SEDs require modification to fit the observed 70 μm counts while maintaining consistency with the counts at other wavelengths. The enhanced 70 μm data would also be used to constrain the far-infrared properties of BzK galaxies, X-ray sources, radio galaxies, as well as other 24 μm and IRAC populations of high-redshift sources being uncovered by Spitzer. Having a large set of optimally, uniformly reduced 70 μm data is crucial to address these and other similar questions.

Currently, the Spitzer 70 μm archived data are under-utilized by the public. Unlike MIPS 24 μm and IRAC, most MIPS-Ge science programs have relied on just a few experts for their data reduction. Having catalogs and enhanced mosaic images available to the general user would trigger much wider usage of these data, overcoming the typical user's concern about the tractability of the Ge data.

6.3. Searches for Galaxy Clusters at z > 1

Spitzer data, when combined with wide-field optical/near-IR imaging, is a powerful tool for finding distant galaxy clusters in the redshift range z $\sim 1-2$, the so-called "cluster desert". This range is particularly important as it spans the gap between the clusters of galaxies found in optical surveys (e.g. the RCS, Gladders *et al.* [24]) and the non-virialized high-z protoclusters found, e.g., though Lyman break galaxy searches at z $\sim 2-5$. It is thus the redshift range at which we expect to see clusters virialize, and the cluster red sequence and morphology-density relation to be established. However, until recently, few clusters had been found in the cluster desert. The low areal density of rich, high-z galaxy clusters makes them hard to find in both ground-based near-infrared surveys, and space-based X-ray surveys. Spitzer/IRAC, with its highly efficient mapping AOT and low background is an excellent cluster-hunting tool. Using Spitzer data, in combination with the ground-based 9 deg^2 NOAO deep wide survey, (Stanford *et al.* [25]) recently discovered a z $= 1.41$ galaxy cluster, the second highest redshift cluster known in the cluster desert. The ongoing Sparcs project (Wilson *et al.* [26]) uses the 50 deg^2 SWIRE survey to search for clusters, using $z' - [3.6]$ colors to identify the cluster red sequence. We expect that cluster surveys will constitute a significant part of the archive use. Including all the wide area surveys such as SWIRE, the IRAC shallow survey and smaller surveys such as the XFLS and COSMOS, about 70 deg^2 in total are available for this work, much of which already has high quality, ground-based imaging data. Although some surveys will already be complete before the end of the cryogenic mission, better cluster finding algorithms will continue to be developed, and better ground-based data will become available to help with photometric redshift estimates. High-z cluster finding is also well-suited to the warm mission, as IRAC channels 1 and 2 are most useful for these surveys, and the warm mission archive will contain a large amount of suitable data.

6.4. Searches for Dust Obscured AGN and Quasars

AGN and quasars obscured by dust make up 1/2-3/4 of the total AGN population, and dominate the 24 μm source counts at $S_{24} > 1$mJy (e.g. Brand *et al.* [27]). The areal density of these objects is ≈ 100 deg^{-2} in shallow-medium Spitzer surveys, thus, in the total area of Spitzer surveys there are about 7000 objects. Efficient mechanisms to identify AGN on the basis of Spitzer broad-band photometry at 3.6-24 μm have been developed for shallow and medium Spitzer surveys (Lacy *et al.* [28], Sajina *et al.* [29], Stern *et al.* [30]). Importantly, Spitzer surveys are able to find the Compton thick AGN that are almost impossible to find in X-ray surveys, yet which comprise about half of the local AGN population, and probably a similar fraction of the hard X-ray background (Gilli *et al.* [31]). Finding large numbers of obscured AGN and quasars will allow us to quantify the fraction of obscured objects, and how this changes with redshift and AGN luminosity. It will also allow us to investigate the relationship between the obscured and unobscured quasar populations.

6.5. Stacking Analyses of Galaxy Populations

Many galaxy populations are well-represented in the wide area Spitzer surveys, and well detected at the shorter Spitzer wavelengths, but typically undetected at longer wavelengths. Dole *et al.* [32] show that stacking 70 μm and 160 μm images at the positions of faint galaxies detected at 24 μm results in statistical detections an order of magnitude fainter than the individual source detection limits in those bands. Eyles *et al.* [33] stack z $\sim$ 6 galaxies detected in IRAC bands 1 and 2 in IRAC bands 3 and 4, and show for the first time that the average z $\sim$ 6 galaxy probably has a sub-solar metallicity. Other examples of other possible uses of this sort of analysis would include: stacking of IRAC channel 4 and MIPS data for Lyman break galaxies to establish their AGN activity (using 8 and 24 μm data) and star formation activity (using 70 and 160μm data), and stacking of 70 and 160 μm data on quasars detected at 24 μm to investigate their far-IR SED as a function of quasar luminosity and redshift.

7. EXAMPLES OF LARGE ARCHIVAL PROJECTS - SOLAR-SYSTEM SCIENCE

Asteroids are remnants of the early solar system and provide clues to its origin and evolution. Unlike planets and satellites, they are not subject to appreciable geological evolution, have remained relatively pristine since their formation and can provide insight into the primordial solar nebula. Their current surface characteristics and size distributions serve as markers of the physical and orbital evolution and of the effects of long-term space weathering on objects within our solar system. As asteroids are routinely present in Spitzer observations near the ecliptic plane, characterizing their prevalence is of interest to non-planetary astronomers as well, who desire to distinguish nearby objects in their mid-infrared data from extrasolar sources they are studying.

The solar system's Main Belt asteroid (MBA) population is characterized via Size Frequency Distribution functions (SFD), taxonomies and surface properties. To date, this work has been carried out using primarily ground-based data. Models such as the Statistical Asteroid Model (Tedesco *et al.* [34]), are limited by observational constraints to defining properties for asteroids with diameters $>$ 1 km. Ground-based observations

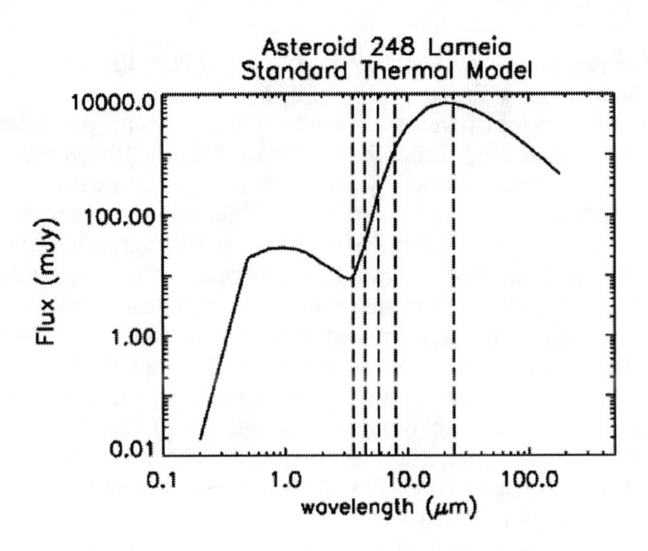

FIGURE 7. A typical main belt asteroid spectral energy distribution. The vertical dashed lines indicate the IRAC 3.6, 4.5, 5.8, and 8.0 μm and the MIPS 24 μm detection bands. The IRAC 8 μm and MIPS 24 μm detectors are well suited to detect MBA thermal emission. For warmer temperatures, the SED is shifted to shorter wavelengths, allowing for detection of NEOs.

are biased towards higher albedo, larger MBAs. Number counts based on traditional surveys are vulnerable to selection effects from intense study of known asteroids, leading to 'clustering' of new discoveries near previously known objects. Mid-infrared measurements with Spitzer have provided, for the first time, a detailed look at the sub-km MBA population. In addition, the Spitzer archive provides a wealth of information about larger, serendipitously observed known asteroids. Spitzer is highly sensitive and can measure thermal emission from previously undetected, small asteroids. As shown in Fig. 7, where a sample spectral energy distribution (SED) for a typical X-type MBA is provided, peak thermal flux is emitted within the Spitzer band passes, making these objects ideal candidates for study in the mid-infrared.

Spitzer asteroid surveys have been carried out using targeted observing programs as well as with archival surveys. Although both types of programs have examined only a small portion of the sky, they clearly demonstrate the feasibility of using IRAC and MIPS to characterize MBAs. The number of objects found, particularly out of the ecliptic plane, where smaller asteroids appear to be prevalent (Xenos *et al.* [35]), is far greater than predicted by empirically-based models such as the Statistical Asteroid Model (Tedesco *et al.* [34]). The First Look Survey Ecliptic Plane Component (Meadows *et al.* [36]) scanned portions of the sky in and near the ecliptic plane and demonstrated Spitzer's sensitivity to MBAs. The survey found double the number of asteroids expected at $\beta = +5°$. Analysis of data in the SWIRE XMM-LSS field (Ryan and Woodward [37]) at $\beta = -17°$ found 39 asteroids in the field and only 11 of these asteroids have ground based optical detections. In addition, Xenos *et al.* [35] have found previously unknown asteroids as far away from the ecliptic plane as $\beta = 26°$.

Deep Spitzer observations over multiple observing epochs are particularly well-suited for examining the scale height behavior and size and albedo distributions of previously unknown, small asteroids. Bhattacharya *et al.* [38] have examined the S-COSMOS field and found over a hundred asteroids in a 0.17 deg^2 field. Other programs that may provide a wealth of information include SWIRE and TAUPROP. Additional data can be found in multi-epoch GO programs and in future Legacy programs. The studies described above have worked with small data sets and relied on a significant level of non-automated identification of asteroids. Studies of the archive as a whole will require automated software, such as that currently being developed by Trilling [39] (PID 30213).

Asteroid observations taken throughout Spitzer's cryogenic and warm mission will also enhance the scientific return of future projects such as Pan-STARRS. This ground-based project will identify up to tens of missions of asteroids, a significant fraction of which will be new. The Spitzer archive will provide linkages for MBAs observed during the cryo mission and for Near Earth Objects (NEOs) observed during the warm mission, allowing for further characterization of a large number of small bodies. Continuation of these efforts through the warm mission and beyond, as well as extending their technique to identifying unknown asteroids, is critical.

The solar elongation constraints inherent in Spitzer observations limits asteroidal motions to a range of approximately 10-40 arcseconds per hour. Thus 2 epochs of data are sufficient to detect motion on the sky by the asteroids in survey data. Known asteroids can be identified using a single epoch, if their orbits are well defined. For surveys with large area coverage it is also possible to link IRAC and MIPS observations of known asteroids obtained with a one-month timescale. Use of only the SCOSMOS, SWIRE and TAUPROP data constitutes 350 Gigabytes of data over 56 square degrees which will yield an estimated asteroid catalog of 1000-2000 objects. This will double the number of asteroids for which there are known albedos and diameters based on IRAS detections and extend the size distribution past the IRAS reliability limit of 10 km.

8. SUMMARY

The programs outlined in the previous section are by no means an exhaustive description of the possible archival programs that will likely be undertaken using the Spitzer cryo and warm mission data. In particular, such a large archive lends itself to serendipitous discoveries of rare and unique objects. The archive and interfaces to it are being constructed to facilitate use by Spitzer experts as well as those new to the mission. The archive programs approved in the first four cryogenic cycles can be found on the SSC Web Pages at http://ssc.spitzer.caltech.edu/approvdprog/.

ACKNOWLEDGMENTS

This work is based (in part) on observations made with the Spitzer Space Telescope, which is operated by the Jet Propulsion Laboratory, California Institute of Technology under contract with NASA. This publication makes use of data products from the Two Micron All Sky Survey, which is a joint project of the University of Massachusetts and the Infrared Processing and Analysis Center/California Institute of Technology, funded by the National Aeronautics and Space Administration and the National Science Foundation.

REFERENCES

1. Deguchi, S., Nakashima, J., Kwok, S., and Koning, N., 2007, ApJ, 664, 1130
2. Song, I. *et al.*, 2005, Nature, 436, 363
3. Kalas, P., Graham, J. R., Beckwith, S. V. W., Jewitt, D. C., and Lloyd, J. P., 2002, ApJ, 567, 999
4. Gaustad, J. and van Buren, D. 1993, PASP, 105, 1127
5. Stauffer, J.R. *et al.*, 2007, ApJS, in press (astroph 0704:1832)
6. Beichman, C., Neugebauer, G., Habing, H., Clegg, P., Chester, T., 1988, IRAS Catalog and Explanatory Supplement
7. Moshir, M., Kopman, G., Conrow, T., 1992, IRAS Faint Source Survey and Explanatory Supplement
8. Rebull, L. *et al.*, 2007, ApJS, 171, 447
9. Román-Zúñiga, C. G., Lada, C. J., Muench, A., and Alves, J. 2007, astro-ph/07043203
10. Jiang, B., Gao, J., Omont, A., Schuller, F., and Simon, G., 2006, A&A, 446, 551
11. Weingartner, J. C., and Draine, B. T. 2001, ApJ, 548, 296
12. Indebetouw, R. *et al.*, 2005, ApJ, 619, 931
13. Robitaille, T. P, Whitney, B. A., Indebetouw, R., and Wood, K. 2007, ApJS, 169, 328
14. Marshall, D. J., Robin, A. C., Reylé, C., Schultheis, M., and Picaud, S. 2006, A&A, 453, 635
15. Simon, R., Jackson, J. M., Rathborne, J. M, and Chambers, E. T., 2006, ApJ, 639, 227
16. Bensch, F., Stutzki, J., Ossenkopf, V., 2001, A&A, 366, 636
17. Falgarone, E., *et al.*, 2005 in Star Formation in the Interstellar Medium, ASPC, 323, 185
18. Burgdorf, M. *et al.*, 2005, AdSpR, 36, 1050
19. Ingalls, J. *et al.*, 2004, ApJS, 154, 281
20. Kasliwal, M. M., Charmandaris, V., Weedman, D., Houck, J. R., Le Floc'h, E., Higdon, S. J. U., Armus, L., and Teplitz, H. I. 2005, ApJL, 634, L1
21. Takagi, T. and Pearson, C. P. 2005, MNRAS, 357, 165
22. Teplitz, H. I., Charmandaris, V., Chary, R., Colbert, J. W., Armus, L., and Weedman, D. 2005, ApJ, 634, 128
23. Frayer, D. *et al.*, 2006, ApJL, 647, L9
24. Gladders, M. *et al.*, 2007, ApJ, 655, 128
25. Stanford, S.A. *et al.*, 2005, ApJL, 634, L129
26. Wilson, G. *et al.*, 2007, To appear in the proceedings of "Infrared Diagnostics of Galaxy Evolution", 2007 (astroph 0604289)
27. Brand, K. *et al.*, 2006, ApJ, 644, 143
28. Lacy, M. *et al.*, 2004, ApJS, 154, 166
29. Sajina, A., Lacy, M., Scott, D., 2005, ApJ, 621, 256
30. Stern, S.A. *et al.*, 2005, ApJ, 631, 163
31. Gilli, R., Comastri, A., Hasinger, G., 2007, A&A, 463, 79
32. Dole, H. *et al.*, 2006, A&A, 451, 417
33. Eyles, W., Bunker, A. J., Ellis, R. S., Lacy, M., Stanway, E. R., Stark, D. P., and Chiu, K., 2007, MNRAS, 374, 910
34. Tedesco, E. T., Cellino, A., and Zappalá, V., 2005, AJ, 129, 2869
35. Xenos, S., Bhattacharya, B., Meadows, V., and Puget, J., 2006, DPS 38, 58.03
36. Meadows, V.S. *et al.*, 2004, ApJS, 154, 469
37. Ryan, E. and Woodward, C., 2006, DPS, 38, 5914
38. Bhattacharya, B. Salvato, M., Aussel, H., Ilbert, O., Sanders, D. B., and Scoville, N., 2006, DPS 38, 59.19
39. Trilling, D., Spitzer 30213 abstract, http://ssc.spitzer.caltech.edu/geninfo/ar/abs-ar3/30213.txt

4. – SOLICITED WHITE PAPERS FROM THE WORKSHOP STEERING COMMITTEE

Observations of Extrasolar Planets During the non-Cryogenic Spitzer Space Telescope Mission

Drake Deming*, Eric Agol†, David Charbonneau**, Nicolas Cowan†, Heather Knutson** and Massimo Marengo‡

**NASA's Goddard Space Flight Center, Planetary Systems Laboratory, Code 693, Greenbelt, MD 20771, USA*
†Department of Astronomy, University of Washington, Box 351580, Seattle, WA 98195-1580, USA
***Department of Astronomy, Harvard University, 60 Garden St., MS-16, Cambridge, MA 02138, USA*
‡Harvard-Smithsonian Center for Astrophysics, 60 Garden St., MS-45, Cambridge, MA 02138, USA

Abstract. Precision infrared photometry from Spitzer has enabled the first direct studies of light from extrasolar planets, via observations at secondary eclipse in transiting systems. Current Spitzer results include the first longitudinal temperature map of an extrasolar planet, and the first spectra of their atmospheres. Spitzer has also measured a temperature and precise radius for the first transiting Neptune-sized exoplanet, and is beginning to make precise transit timing measurements to infer the existence of unseen low mass planets. The lack of stellar limb darkening in the infrared facilitates precise radius and transit timing measurements of transiting planets. Warm Spitzer will be capable of a precise radius measurement for Earth-sized planets transiting nearby M-dwarfs, thereby constraining their bulk composition. It will continue to measure thermal emission at secondary eclipse for transiting hot Jupiters, and be able to distinguish between planets having broad band emission *vs.* absorption spectra. It will also be able to measure the orbital phase variation of thermal emission for close-in planets, even non-transiting planets, and these measurements will be of special interest for planets in eccentric orbits. Warm Spitzer will be a significant complement to Kepler, particularly as regards transit timing in the Kepler field. In addition to studying close-in planets, Warm Spitzer will have significant application in sensitive imaging searches for young planets at relatively large angular separations from their parent stars.

Keywords: Spitzer Space Telescope, infrared astronomical observations, extrasolar planets
PACS: 95.85.Hp, 97.82.Cp, 97.82.Jw, 97.82.Fs

1. INTRODUCTION

The Spitzer Space Telescope (Werner *et al.* [1]) was the first facility to detect photons from known extrasolar planets (Charbonneau *et al.* [2], Deming *et al.* [3]), inaugurating the current era wherein planets orbiting other stars are being studied directly. Cryogenic Spitzer has been a powerful facility for exoplanet characterization, using all three of its instruments. Spitzer studies have produced the first temperature map of an extrasolar planet (Knutson *et al.* [4]), and the first spectra of their atmospheres (Grillmair *et al.* [5], Richardson *et al.* [6]). Spitzer will continue to study exoplanets when its store of cryogen is exhausted. 'Warm Spitzer' (commencing $\sim$ spring 2009) will remain at T $\sim$ 35K (passively cooled by radiation), allowing imaging photometry at 3.6 and 4.5 μm, at full sensitivity. The long observing times that are projected for the warm mission will facilitate several pioneering exoplanet studies not contemplated for the cryogenic

CP943, *The Science Opportunities for the Warm Spitzer Mission Workshop,*
edited by L. J. Storrie-Lombardi and N. A. Silbermann

mission.

2. EXTRASOLAR PLANETS IN 2009

Currently over 200 extrasolar planets are known, including 22 transiting planets (17 orbiting stars brighter than V=13). Some of these have been discovered by the Doppler surveys, but an increasing majority of the transiting systems are being discovered by ground-based photometric surveys. However, the Doppler surveys remain an efficient method to find hot Jupiters, and surveys such as N2K (Fischer *et al.* [7]) continue to be a productive source of both transiting and non-transiting close-in exoplanets. The discovery rate from the photometric surveys is accelerating, because these teams have learned to efficiently identify and cull their transiting candidates, and quickly eliminate false positives. Several transit surveys (HAT, TrES, and XO) recently announced multiple new giant transiting systems (Burke *et al.* [8], O'Donovan *et al.* [9], Johns-Krull *et al.* [10], Mandushev *et al.* [11], Torres *et al.* [12]), and a Neptune-sized planet has been discovered transiting the M-dwarf GJ 436 (Gillon *et al.* [13]). We estimate that the number of bright (V<13) stars hosting transiting giant planets will increase to $\sim$100 in the Warm Spitzer time frame.

The discovery of transits in GJ 436b has stimulated interest in finding more M-dwarf planets, both by Doppler surveys (Butler *et al.* [14]), and using new transit surveys targeted at bright M-dwarfs. It is reasonable to expect that $\sim$10 transiting hot Neptunes will be discovered transiting bright M-dwarf stars by the advent of the warm mission. Moreover, the Doppler surveys are finding planets orbiting evolved stars (Johnson *et al.* [15]). The greater luminosity of evolved stars can potentially super-heat their close-in planets and facilitate follow-up by Warm Spitzer at 3.6 and 4.5 μm.

3. PHOTOMETRY USING WARM SPITZER

Warm Spitzer has a particularly important role in follow up for bright transiting exoplanet systems, as well as non-transiting systems, because in 2009 it will be the largest aperture general-purpose telescope in heliocentric orbit. Heliocentric orbit provides a thermally stable environment, and it allows long periods of observation, not blocked by the Earth. Although Kepler will have a greater aperture than Spitzer, Kepler will be locked-in to a specific field in Cygnus, so it cannot follow-up on the numerous bright transiting systems that will be discovered across the sky.

The thermally stable environment of heliocentric orbit has proven to be a boon for precision photometry from Spitzer. For example, the recent Spitzer 8 μm observations of the HD 189733b transit reported by Knutson *et al.* [4], illustrated in Fig. 1, are among the most precise transit observations ever made. These investigators measured the planet-to-star radius ratio for HD 189733b as 0.1545 ± 0.0002, corresponding to a precision of $\pm$ 90 km in the radius of the giant planet, and they also measured the orbital phase variation of the planet's thermal emission.

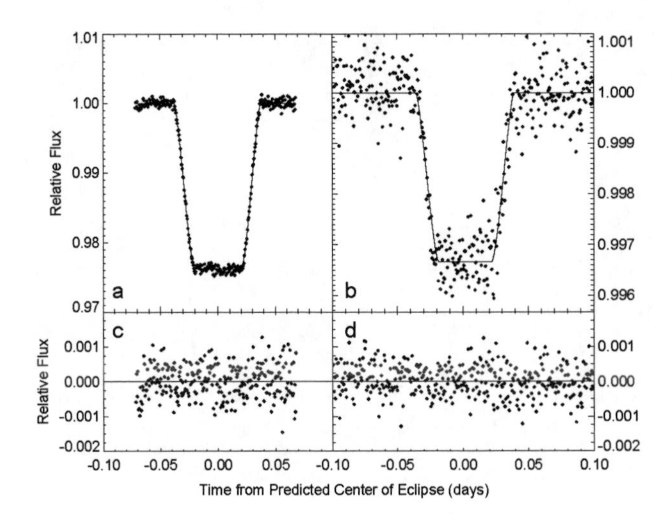

FIGURE 1. Transit (left) and a 60σ secondary eclipse detection (right) of HD 189733b at 8 μm using a continuous 33-hour Spitzer photometry sequence (Knutson *et al.* [4]).

4. MASS-RADIUS RELATIONS

Spitzer's precision for transits derives not only from its stable thermal environment, but also from the lack of stellar limb darkening in the IR. Without limb darkening, the transit becomes extremely 'box-like', with a flat bottom (Richardson *et al.* [16], Knutson *et al.* [4], see Fig. 1). The IR transit depth yields the ratio of planet to stellar area simply and directly, without the added uncertainty of fitting to limb-darkening. Spitzer is now the facility of choice for transiting planet radius measurements. A Warm Spitzer transit program - exploiting the bright stellar flux at 3.6 and 4.5 μm - could significantly improve our knowledge of the mass-radius relationship, and clarify differences in bulk composition, for all but the faintest hot Jupiter systems. Figure 2 shows the mass-radius relation for several of the transiting giant planets (Charbonneau *et al.* [17]). The mass-radius relation encodes fundamental information on the global structure of these planets. For example, HD 149026b is inferred to have a heavy element core of at least 70 Earth masses, based on the small radius for its mass (Fig. 2, and Sato *et al.* [18]). This information is crucial to our understanding of planet formation, e.g., by the core accretion and gravitational instability mechanisms (Lissauer & Stevenson [19]). The scientific utility of these measurements will be maximized if all transiting exoplanet radii are measured to high precision, in a mutually consistent manner. Moreover, as the Doppler and transit surveys discover Neptune to Earth-sized planets orbiting M-dwarfs, the highest precision photometry will be needed to measure their radii to a precision sufficient to constrain their interior structure.

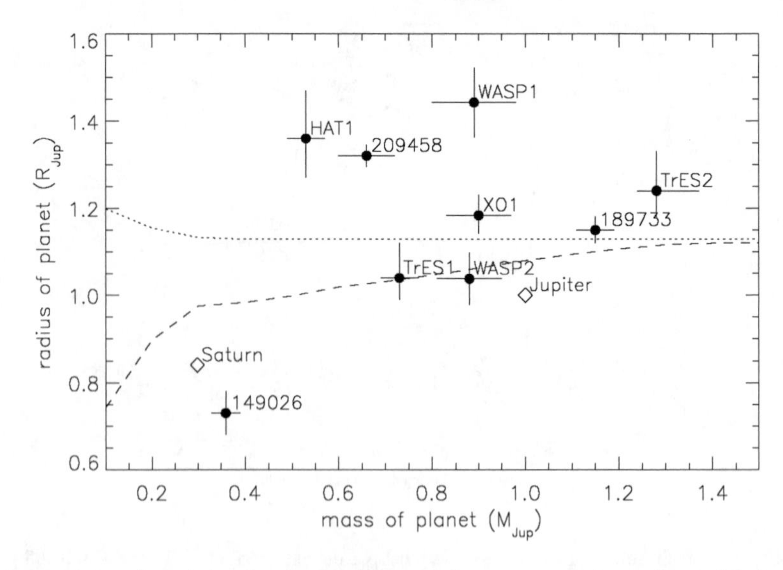

FIGURE 2. Mass-radius relation for giant transiting exoplanets, compared to Jupiter and Saturn (Charbonneau *et al.* [17]). The lines show the theoretical relations (Bodenheimer *et al.* [20] for planets having no core (dotted) and a 20 Earth-mass solid core (dashed).

4.1. Spitzer vs. Ground-Based Photometry

Ground-based photometry in the z-band is achieving sub-milli-magnitude levels of precision in many cases (Winn *et al.* [21]), and can determine the radii of some transiting giant planets to error limits imposed by astrophysical uncertainty in the stellar mass. The most favorable systems for ground-based observation are those occurring in fields with numerous nearby reference stars of comparable brightness. Planets transiting bright, spatially isolated, stars are not as favorable for ground observation. Moreover, as the radius of the transiting planet decreases, greater photometric precision is needed to reach the limits imposed by uncertainty in the stellar mass. Nearby M-dwarfs have flux peaks longward of the visible and z-band spectral regions, and they often lack nearby comparison stars of comparable infrared brightness. Neptune- to Earth-sized planets orbiting nearby M-dwarfs will therefore require infrared space-borne photometry for the best possible radius precision. Figure 3 illustrates a single transit of a 1-Earth radius planet across an M-dwarf, observed by Spitzer at 8 μm. We simulated this case by re-scaling a real case: Spitzer's recent photometry of GJ 436b (Deming *et al.* [22], Gillon *et al.* [23]). Spitzer's nearly photon-limited precision detects this Earth-sized planet to 7σ significance in a single transit.

Although Fig. 3 is based on Spitzer observations at 8 μm, the photon-limit for observations during the warm mission (e.g., at 4.5 μm) will be even more favorable, simply because stars are brighter at the shorter wavelength. Stellar photometry at the wavelengths used by the warm mission is affected by a pixel phase effect in the IRAC instru-

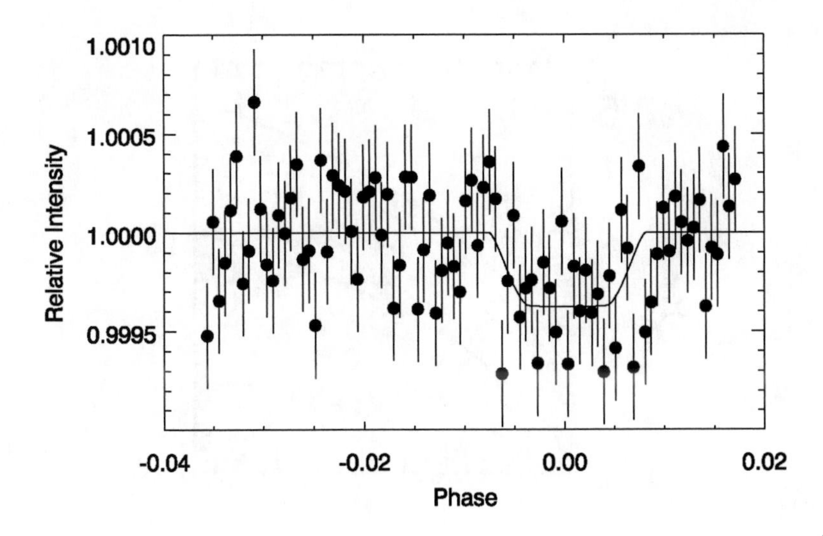

FIGURE 3. A transit of a 1 Earth radius planet across an M-dwarf, simulated by rescaling Spitzer 8 μm observations of GJ 436b (Deming *et al.* [22])

ment (Reach *et al.* [24]), but that can be successfully corrected by decorrelation (Charbonneau *et al.* [2]), and recent results have demonstrated secondary eclipse detections with a precision of $\sim 10^{-4}$ (Charbonneau *et al.* [2], Knutson *et al.* [25]). The pixel phase effect should be correctable to even greater precison using the large data sets contemplated for the warm mission.

5. NEW TYPES OF TRANSITING PLANETS

Ongoing Doppler and transit monitoring of known hot Jupiters can detect subtle deviations from Keplerian orbits (Charbonneau *et al.* [26]), indicating the presence of additional planets, e.g., 'warm Jupiters' in longer period orbits, or terrestrial mass planets in low order mean motion resonances. The likely co-alignment of orbital planes increases the chance those planets will also transit, and intensive radial velocity monitoring could constrain the transit time for giant planets. Warm Spitzer will be a sensitive facility for confirming those transits, and extending the mass-radius relation (Fig. 2) to planets in more distant orbits, and even to close-in terrestrial planets. Even lacking specific indications from Doppler measurements, searches for close-in terrestrial planets in low order mean motion resonances with known giant transiting planets (Thommes [27]) are warranted using Warm Spitzer. These searches could be combined with radius and transit timing measurements for the giant planets, in the same observing program.

For stars not known to host a hot Jupiter, ongoing Doppler surveys and space-borne transit surveys (e.g., COROT) will find more transiting planets, extending to Neptune

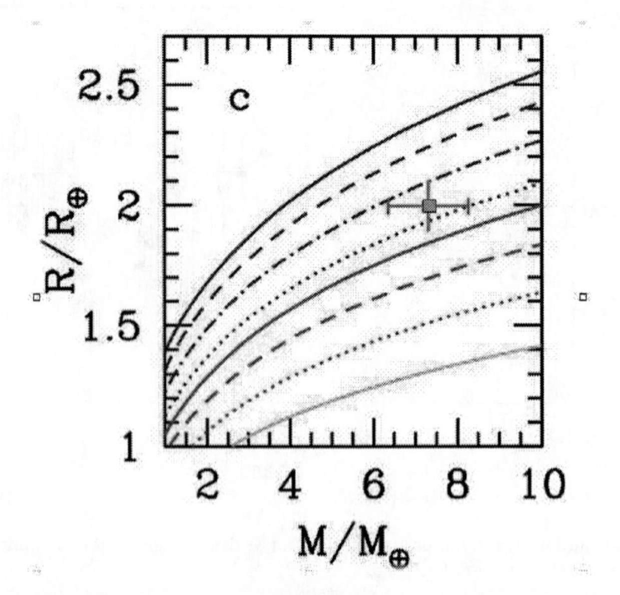

FIGURE 4. Mass-radius relations for solid exoplanets of various compositions, from Seager *et al.* [28]. Blue represents water ice planets, red are silicate planets, and green is a pure iron planet. The magenta point is a hypothetical observation of a hot Earth transiting an M-dwarf at 50 pc, observed by Warm Spitzer at 4.5 μm. The horizontal error bar is the mass error from Gliese 876d (Rivera *et al.* [29]); the vertical (radius) error bar is calculated for Warm Spitzer, not including error in the stellar radius.

mass and below. Spitzer transit measurements can precisely determine the radii of small planets. Exoplanet radius measurements and transit searches are particularly appropriate for Warm Spitzer, because: a) stars are bright in the 3.6 and 4.5 μm bands, while limb darkening is still absent, b) stellar activity is muted at IR wavelengths, and c) longer observing times are congruent with the goal of simplified operations in the extended mission. Figure 4 shows a potential example of a precise radius (allowing precise density) determination for a hot super-Earth, compared to the mass-radius relation for solid exoplanets of various composition (Seager *et al.* [28]). In this case, the Spitzer radius is sufficiently precise ($\pm$ 0.1 Earth radii) to constrain the bulk composition of this solid exoplanet by comparison to the Seager *et al.* [28] models.

6. THERMAL EMISSION AT 3.6 AND 4.5 μm

6.1. Absorption vs. Emission Spectra

The secondary eclipse of a transiting planet has the greatest depth at Spitzer's longest wavelengths. However, the eclipses are quite detectable at Spitzer's shortest wavelengths, because these are close to the peak of the Planck function at the temperatures of transiting planets. Secondary eclipse photometry using Warm Spitzer can therefore continue to define the brightness of exoplanets at 3.6 and 4.5 μm for new bright transiting systems. This will be especially valuable if complemented by ground-based detections exactly at the expected spectral peaks at 2.2 and 3.8 μm, which is believed to be feasible (Snellen and Covino [30], Deming *et al.* [31]). The predicted spectrum of a hot Jupiter exoplanet is illustrated in Fig. 5, from Charbonneau *et al.* [2]. Normally, the broad band spectrum is expected to be shaped by water vapor absorption. However, recent Spitzer photometry of HD 209458b (Knutson *et al.* [25]) indicates much better agreement with a spectrum wherein the water bands appear in *emission*, and can only be explained by the presence of a thermal inversion at high altitude (Burrows *et al.* [32]). The nature of the high altitude absorber needed to create this inversion is unknown, and may be connected to the specific properties of the system, such as the planet's level of irradiation or surface gravity. A comprehensive survey of the bright transiting systems would make it possible to search for correlations between the presence of a temperature inversion and other properties of the systems, thus providing insight into the origin of the inversions. Such a survey would also identify the best systems for spectroscopic follow up by JWST.

The signature of such an inversion is easily observed in the 3.6 and 4.5 μm Spitzer bandpasses. Standard models for atmospheres without temperature inversions predict that the planets will appear brighter at 3.6 μm than at 4.5 μm (see Fig. 5) due to the presence of water absorption bands at wavelengths longer than 4.5 μm. With a thermal inversion the relative brightnesses in the two bands are reversed, as the 3.6 μm flux is suppressed and the 4.5 μm flux is correspondingly enhanced by the presence of water emission at the longer wavelengths.

6.2. Temporal Variations Due to Dynamics

The cycle of variable stellar heating caused by planetary rotation can drive a lively dynamics in hot Jupiter atmospheres (Cooper & Showman [33]). In turn, the atmospheric dynamics will produce a spatially and temporally varing temperature field, and the temperature fluctuations are expected to be of large spatial scale for close-in planets (Rauscher *et al.* [34]). For the temperatures found in hot Jupiter atmospheres ($T \sim 1200K$), the Planck function at Warm Spitzer wavelengths varies strongly with temperature. Consequently, temporal variations in thermal emission from close-in planets should be readily observable during the warm mission. The secondary eclipse depth of a given transiting system can vary, and observations of multiple eclipses could yield key insights into the atmospheric physics. Observations extended over a full orbit - even for non-transiting systems - can potentially reveal variations in thermal emission correlated with orbit phase (Cowan, Agol and Charbonneau [35]). Orbital phase variations can reflect the changing viewing geometry, but can also be caused by strong atmospheric dynamics in response to the variable stellar forcing that is characteristic of eccentric orbits.

Extrasolar planets have more eccentric orbits on average than do the planets of our own solar system. In some cases, their eccentricity extends to strikingly high values. For example, HD 80606b has an eccentricity of 0.93 (Naef *et al.* [36]). During its close periastron passage, it receives a stellar flux more than 1000 times greater than

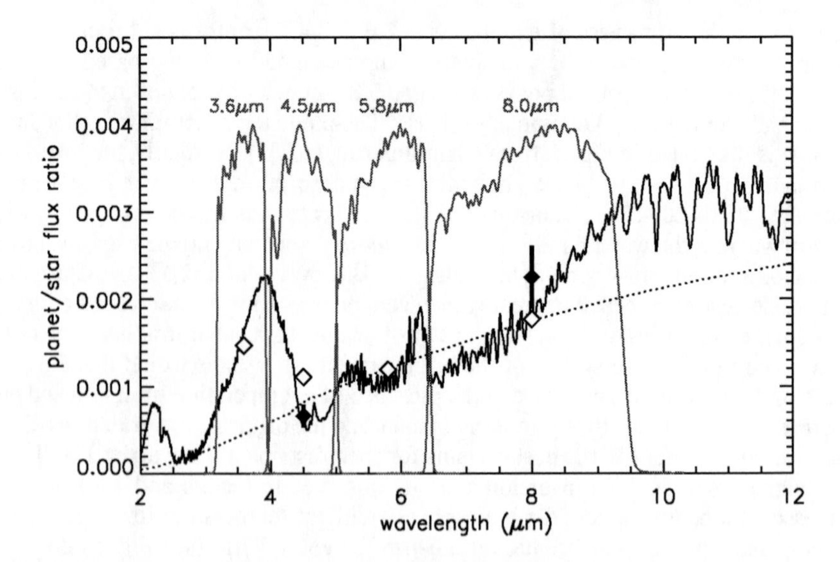

FIGURE 5. Predicted specturm of a hot Jupiter exoplanet, shown in comparison to observations of TrES-1 by Charbonneau *et al.* [2] (solid diamonds). The expected contrast (planet divided by star) for the IRAC bands is shown by open diamonds. Note the higher contrast expected at 3.6 μm compared to 4.5 μm, indicative of a water absorption spectrum. These contrast values will be reversed for a water emission spectrum. The dotted line is a blackbody spectrum.

the flux received by Earth from the Sun. This strong flux will cause a rapid heating of the planet's atmosphere, and its time dependence encodes crucial information on the radiative time scale, and thus the composition, of the planet's atmosphere (Langton and Laughlin [37]). This rapid heating of HD 80606b and similar systems may be observable at 3.6 and 4.5 μm. In this regard it is interesting to note that the exoplanet HD 185269b orbits a sub-giant star, having greater than solar luminosity, in a close orbit (6.8 day period), with an eccentricity of 0.3 (Johnson *et al.* [38]). The resultant strong variation in stellar heating over the orbit will force a corresponding variation in the planet's thermal emission, that should correlate with orbital phase. Recently, the transiting planets XO-3 and HAT-P-3 have also been found to have a significant eccentricity (Johns-Krull *et al.* [10], Torres *et al.* [12]), opening the possibility to also measure the spatial distribution of time dependent heating on the planet's disk (Williams *et al.* [39], Knutson *et al.* [4]). Given the high precision possible from Spitzer, it may be possible to observe all of these effects at 4.5 μm using Warm Spitzer.

7. TRANSIT TIMING

There is considerable recent interest in the indirect detection of extrasolar terrestrial planets via their perturbations to the transit times of giant transiting planets (Agol *et al.* [40], Holman and Murray [41], Steffen and Agol [42], Agol and Steffen [43]). Spitzer

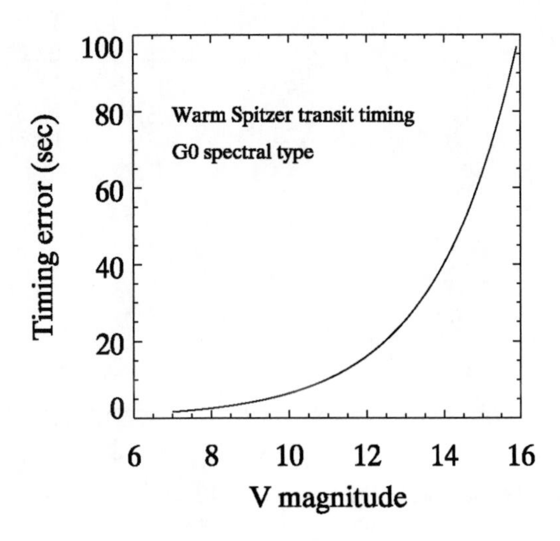

FIGURE 6. Calculation of the transit timing error (1σ) for Warm Spitzer observing giant planet transits at 4.5 μm, as a function of stellar brightness.

transit photometry during the warm mission is an excellent way to make precise transit timing measurements. The lack of IR limb darkening is again a significant advantage, because it results in very steep ingress and egress curves, producing excellent timing precision. Knutson *et al.* [4] found a timing precision for the HD 189733b transit of 6 seconds. Hubble transit timing errors range from 10 to 50 seconds (Agol and Steffen [43]), although a 3 second timing precision was recently achieved for HD 189733b by Pont *et al.* [44]. However, Spitzer has the advantage that the lower contrast of star spots and plage in the IR as compared to visible wavelengths minimizes systematic errors due to stellar activity noise. Also, Spitzer's heliocentric orbit permits continuous measurements before, during, and after transit - unlike Hubble where blocking by Earth interrupts transits. Moreover, Spitzer transit timing precision should be even better at the shorter wavelengths available for the warm mission, because stars are brighter at shorter wavelengths, and limb darkening remains negligible.

The continued success of the ground-based transit surveys, the advent of COROT, and the upcoming launch of Kepler, will provide a wealth of targets for Warm Spitzer follow-up. We have calculated the transit timing precision by Warm Spitzer at 3.6 and 4.5 μm, for solar-type stars at different distances (Fig. 6). This calculation is consistent with the Knutson *et al.* [4] result, and it projects a Spitzer timing precision of better than 40 seconds down to the faint end of Kepler's range at V=14. This is sufficient to detect perturbations by terrestrial planets well below one Earth mass in resonant orbits (Agol et al. 2005), or to $\sim 10\,(150\ \text{days}/P_1)$ Earth masses for $P_2/P_1 < 4$, where $P_{1,2}$ are the periods of the transiting and perturbing planets (Holman and Murray [41], Agol *et al.* [40]). The Warm Spitzer mission will begin at about the same time that Kepler begins to discover

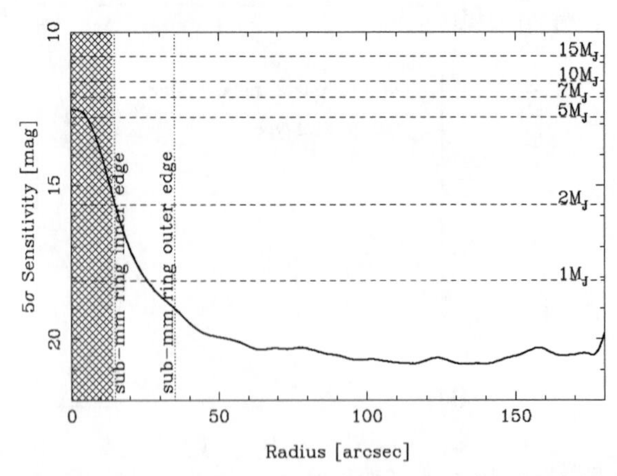

FIGURE 7. Residual noise, after PSF subtraction, of 4.5 μm ε Eridani radial profile (Marengo *et al.* [51]). Dashed lines are the predicted fluxes (from Burrows, Sudarsky, and Lunine [49]) of 1 Gyr old planets with mass from 1 to 15 M_J. Dotted lines enclose the region where the sub-millimeter debris disk around this star is located.

multiple new giant transiting planets (February 2009 launch). The observing cadence of the Kepler mission is not optimized for transit timing (Basri *et al.* [45]), so transit timing observations in the Kepler field by Warm Spitzer could leverage and enhance Kepler's science return.

8. DIRECT IMAGING

Although the bulk of Spitzer's results for exoplanets have relied on time series photometry and spectroscopy, Spitzer's high sensitivity in imaging mode is also important for exoplanet imaging studies. Radial velocity surveys have reached the precision required to detect planets only in the last 10 years, so little is known about the frequency of exoplanets and other low mass companions at distances greater than $\sim$ 5 - 10 AU. This is unfortunate because determining the presence of planetary mass bodies in the periphery of known exoplanetary systems has important implications for their evolution. These include studying the dynamical "heating" of the orbits in the system, which may result in higher eccentricities (or even expulsion) of some components, or in enhanced collision rates between the bodies in extrasolar Kuiper belts, which may be responsible for the formation of transient debris disks.

Imaging can in principle fill this observational gap. Three objects of near planetary mass have already been detected within 300 AU from the primary around the brown dwarf 2M1207 (Chauvin *et al.* [46]) and the stars GQ Lup (Neuhauser *et al.* [47]) and AB Pic (Chauvin *et al.* [48]) with ground based adaptive optics observations. Other optical and near-IR searches from the ground and from space, have so far produced negative results.

Warm Spitzer may provide a significant contribution in this arena, as the two surviving IRAC bands are particularly suited for the detection of cool extrasolar planets and brown dwarf companions (T and the so-called Y dwarfs). A gap in molecular opacities of giant planets near 4.5 μm allows emission from deep, warmer atmospheric layers to escape: giant planets are very bright at this wavelength (Burrows, Sudarsky, and Lunine [49]). A strong methane absorption band strongly depresses the planetary flux at 3.6 μm: as a result, the IRAC [3.6]-[4.5] color of planetary mass bodies is expected to be unique, and allow for their identification among background objects in the field. Model atmosphere calculations predict that a 1 Gyr old, 2 M_J planet around a star 10 pc from the Sun will have a 4.5 μm magnitude of $\sim$ 18. Such planets are detectable today with IRAC provided that the diffracted light from the central bright star can be removed at the planet's image location. The stability of Spitzer's optical and pointing systems assures that the stellar Point Spread Function (PSF) is highly reproducible, allowing much fainter nearby sources to be identified in the PSF wings using differential measurements.

This search has already been carried out for the debris disk star ε Eridani, which is also home of a Jovian class radial velocity planet orbiting the star at 3.4 AU (Hatzes *et al.* [50]). The search has set stringent limits for the mass of external planetary bodies in the system (including the area occupied by the debris disk, Marengo *et al.* [51]), and demonstrated that this technique is sensitive to the detection of planets with mass as low as 1 M_J (Fig. 7) outside the 14 arcsec radius (50 AU) where the IRAC frames are saturated. The search radius can be reduced to $\sim$ 5 arcsec or less by using shorter frame times available in IRAC subarray mode. A pilot search of 16 nearby stellar systems is being conducted in Spitzer cycles 3 and 4. These programs will identify possible candidates based on their [3.6]-[4.5] colors, which will need to be verified by second epoch observations during the warm mission, to detect their common proper motion with the primary.

The Spitzer warm mission will provide the opportunity to extend the search of planetary mass companions through imaging techniques to a large number of systems in the solar neighborhood, probing a search radius from $\sim$ 10 to 10,000 AU around stars within 30 pc from the Sun. This search will be sensitive to masses as low as a few Jupiter masses, depending on the age and distance of the systems. These observations will be complementary to ground based radial velocity and imaging searches with adaptive optics systems, given the larger field of view and higher sensitivity of IRAC, in a wavelength range where the required contrast ratio (as low as 10^{-5} of the parent star flux) is more accessible than in the optical and near-IR.

ACKNOWLEDGMENTS

We thank the Spitzer Science Center for the opportunity to consider and discuss the potential for exoplanet science during the warm mission. We are grateful to Josh Winn and Andy Gould for helpful conversations and remarks regarding the relative merits of of ground-based vs. space-borne photometry. We also acknowledge informative conversations with Greg Laughlin on the effects of heating in eccentric orbits.

REFERENCES

1. Werner, M. W., *et al.*, 2004, ApJS, 154, 1.
2. Charbonneau, D., *et al.*, 2005, ApJ, 626, 523.
3. Deming, D. *et al.*, 2005, Nature, 434, 740.
4. Knutson, H., *et al.*, 2007a, Nature, 447, 183.
5. Grillmair, C. *et al.*, 2007, ApJ 658, L115.
6. Richardson, L. J. *et al.*, 2007, Nature 445, 892.
7. Fischer, D., *et al.*, 2005, ApJ 620, 481.
8. Burke, C., *et al.*, 2007, submitted to ApJ, (astro-ph/0705.0003).
9. O'Donovan, F. T., *et al.*, 2007, ApJ 663, L37.
10. Johns-Krull, C., *et al.*, 2007, BAAS 39, 096.05.
11. Mandushev, G., *et al.*, 2007, ApJ, in press, (astro-ph/0708.0834).
12. Torres, G., *et al.*, 2007, ApJ, 666, L121.
13. Gillon, M., *et al.*, 2007a, A&A, 472, L13.
14. Butler, R. P., *et al.*, 2004, ApJ 617, 580.
15. Johnson, J. A., *et al.*, 2007, ApJ, 655, 785
16. Richardson, L. J. *et al.*, 2006, ApJ 649, 1043.
17. Charbonneau, D., *et al.*, 2006, ApJ 636, 445.
18. Sato, B., *et al.*, 2005, ApJ 633, 465.
19. Lissauer, J. J. and Stevenson, D. J., 2007, in Protostars and Planets V, (editors D. Jewitt and B. Reipurth), 591.
20. Bodenheimer, P., Laughlin, G., and Lin, D. N. C. 2003, ApJ 592, 555.
21. Winn, J. N., *et al.*, 2007, AJ 133, 1828.
22. Deming, D., *et al.*, 2007a, ApJL, in press, (astro-ph/0707.2778).
23. Gillon, M., *et al.*, 2007b, A&A 471, L51.
24. Reach, W. T., *et al.*, 2005, PASP 117, 978.
25. Knutson, H., *et al.*, ApJ, submitted.
26. Charbonneau, D., *et al.*, 2007, in Protostars and Planets V, (editors D. Jewitt and B. Reipurth), 701.
27. Thommes, E. W., 2005, ApJ 626, 1033.
28. Seager, S., *et al.*, 2007, ApJ, in press (astro-ph/0707.2895).
29. Rivera, E. J., *et al.*, 2005, ApJ 634, 625.
30. Snellen, I. A. G., and Covino, E., 2006, MNRAS 375, 307.
31. Deming, D. *et al.*, 2007, MNRAS, 378, 148.
32. Burrows, A. *et al.*, ApJL, in press.
33. Cooper, C. S., & Showman, A. P. 2005, ApJ, 629, L45.
34. Rauscher, E., *et al.*, 2007, ApJ, 662, L115.
35. Cowan, N. B., Agol, E., and Charbonneau, D., 2007, MNRAS, 379, 641.
36. Naef, D., *et al.*, 2001, A&A 375, L27.
37. Langton, J., and Laughlin, G., 2007, ApJ 657, L113.
38. Johnson, J. A., *et al.*, 2006, ApJ 652, 1724.
39. Williams, P. K. G., *et al.*, 2006, ApJ 649, 1020.
40. Agol, E., *et al.*, 2005, MNRAS 359, 567.
41. Holman, M. J., and Murray, N. M., 2005, Science 307, 1288.
42. Steffen, J., and Agol, E., 2005, MNRAS 364, L96.
43. Agol, E., and Steffen, J., 2007, MNRAS 374, 941.
44. Pont, F., *et al.*, 2007, A&A, submitted (astro-ph/0707.1940).
45. Basri, G., Borucki, W. J., and Koch, D. 2005, New Astron. Rev. 49, 478.
46. Chauvin, G. *et al.*, 2004, A&A 425, L29.
47. Neuhauser, R. *et al.*, 2005, A&A 345, L13.
48. Chauvin, G. *et al.*, 2005, A&A 438, L29.
49. Burrows, A., Sudarsky, D., and Lunine, J. L. 2003, ApJ 596, 587.
50. Hatzes, A. P. *et al.*, 2000, ApJ 544, L145.
51. Marengo, M. *et al.*, 2006, ApJ 647, 1437.

The Warm Spitzer Mission: Opportunities to Study Galactic Structure and the Interstellar Medium

R. A. Benjamin*, B. T. Draine†, R. Indebetouw**, C. J. Lada‡, S. R. Majewski**, I. N. Reid§ and M. F. Skrutskie**

*Dept. of Physics, University of Wisconsin-Whitewater, Whitewater, WI 53190, USA
†Princeton University Observatory, Peyton Hall, Princeton, NJ 08544, USA
**Department of Astronomy, University of Virginia, Charlottesville, VA 22903, USA
‡Harvard-Smithsonian Center for Astrophysics, 60 Garden Street, Cambridge, MA 02138, USA
§Space Telescope Science Institute, Baltimore, MD 21218, USA

Abstract. The Spitzer Space Telescope Warm Mission capability, specifically broadband IRAC imaging at 3.6 and 4.5μm, provides a unique opportunity to probe the content and structure of the Milky Way's disk, particularly along lines of sight that suffer significant interstellar obscuration. This white paper examines the factors that favor and constrain the application of a warm Spitzer to Galactic structure and interstellar medium science. Although the paper briefly discusses some specific survey choices, its primary purpose is to highlight the value of an extended Spitzer mission and outline the factors that might influence the construction of a large proposal focused on Galactic science in the future.

Keywords: Spitzer Space Telescope, infrared astronomical observations, ISM, stars, Galaxy:structure, Galaxy:stellar content
PACS: 95.85.Hp,98.20.-d,98.35.-a,98.38.-j,98.38.Jw

1. INTRODUCTION

There is only one galaxy in the sky for which we can truly obtain a three dimensional view of its internal structure: the Milky Way. Our view from within the Galaxy removes many of the factors that limit the study of distant galaxies. With the high spatial resolution resulting from proximity one can resolve different stellar populations and spatial components of the Milky Way that appear only in the integrated light of other galaxies. Historically, the two principal difficulties in making progress in Galactic stellar structure have been the large area of the sky that must be covered uniformly and the presence of significant interstellar obscuration. The Spitzer/IRAC Warm Mission offers a unique solution to these issues.

Making the case, however, for extending Spitzer's Galactic plane coverage to the entire Galactic disk with warm mission observations is more difficult than it might seem. This difficultly does not arise from the shortcomings of Spitzer's warm capabilities, but instead arises from considering the *unique* contributions possible with Spitzer in the context of the wealth of existing, and soon to be available, space and ground-based survey data. For example:

- Visible and near-infrared wavelengths – SDSS and 2MASS have already provided

CP943, *The Science Opportunities for the Warm Spitzer Mission Workshop,*
edited by L. J. Storrie-Lombardi and N. A. Silbermann

detailed star count and population statistics. 2MASS, in particular, delivers diagnostics deep in the Galactic plane. Unlike Spitzer IRAC Bands 1 and 2, visual and near-infrared broadband colors are sensitive to stellar spectral and luminosity class, so these surveys support both star counts studies and true 3-dimensional reconstruction using selected populations as crude standard candles. 2MASS, for example, can detect and classify red clump stars at distances of $\sim$10kpc. On the other hand, Spitzer's IRAC Bands 1 and 2 fall in the Rayleigh-Jeans portion of most stellar spectral energy distributions, thus most stars have the indistinguishable IRAC colors and population analysis is not possible. Spitzer, nevertheless has a unique advantage over these "short" wavelength surveys in its ability to penetrate interstellar extinction. Combined with the available near-infrared data, Spitzer's warm capabilities provide a means of establishing the extinction to individual stars, and thus can be leveraged to inform a near-infrared view of the Milky Way largely free of Galactic extinction.

- Mid-infrared wavelengths – The Wide Field Infrared Survey Explorer (WISE) will launch in late 2009 and provide full-sky coverage at 3.3, 4.7, 12, and 23 μm during a 7-month primary mission. The two short-wavelength WISE bands effectively duplicate the available IRAC warm mission spectral bandpasses with 120 and 160 μJy (5σ) sensitivity in unconfused regions - essentially the same sensitivity as for GLIMPSE-style coverage. WISE, however, will have a 6 arcsecond FWHM PSF at 3.3 and 4.7 μm compared with 2 arcseconds for IRAC bands 1 and 2, giving Spitzer a substantial advantage in source confused regions.

In the context of these existing and future capabilities, the Spitzer warm mission still provides unique scientific opportunity to extend our knowledge of the structure of the Milky Way. Two significant examples are:

- Although WISE will survey the entire sky, Spitzer enjoys greater spatial resolution and can surpass WISE's sensitivity in modest integration time. Source confusion then defines the regions in which Spitzer will be most effective and, conversely, those in which WISE coverage alone may be sufficient for illuminating Galactic structure.
- Many regions of the Galactic plane suffer substantial extinction even at near-infrared wavelengths. At low Galactic latitude some of the most interesting features of the Milky Way's disk (e.g. bars, interacting dwarf satellites, and spiral arms) lie hidden behind the dust and are inaccessible or poorly defined in near-infrared surveys. With its ability to penetrate extinction, and to obtain higher sensitivity at higher spatial resolution than WISE, Spitzer has a unique niche to ally with near-infrared surveys such as 2MASS and UKIDSS to lift the veil of extinction in regions where extinction impairs the near-infrared surveys.

2. COMPARISONS OF PAST, PRESENT, AND FUTURE SURVEYS

Figure 1 shows the point source sensitivities and wavebands covered by several generations of infrared surveys. Since the ground-breaking IRAS survey, infrared surveys have delivered ever increasing sensitivity and improved angular resolution. Included in

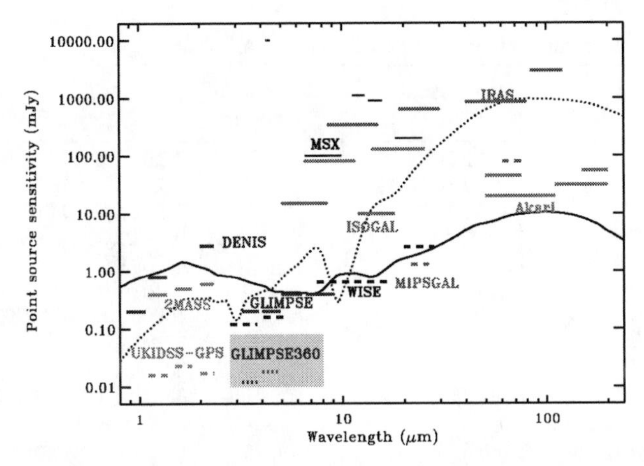

FIGURE 1. A comparison of GLIMPSE and GLIMPSE360 sensitivity limit to the sensitivity of other ground and space-based infrared surveys. This comparison shows the good match in sensitivity between GLIMPSE and 2MASS, and GLIMPSE360 and UKIDSS. The curves show model spectra of Whitney et al. [1] for a 1 L_o T Tauri star at a distance of 0.7 kpc (solid) and a deeply embedded 1 L_o protostar at a distance of 0.6 kpc(dotted).

this plot are the point source sensitivity of GLIMPSE, a shallow, but large area survey which consisted of (at least) two 2 second integrations over the galactic longitude range $|l| < 65°$ and $|b| < 1°$, with some selected vertical extensions at selected longitudes (See Figure 2). The depth of the four-band GLIMPSE survey is well matched to the three-band near-infrared sensitivity of 2MASS. This combined seven-band data set is especially powerful for analyzing Galactic stellar structure and identifying different classes of sources, such as carbon stars, AGB stars, red clump giants, Wolf-Rayet stars, etc.

Figure 1 also shows the point source sensitivity of a potential Warm Spitzer program, hereafter GLIMPSE360. This survey would be deeper than GLIMPSE, consisting of 12 second dithered HDR mode observations with two epochs of coverage, similar to the strategy used for the SAGE legacy project. This style of coverage takes approximately twice as long as GLIMPSE,[1] but goes significantly deeper. Like the GLIMPSE/2MASS combination, the combination of GLIMPSE360 data with data from the Northern hemisphere UKIDSS-GPS (UKIRT Infrared Deep Sky Survey-Galactic Plane Survey), will provide a five band combination to fainter than 17.5 mag in all bands.[2] It should also be noted that most of the UKIDSS-GPS survey region (see Table 1) has yet to be comprehensively covered by Spitzer in any wavelength range.

---

[1] The GLIMPSE style observing strategy covers 0.55 square degrees/hour, while GLIMPSE360 style averages 0.227 square degrees/hour.

[2] The GLIMPSE360 5σ limits would be 18.4 and 17.5 magnitude in the 3.6 and 4.5 μm IRAC bands. The limiting magnitudes for UKIDSS-GPS are 20, 19.1, and 19.0 for J, H, and K bands respectively. See `www.ukidss.org` for more information.

TABLE 1. Summary of Infrared Galactic Plane and All-Sky Surveys

	Wavebands (μm)	Resolution (arcsec)	Sensitivity*
DENIS	0.97,1.22,2.16	1-3	0.2,0.8,2.8 mJy
2MASS	1.22,1.65,2.16	2	0.4,0.5,0.6 mJy
UKIDSS-GPS	1.2, 1.6, 2.2	0.5	0.016,0.023, 0.017 mJy
GLIMPSE	3.6,4.5,5.8,8.0	≤ 2	0.2,0.2,0.4,0.4 mJy
GLIMPSE360	3.6,4.5	≤ 2	0.012, 0.018 mJy
WISE	3.3, 4.7, 12, 24	6-10	0.12, 0.16, 0.65, 2.6 mJy
MSX	4.1,8.3,12,14,21	18.3	10000,100,1100,900,200 mJy
ISOGAL†	7,15	6	15,10 mJy
Akari**	9.1,18.3,62.5,80,155,175	5–44	20-100 mJy
IRAS	12,24,60,100	25–100	350,650,850 3000 mJy
Hershel/Hi-Gal‡	60-600	4 – 40	~100 mJy
COBE/DIRBE§	1.25–240	0.7°	0.01-1.0 MJy sr^{-1}

* 5σ point source sensitivity
† Survey covered non-continuous regions of Galactic plane
** Formerly Astro-F
‡ Galactic plane survey still being formulated
§ DIRBE photometric bands are 1.25, 2.2, 3.5, 4.9, 12, 25, 60, 100, 140, and 240 μm. We report the diffuse flux sensitivity rather than point source sensitivity due to the large beam size.

Also important to a Warm Spitzer Galactic plane survey is the coming deluge of complementary longer wavelength data from MIPSGAL, WISE, Akari, and Herschel (for which a galactic survey, Hi-Gal, is currently being developed; Molinari et al. [2]). While shorter wavelengths characterize stars and embedded sources, these longer wavelength bands are vital for the detection of circumstellar or interstellar dust. For YSOs, for example, data at all near- and mid-infrared bandpasses are necessary to provide meaningful constraints on the viewing geometry, stellar mass, accretion rate and disk geometries of these objects (Robitaille et al. [3]).

Angular resolution, and the related issue of source confusion, further distinguish existing and future surveys. Table 2 shows that the latitude-averaged confusion limit for GLIMPSE is approximately 0.5 to 1.0 magnitudes brighter than the sensitivity limit even for this shallow survey. Toward the Galactic center confusion abates rapidly with increasing Galactic latitude, and slowly (13.3 to 13.6 mag) with increasing longitude. Confusion estimates based on source counts from 2MASS, UKIDSS and previous IRAC observations will help inform the design of any large-area warm Spitzer survey.

For reference, if 6000 hours of warm mission time were devoted to a large-area Galactic plane survey, that allocation could enable:

- a GLIMPSE-style (two 2-second exposures for each sky location) survey of 3300 square degrees. Spread uniformly in longitude (resurveying the 388 square degrees already covered by GLIMPSE, GLIMPSE2, and GLIMPSE 3D), this survey would cover a latitude range of $|b| < 4.6°$.
- a GLIMPSE360-style (two 12-second exposures for each sky location) survey of 1360 square degrees, equivalent to a full-plane survey with $|b| < 1.9°$.

TABLE 2. GLIMPSE Catalog Information

IRAC band (μm)	S_0* (Jy)	A_λ/A_K†	m_{sens} (mag)	m_{sat} (mag)	m_{conf}** (mag)
3.55	277.5	0.56 ± 0.06	14.2	7.0	13.3–13.6
4.49	179.5	0.43 ± 0.08	14.1	6.5	13.3–13.6
5.66	116.5	0.43 ± 0.10	11.9	4.0	11.7–12.3
7.84	63.13	0.43 ± 0.10	9.5	4.0	11.0–12.4

* Vega isophotal wavelengths and IRAC zero magnitude flux from Reach et al. (2005)

† Extinction from Indebetouw et al. [4]

** Confusion limit varies over the longitude range $|l| = 10°$ to $|l| = 65°$.

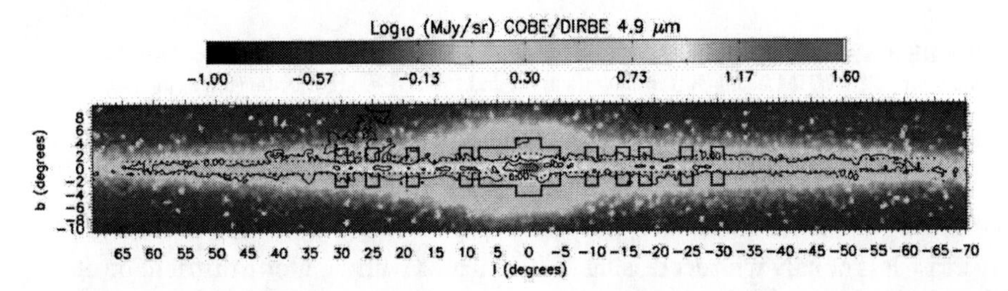

FIGURE 2. COBE/DIRBE 4.9 micron map of the Galactic plane. Irregular contours show regions of $A_K > 1$ ($A_V \sim 10$) as inferred from Dame CO maps. Straight lines show the area covered by GLIMPSE (dotted) and GLIMPSE-3D (solid) survey.

3. LIMITATIONS IMPOSED BY GALACTIC PLANE CONFUSION

Confusion plays a significant role in defining the characteristics and priorities for a warm Spitzer galactic structure campaign, especially when making direct comparisons with the capabilities of the upcoming WISE mission. In particular, the higest priority lines of sight for a Spitzer warm survey are those in which Spitzer's superior source confusion limit (2″ FWHM vs. 6″ FWHM for WISE) yields significant advantage for Spitzer. One, however, must also account for directions in which confusion noise will thwart Spitzer's advantage of being able to improve sensitivity with longer integrations.

Estimating confusion noise for any direction requires either observations that extend to or below the flux limits under consideration or extrapolation of source counts at brighter flux limits with some knowledge of the power law coefficient, α, of the $\frac{\log N}{\log S}$ relation. Lacking the former for most lines of sight in the Galaxy, we have used the MSX 8 μm source counts as the normalization for the confusion noise estimate. Since stars have Rayleigh-Jeans spectral energy distributions at mid-infrared wavelengths we extend these confusion predictions to wavelengths other than 8μm by scaling by λ^2.

The distribution of Galactic stars is non-Euclidean. In the outer Milky Way, counts

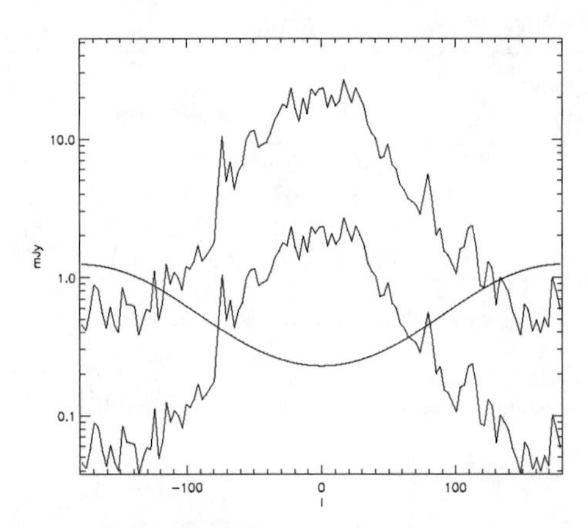

FIGURE 3. A plot of Galactic plane source confusion vs. longitude estimated from MSX 8μm source counts for a 2″ FWHM beam (lower jagged curve) and for a 6″ FWHM beam (upper jagged curve). The curve has been scaled to be representative of IRAC Band 1 assuming the primary source of confusion has a Rayleigh-Jeans spectral energy distribution. The smooth curve represents a K3III star at a galactocentric distance of 20 kpc - a red clump star in the outer Galactic disk.

increase less rapidly with decreasing flux compared with a uniform distribution of targets due to the decline in disk surface density with distance. Model estimates (T. Jarrett, personal communication)) and empirical measurements (based on 2MASS) of the $\frac{\log N}{\log S}$ power law coefficient yield a slope of $\alpha \sim -0.88$ - shallower than the Euclidian case. We have used this coefficient for our estimate of source confusion noise at the Spitzer/WISE flux density levels. In fact, at these flux density levels extragalactic populations begin to dominate over stars while at the same time Galactic star counts begin to decline in the outer disk leading to some uncertainty in the proper value of α for estimates of the confusion limit. The confusion analysis that follows should be considered uncertain to about a factor of two pending direct measurement of in-plane confusion in the outer Milky Way.

In the inner Milky Way confusion dominates and is well estimated by GLIMPSE at the flux levels of interest. Overall the plots suggest that there are two distinct confusion regimes of interest:

- *The inner Milky Way ($|l| < 90°$):* Spitzer and WISE are both confused near the plane over this range of longitudes. Spitzer enjoys nearly an order of magnitude better sensitivity relative to WISE owing to its better FWHM. At the same time, Spitzer achieves sensitivity better than the confusion limit in the two 2-second exposures characteristic of the original GLIMPSE survey. Gains from increasing the base exposure time to 12-seconds will be limited (except at high galactic latitude).
- *The outer disk ($|l| > 90°$):* WISE remains source confused at these longitudes. Spitzer again reaps significant gains over WISE in GLIMPSE-like exposure times, but more importantly Spitzer can take full advantage of pairs of 12-second frames.

4. GALACTIC SCIENCE IN THE WARM SPITZER MISSION

Three principal scientific goals motivated the GLIMPSE survey. Much of the science that will be accomplished with a Warm Spitzer mission will be extensions of these results to cover the whole plane to deeper magnitudes and higher latitudes. As of June 2007, data from the GLIMPSE project has been used in a total of 27 papers by members of the GLIMPSE team and 41 papers by the community at large. The principal goals of GLIMPSE were:

1. *To provide a uniform stellar census of the Galaxy.* Analysis of the GLIMPSE point source catalog (Benjamin et al. [5], Benjamin et al. in prep.) has provided convincing confirmation of two previously under-appreciated features of Galactic structure: (1) the presence of a long ($R = 4.5$ kpc), thin, stellar bar oriented at 45° with respect to the Sun-Galactic center line, *in addition to* the shorter, thicker bar oriented at $\sim 20°$ characterized by COBE and microlensing studies (Nishiyama et al. [6], Cabrera-Lavers et al. [7]), (2) the lack of stellar counterpart for the Sagittarius arm (Drimmel [8]) together with a convincing detection of the Centaurus spiral arm tangency, supporting the hypothesis that the Milky Way is a two-armed spiral in mass density (traced by mid-infrared light) and a four-armed spiral in star formation. GLIMPSE data are also being used to characterize the scalelength of the stellar disk, the structure the Galactic bulge, and (to a limited extent) the vertical structure of the disk, bar, and bulge.
2. *To provide a unbiased survey of star formation in the inner galaxy.* The original GLIMPSE survey includes a uniform sample of thousands of star formation regions and stellar clusters, a large fraction of which were previously uncharacterized (Mercer et al. [9, 10], Churchwell et al. [11, 12]). Many of these star formation regions are close enough that the stellar clusters can be resolved into individual stars and protostars. The spatial and mass distribution of lower mass stars can be studied in the more nearby star formation regions (Churchwell et al. [13], Povich et al. [14], Indebetouw et al. in prep.), and the 4.5 μm band has allowed the detection of a few dozen new outflows/jets in star forming regions (Watson et al. [15]). This unbiased sample of star formation allows a study in the variation in star formation properties, i.e., cluster density, initial mass function, gas content, in a wide range of Galactic environments.
3. *To discover new objects hidden behind the dust and new classes of objects that are bright in mid-infrared wavelengths.* At the outset of the GLIMPSE survey, it was realized that there was likely to be a large component of serendipitous discoveries since high angular resolution mid-infrared observations would allow one to see through regions of the Galaxy that were obscured even at near-infrared wavelengths. Some of the surprises of GLIMPSE so far include (1) a new globular cluster (Kobulnicky et al. [16]), (2) an overdensity of bright ($> 3L_*$) galaxies associated with the Great Attractor (Jarrett et al. [17]), (3) identification of infrared counterparts for X-ray bright sources (Reynolds et al. [18]), and (4) mid-IR bright bowshocks surrounding single stars both in star formation regions and the field (Benjamin and GLIMPSE [19]).

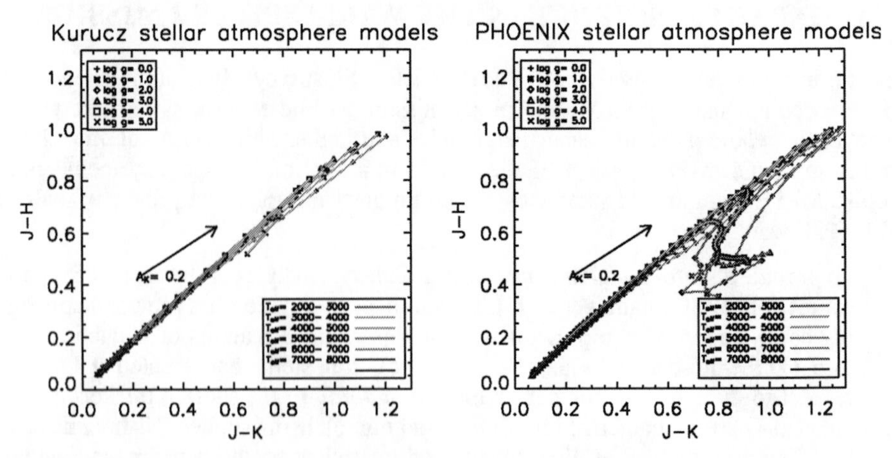

FIGURE 4. Infrared color-color plot showing the near-infrared colors as a function of T_{eff} and log g for two different stellar atmosphere models. Note the degeneracy between temperature and extinction for near-infrared colors.

4.1. Identifying Stellar Sources with Near/Mid-Infrared Data

The Galactic stellar structure goals of a Warm Spitzer mid-infrared Galactic plane survey are similar to the goals of near-infrared investigations. *However, the combination of Spitzer and 2MASS/UKIDSS-GPS data allows for two key improvements over near-infrared data alone:*

- **More sources:** Since $A_L = 0.06A_V = 0.6A_K$, for resolution/sensitivity matched Galactic plane surveys, mid-infrared data will yield a greater number of sources and any investigation using star counts with be far less limited by patchy extinction. For example, the 99.5% reliable GLIMPSE point source catalog yields approximately 50% more sources than the full 2MASS catalog over the same region. Since this ratio changes as a function of Galactic longitude, uncorrected near-IR and mid-IR star counts would give significantly different estimates of the radial scalelength of the Galactic exponential disk.
- **Accurate extinction corrections:** Figure 4 shows that, in theory, near-infrared colors (e.g J-K) are a strong diagnostic of spectral type/temperature. In practice, application of near-infrared colors for this purpose is quite effective in directions where interstellar extinction is low ($A_V < 2$ and thus $A_K < 0.2$). However, within the Galactic plane, even in the J, H, and K bands, extinction becomes significant and skews the ability to distinguish stellar temperature (Figure 5).
  The addition of mid-infrared to near-infrared data allows for accurate extinction correction of near-infrared data. Estimation of extinction correction is extremely difficult at visible wavelengths because the wavelength dependence strongly depends on the dust density/composition. In the near/mid infrared this wavelength dependence is much reduced (Indebetouw et al. [4], but see Flaherty et al. [20]). With near-infrared data alone, however, the reddening and temperature loci are par-

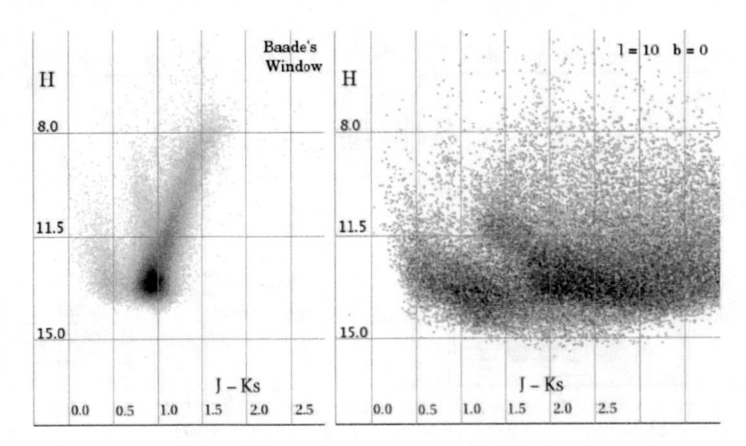

FIGURE 5. A comparison of 2MASS colors in an relatively low extinction region near the Galactic plane (Baade's Window - left) and a region at $l = 10$ where extinction is high (right). In Baade's Window stellar populations are evident, particularly the giant branch and asymptotic giant branch (with the red clump at the base of the giant branch - just above the 2MASS confusion limit). In the plot at right interstellar extinction skews the stellar populations. Reliable dereddening can restore stellar colors to enable population analysis.

allel in color-color space (see Fig. 4). The addition of the first two IRAC bands, however, provides measurement on the Rayleigh-Jeans tail of the stellar blackbody curve, so that (to first order) all near-mid IR colors should be zero. By picking an infrared color that remains fixed for a large range of stellar temperatures expected in the sample, e.g., H-[4.5] (see Fig. 6), one can accurately correct stellar fluxes for extinction (Nidever et al., in prep) and recover the temperature/spectral type information inherent in the near-infrared fluxes.

A Spitzer warm galactic plane survey would provide a powerful tool for dereddening surveys such as 2MASS and UKIDSS at low galactic latitudes, enabling population based (i.e. standard candle) reconstruction of Galactic structure and thus permitting the construction of a three-dimensional view of the Milky Way. Such a survey would be directly applicable to 2MASS and UKIDSS, and ultimately would serve as a legacy for application to any low-latitude near-infrared survey with overlapping sensitivity.

4.2. Galactic Stellar Structure Goals

Given the availability of near-infrared surveys, the highest priority focus for a Spitzer Warm Mission galactic plane survey should be those direction that suffer the greatest extinction. Given the utility of mid-infrared data for de-reddening and luminosity classification, coverage beyond the highest extinction regions is warranted as well, but of lower priority. A list of goals for Galactic stellar structure are as follows:

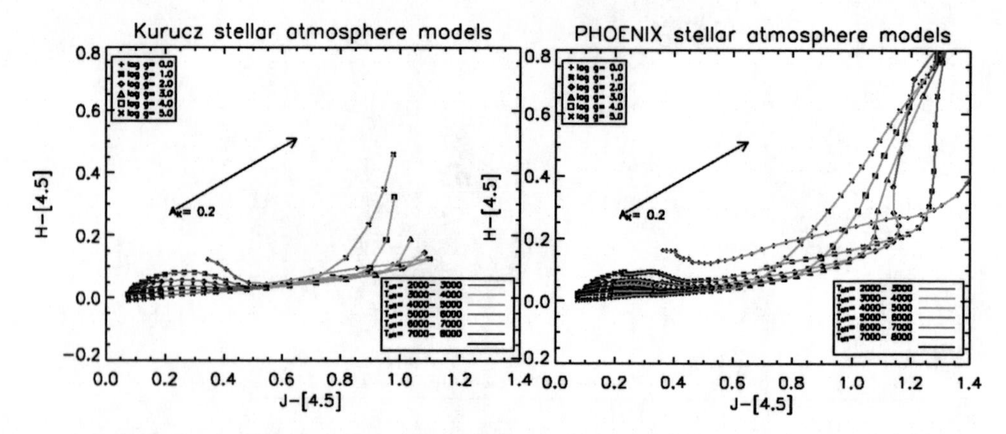

FIGURE 6. Infrared color-color plot showing how mid-infrared colors can be used to break the degeneracy between temperature, extinction, and (to a lesser extent) log g for ordinary stellar sources.

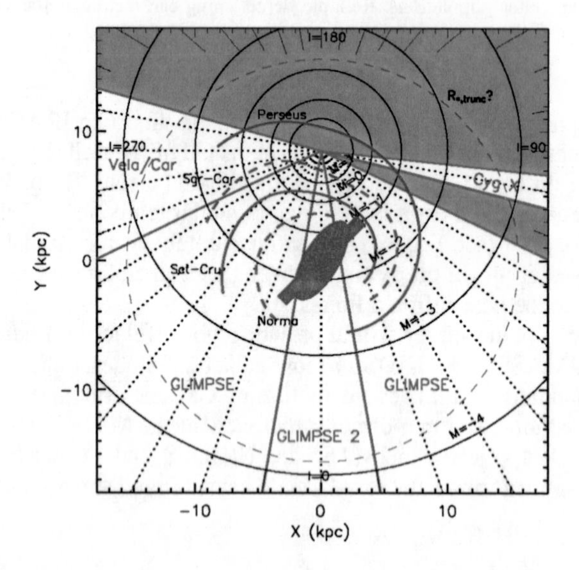

FIGURE 7. The Galactic Plane showing the areas covered by GLIMPSE, GLIMPSE2, Vela- Carina Survey (PI: Majewski), and Cygnus-X survey (PI: Hora). Solid circles indicate the GLIMPSE detection distances for objects with absolute magnitude M, assuming no extinction. The approximate positions of Galactic spiral arms (Taylor and Cordes [21]) are indicated in red. The central oval and bar represents the approximate extent of the central triaxial bulge/bar (Gerhard [22], Cole and Weinberg [23]) and the "Long" bar (Hammersley et al. [24], Benjamin et al. [5]). The expected truncation radius for the Galactic stellar disk is also shown with a dashed red line.

1. *The vertical structure of the stellar bar and disk:* Our own Galaxy is the *only* edge-on spiral galaxy in the universe in which we currently can hope to study the vertical distribution of stellar populations as a function of position within the Galaxy. The inner disk, bar, and bulge lie in the directions where interstellar extinction is at a maximum, complicating our ability to constrain scaleheights of the stellar components tracing these structures. To this end, GLIMPSE-3D (PI: Benjamin) was approved to add vertical coverage at selected longitudes. However, constraints on observing time precluded uniformly covering the latitude range $1 < |b| < 3°$ (see Figure 2). *Experience shows that in the presence of patchy extinction, even in the mid-infrared, uniform coverage is vital to constrain galactic structure parameters.* To probe the thin disk, vertical coverage should extend to at least two scale heights at the distance of Galactic center which is approximately $|b| = 3°$. Using COBE/DIRBE data, Binney et al. [25], for example, found evidence for the stellar disk scale height increasing from $z_o = 97$ to 220 pc ($0.69 - 1.57°$ at a distance of R=8 kpc). Assuming the thick disk maintains a constant thickness of z_{th} =1060 pc (Cabrera-Lavers et al. [26]), latitude coverage of 7.6° would be needed to cover a single scaleheight. Finally, models of stellar bars are still under rapid development, but models of Debattista et al. [27], for example, predict a scaleheight of $\sim$ 500 pc (3.6°).
2. *The edge of the stellar disk of the Galaxy:* The stellar disk of $\sim$ 80% of galaxies exhibit an outer truncation radius at approximately four scalelengths (van der Kruit and Searle [28]). Using stellar disk scalelength derived from GLIMPSE data (Benjamin et al. [5]), this result predicts that the Milky Way's truncation should lie at a galactocentric radius of 15.6 ± 2.4 kpc, putting our Sun about halfway between the center and edge of the Galaxy. If the thin disk does not flare at large radii (due to lower surface gravity), the edge of the thin disk would have the same angular height (1.57°) as at the Galactic center. A search for a deficit of red giant branch and red clump stars towards the outer galaxy could be used to constrain this outer truncation. Note, in Fig. 7, however, that the entire outer Galaxy is obscured by the Perseus arm; an accurate determination of the stellar edge of the Galaxy requires the combination of near/mid IR data to provide a reliable extinction correction.
3. *A survey of low latitude star clusters:* Star clusters are a particularly valuable commodity in stellar populations research because they are the only disk tracer for which reliable ages can be derived. This feature not only makes it possible to explore such attributes of the disk as the age-metallicity relation, age-velocity dispersion relation (where velocity dispersion is intimately correlated to the scaleheight of a stellar population through Poisson's equation), and the disk star formation history, but also to explore the dynamical evolution of the clusters themselves. Unfortunately, proper age-dating (via isochrone fitting) requires accurate distances to the clusters. For the typical globular cluster, a 10% uncertainty in the distance corresponds to about a 20% error in the age. A similarly small uncertainty in the extinction translates to a similar error in the ages, and one of the primary problems with studying low latitude clusters is the degeneracy of age, distance, and reddening/extinction in the isochrone fitting. The use of Spitzer+2MASS data can break this degeneracy and lead to unambiguous isochrone age-dating.

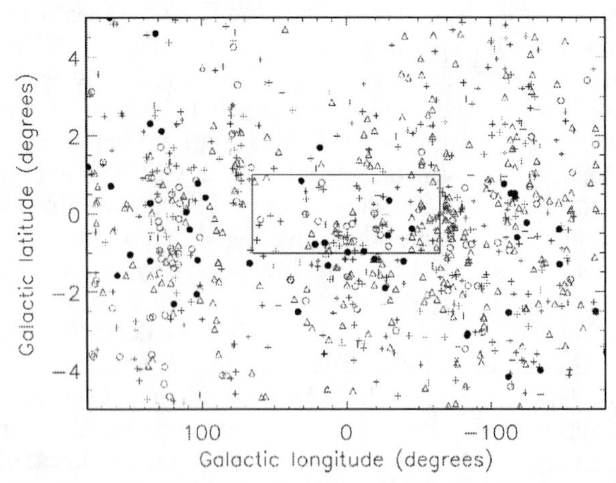

FIGURE 8. Location of open clusters within five degrees of the Galactic plane with the approximate GLIMPSE boundaries shown. Symbol type indicates the distance to the cluster: 0-5 kpc (open triangle), 5-10 kpc (open circle), 10-20 kpc (solid circle), and unknown distance (plus sign).

While a large fraction of star formation is tightly constrained to the inner Galactic plane, the majority of known open clusters are more distributed in both longitude and latitude, but generally still highly extinguished. In addition, the disk and bulge globular cluster population is broadly dispersed, but also lies behind significant dust. Only 2 globular clusters fall within the GLIMPSE survey, whereas a full quarter of known ~ 150 Galactic globular clusters lie within $|b| < 5$, and more than half of the known clusters lie behind more than $A_V = 1$ magnitude of dust, including the bulk of the disk and bulge clusters. Out of the 1100 known open clusters (taken from the WEBDA database), only 101 lie within the area currently covered by GLIMPSE data. A full plane galactic survey covering $|b| < 2°, 3°$ or $4°$ would increase this sample size to 529, 679, and 783 open clusters, respectively. Further strategic IRAC pointing, particularly on all highly reddened globular clusters, would provide substantial, definitive leverage on the age distribution of open and globular clusters. It is also likely that with a deep, general Galactic plane survey, where star-by-star dereddening would be enabled, new open clusters and stellar associations will be identified.

4. *The stellar spiral structure of the Galaxy:* One of the pleasant surprises of the GLIMPSE survey has been the detection of an enhancement in star counts in the direction expected for the Centaurus (or Crux) spiral arm tangency, combined with the lack of detection of the first quadrant tangency for the Sagittarius arm (Benjamin et al. [5]). This supports the Drimmel and Spergel [29] hypothesis that the Galaxy is a two-armed spiral galaxy in mass.[3] Given the positive identification

[3] In this hypothesis, the Perseus and Scutum-Crux spiral arms are traced by the old stellar mass, while the Sagittarius-Carina and Norma arms, while still star formation structures, are not associated with significant overdensities of mass.

of the stellar spiral structure of the Galaxy, it would be extremely desirable to map the vertical and azimuthal structure using stellar populations that trace both the mass (such as red clump giants), and the star formation. Since spiral arms are intrinsically regions of high gas density, the reduced extinction available at mid-infrared wavelengths is essential for such work.

5. *Galactic Satellites and Tidal Streams:* From high latitude, large-area photometric surveys like 2MASS and the Sloan Digital Sky Survey, it is now clear that the halo of our Galaxy is highly substructured, including networked streams of tidal debris from disrupting satellites (e.g., Majewski et al. [30], Belokurov et al. [31], Grillmair [32]). A number of the newly identified, diffuse substructures — such as the Monoceros Stream (Newberg et al. [33], Yanny et al. [34], Rocha-Pinto et al. [35]), the Triangulum-Andromeda star cloud (Rocha-Pinto et al. [36], Majewski et al. [37]), the Anticenter Stream(s) (Grillmair [32]), and the Hercules-Aquila cloud (Belokurov et al. [38]) — lie at low Galactic latitudes, and may be part of a vast complex of tidal debris accumulating around the disk of the Milky Way.
   The tidal disruption of satellites aligned with the disk has been used to explain a similar, diffuse set of features seen wrapping around the M31 disk (Ibata et al. [39]), and this may be a manifestation of the predictions from the prevailing ΛCDM models that Galactic disks grow inside out from the continued agglomeration of subhalos. Several "nuclei" for such accreted structures have been identified, including Canis Major (Martin et al. [40]) and Argo (Rocha-Pinto et al. [41]), but whether these nuclei as well as the "streams" identified above may actually be a misinterpretation of the heavily reddened, low latitude starcounts, further complicated by the presence of a stellar counterpart of the HI warp has been hotly debated (see below; Rocha-Pinto et al. [41], Momany et al. [42], Bellazzini et al. [43, 43], Martin et al. [44], López-Corredoira [45], Momany et al. [46]).
   What is known is that satellites on inclined, prograde orbits will dyamically couple to disks, suffer orbital decay and orbital plane tilting, and eventually lead to the accretion of these systems along the edge of the disk (Quinn et al. [47]). Eventually these systems will dynamically heat the outer disk as the satellite continues to spiral inward. Theoretically, then, it appears that the dust-obscured, outer disk is perhaps the most interesting place to look for the ongoing process of disk building and subhalo accretion. One key way to probe such structures is to analyze near-infrared CMDs of the stellar overdensities. In regions where extinction is significant, Spitzer data is necessary to provide reliable extinction corrections to the near-infrared fluxes.
   Even satellite systems that are on more polar orbits can have large critical patches obscured by the Zone of Avoidance. For example, a large fraction of the Sagittarius dSph core and its tidal stream lies behind the bulge of our Galaxy, in highly reddened regions, and this has made it difficult to follow with accuracy the shape of the tidal stream (e.g., Majewski et al. [30]), leading to disparate models of the system (e.g., Law et al. [48], Fellhauer et al. [49]). Indeed, the very distance of the Sgr core is still uncertain at about the 20% level in part because of the uncertainty on the reddening.
   Finally, it is estimated that if the distribution of satellites is uniform by latitude, then

approximately 1/3 of all dwarf satellites of the Milky Way will be unaccounted for if we do not accurately probe the Zone of Avoidance (Willman et al. [50]). With the doubling of the known satellite population of the Milky Way just in the past year or two of study of the SDSS footprint (which covers 1/5 of the sky), one can estimate that at least another 40-50 satellites remain to be found around the Galaxy, and more than about 20 probably lie in highly reddened regions, perhaps some very close, according to the Quinn et al. mechanism. Thus, a thorough probe of the low latitude Galaxy with a warm mission IRAC survey may teach us a lot about the structure and continued evolution of the Milky Way and its subhalo system.

6. *The Stellar Warp:* The neutral hydrogen warp of the Galaxy has been well studied, but the relation of this to the stellar warp remains far from clear (Momany et al. 2006). Part of the difficulty in tracing the stellar warp with different stellar populations is the difficulty of identifying stellar populations in near-infrared CMDs that have been smeared out by extinction. In addition, there is ongoing debate on the connection between the warp and tidal streams known to be encircling the disk (e.g., López-Corredoira [45], Bellazzini et al. [43], Momany et al. [46]). Once again, the reddening estimates provided by mid-infrared observations are essential to reconstructing the extinction-free near-infrared CMD's of the outer Galaxy and disentangling the structure(s) there.

In addition to these global stellar structure goals, a full plane survey will allow for a global characterization of whole classes of stellar populations, including AGB stars, Wolf-Rayet stars, and carbon stars. Multi-epoch observations could allow for the identification of several classes of variable stars. Finally, one of the unanticipated uses of GLIMPSE data that will also hold true for a full-plane survey is the use of this data for the identification of infrared counterparts or companions for X-ray and gamma-ray sources resolved by Chandra or XMM; this has already added a few sources to the list of $\sim$ 100 known high mass X-ray binaries.

4.3. Star Formation

The dependence of star formation on physical conditions in galaxies is critically important to understanding the evolution of those galaxies. We must search for any dependencies of the star formation process on large-scale physical conditions in our Galaxy, where star formation regions can be resolved and studied in detail. The average metallicity (measured in ionized gas) drops by at least a factor of 4 from the inner to outer Galaxy (Afflerbach et al. [51]). Theoretically, low metallicity should reduce the ability of molecular clouds to cool and collapse into stars. The cosmic-ray flux (which should also heat molecular clouds and affect their collapse) is also reduced away from the inner Galaxy (Bloemen et al. [52]). The average interstellar radiation field is observed to be more intense in the outer Galaxy (Cox and Mezger [53]) , and the interstellar phase balance shifts significantly in favor of atomic over molecular hydrogen. The radiation field produced near a given mass star which does form will be stronger and harder if the star has lower metallicity, hindering the formation of lower-mass stars in the proximity. The structure and temperature of the HII regions surrounding these more

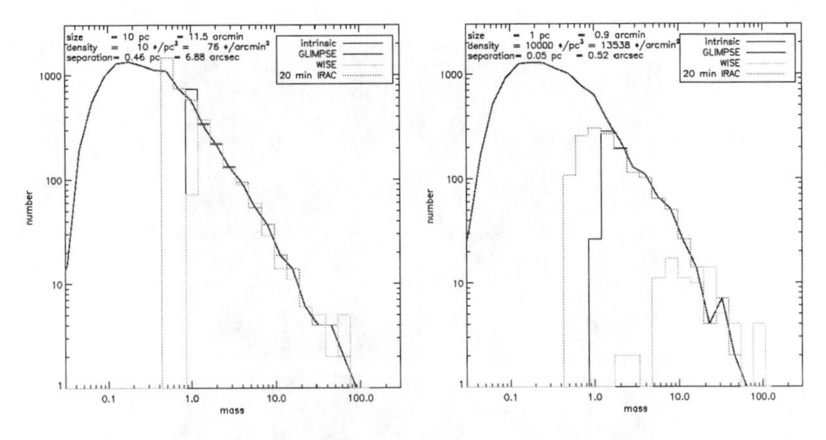

FIGURE 9. Simulated mass functions as "observed" by Spitzer and WISE for two different apparent source densities.

intensely radiating protostars should differ from that around high-metallicity protostars. The different physical conditions in outer Galaxy molecular clouds should also affect the spatial and luminosity distribution and stellar content of star forming regions, the IMF, and the radiation transfer in a nascent cluster, all of which can be studied with Spitzer's warm mission.

A full Galactic plane survey would extend an unbiased study of star formation to the entire plane, providing coverage of the Carina spiral arm tangency and most of the Perseus arm. Using the HII region compilation of Paladini et al. [54], 850 of the 1442 HII regions are in the current GLIMPSE coverage. Full coverage up to $|b| < 2°, 3°$ or $4°$ would increase this sample size to 1250, 1344, and 1393 cataloged HII regions. However, Churchwell et al. [12] have shown many of the star forming regions seen in GLIMPSE are not included in this catalog, so these numbers are lower limits.

Owing to local confusion in clustered regions, Spitzer has a clear advantage over WISE for clustered star formation studies. Figure 9 shows simulated mass functions (with the Kroupa and Weidner [55] functional form) for Galactic clusters, and the mass functions that would be observed with the existing Spitzer/GLIMPSE survey (2.4s exposure time), with IRAC for 20 minutes in a warm mission observation, and with WISE. The observed mass functions are derived from the 3.6μm observed flux density distribution and a polynomial fit to the spectral type-flux density relation (e.g. Bessell and Brett [56]). Poisson noise from the stars and from a uniform background of 200MJy/sr is included – the latter is a best case, since it is known from GLIMPSE that the diffuse background is highly structured, and in practice the faintest sources must be discarded to maintain reliability and exclude false sources extracted from knots in the background. It is clear that in the inner Galaxy and in even modestly crowded clusters (tens of stars per cubic parsec), that Spitzer will improve over WISE. It is less clear that significant improvement can be gained over GLIMPSE in the inner Galaxy and in clusters with significant diffuse background. The greatest impact of a warm mission survey will be in clusters, star formation regions, and regions of high extinction in the

FIGURE 10. Three color composite of outer Galaxy star forming region G124 from a 72 sec integration with IRAC bands 1, 2, and 4.

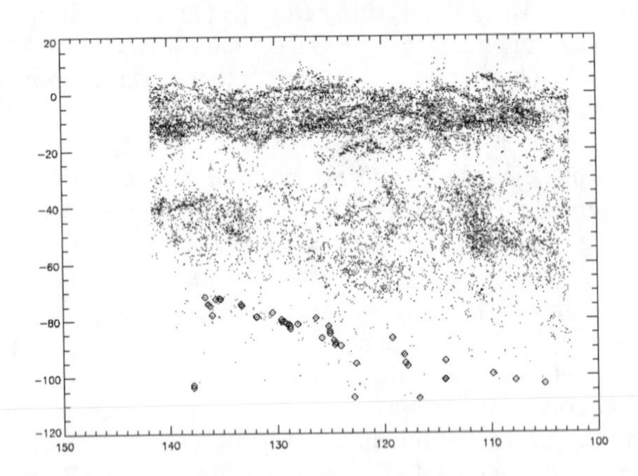

FIGURE 11. Longitude *vs.* V_{LSR} for IRAC/CO associations in the outer Milky Way from Kerton and Brunt [57]. The diamonds highlight the most distant regions in the outer disk.

outer Galaxy that have not yet been observed by Spitzer.

Figure 10 shows an example of an outer Galaxy star formation region, IRAC bands 1,2, and 4, with contours of ^{12}CO at the velocity range of the outer Galaxy cloud. The region was selected from Kerton and Brunt [57] and observed for 72 seconds. These observations are not yet confusion limited – clearly deep observations of outer Galaxy clusters will be useful for measuring the mass function with a large dynamic range in mass, and characterizing the stellar and protostellar population.

One can outline a type of project that would be worth pursuing: Outer Galaxy star formation regions have been identified by correlation of CO clouds with IRAS point sources by Kerton and Brunt [57] among others. Figure 11 shows a region in the outer galaxy with CO/IRAS associations, and the most distant outer galaxy regions marked with diamonds. The Sagittarius spiral arm is clearly visible at lower velocities. Their catalog has 6700 associations, each which could probably be observed with 1-2 IRAC pointings. If one wanted to observe all such regions for two minutes, a ~300 hour program would ensue. Only observing the most distant regions (163 regions with V<-65) for 5 minutes each would require ~15 hours.

4.4. Extragalactic Science

A deep Galactic plane survey may be of great interest to astronomers working outside the field of Galactic astronomy. In particular, a mid-infrared survey will serve to nearly eliminate the "Zone of Avoidance" found in surveys of extragalactic clusters. Several galaxies with $L \sim 3L_*$ have been found in the GLIMPSE survey, with many more appearing in the first analysis of GLIMPSE-3D galaxy. Redshifts have been obtained for some of these galaxies and it appears that these galaxies trace a part of the Great Attractor, a major overdensity of galaxies in the local Universe (Jarrett et al. [17]). The 2MASS extended source distribution indicates that a number of filaments, as well as some large voids, cross the outer plane, including the Local Supercluster and the Perseus-Pisces Supercluster.

5. THE INTERSTELLAR MEDIUM

5.1. Diffuse Interstellar Emission

PAH 3.3um emission (from the C-H stretching mode) makes a significant diffuse contribution in band 1. Emission from PAHs or other ultrasmall grains is thought to also contribute in IRAC band 2. If PAHs are heavily deuterated, as has been suggested (Peeters et al. [58], Draine [59, 60]) then the C-D stretching mode falls in the center of band 2. The 0-0 S(9) 4.695um, 0-0 S(10) 4.410um, and 0-0 S(11) 4.181um lines of H_2 also fall within IRAC band 2; these lines may be excited by outflows from protostars in molecular clouds (Noriega-Crespo et al. [61, 62]). Away from molecular clouds, the H_2 rotational emission should be insignificant, and PAH/ultrasmall grain emission is expected to be the principal source of diffuse emission.

In observations of the emission from other galaxies using IRAC bands 1 and 2, it is nearly impossible to separate the diffuse emission from dust and gas from the (generally much stronger) stellar continuum. However, in observations of the Milky Way, a large fraction of the stellar contribution can be removed by subtracting point sources from Spitzer images of the Milky Way. This has been done by Flagey et al. [63]), who were able to determine the emission in bands 1 and 2 that correlates with diffuse emission (due to PAHs) in IRAC band 4.

During a warm Spitzer mission, deeper observations of Milky Way fields may allow improved subtraction of point sources, due to higher signal/noise and perhaps also improved understanding of the instrumental p.s.f. If the regions in question have been previously imaged with IRAC bands 3 and 4, it may be possible to improve upon the results of Flagey et al. [63] concerning PAH+dust emission into IRAC bands 1 and 2, and how this correlates with diffuse emission in bands 3 and 4. One can also correlate diffuse emission in bands 1 and 2 with other interstellar tracers, such as 21 cm emission.

It is not clear at this time how much of an improvement can realistically be expected in terms of our ability to measure diffuse emission in bands 1 and 2, and this is perhaps best viewed as "added value" for imaging at moderate and high galactic latitudes that will be carried out for other reasons (e.g., deep imaging to detect high redshift objects). Such data can be "mined" to extract the diffuse emission from the Milky Way, which can then be compared to other tracers of the ISM, such as FIR emission (e.g., perhaps comparing with Akari sky maps at 80um , 140um, or 160um), 21cm emission, or CO 1-0 emission. Indeed, it might be of interest to select regions for observation for which imaging data from Akari will be available. [Note that Akari bands N3 and N4 have overlap with IRAC bands 1 and 2.] Prior ISO and MSX imaging would also provide context for IRAC band 1 and band 2 observations.

5.2. Infrared Extinction in Dense Cores

A specific interstellar medium topic that exploits Spitzer's mid-infrared capability to work at wavelengths that minimize interstellar extinction involves the ability to map the extinction profile of dense cores in molecular clouds. Even at near-infrared wavelengths these cores are sufficiently opaque that stars are not visible through the densest regions in deep exposures.

Quantitative knowledge of the detailed structure of dense cores is critical for both setting the initial conditions for star formation theory and directly testing predictions of protostellar collapse models. Extinction measurements toward background field stars are a very powerful, robust and quantitative probe of cloud structure. Deep ground-based near-infrared (JHK) observations obtained on large aperture telescopes enables extinction mapping of nearby dense cores to depths of typically $A_V \sim 30$ magnitudes and in very favorable circumstances extinctions as high as 40 magnitudes can be measured. This depth is often sufficient to probe the structure of starless cores with small or modest density contrasts. However, to penetrate more evolved starless cores on the verge of forming a star and typical protostellar cores requires the ability to reach depths of 75 - 100 magnitudes. Deep Spitzer observations in Bands 1 & 2 are capable of probing such depths.

A Spitzer Warm Mission survey of dense cores could obtain extinction maps of the inner regions of some of a comprehensive list of the most opaque dense cores known in molecular clouds in order to provide robust measurements of their structure which in turn will provide the initial conditions for, and direct tests of, star formation theories. It is in such regions that the early star formation process will exhibit its most significant structural evolution. An observational campaign to obtain deep, pointed band 1 & 2 images of a sample of nearby dense cores of different masses and in different stages of development would provide a quantitative empirical description of core evolution and yield direct tests of collapse calculations.

Because of the much longer integration times required, such a survey would be independent of the shallow Galactic plane surveys described to this point. Typical observations would require integration times of at least ~10 minutes. Cores need to be of order ~ 2 arcminutes in angular size to produce meaningful morphological conclusions. The number of such nearby cores limits such a survey to a few dozen targets. A more extensive survey could be conducted on more distant cores but with limited angular resolution.

5.3. Extinction mapping of Infrared Dark Clouds

Infrared Dark Clouds (IRDCs) are distant, massive and dense molecular clouds that appear dark at mid-infrared wavelengths indicating relatively large opacities ($A_v > 40$ magnitudes). These objects are not typically sites of significant star formation and are consequently thought to be the future sites of cluster formation. Little is known about their physical conditions and infrared extinction mapping could provide robust estimates of some of their most fundamental properties, namely mass and structure.

A warm mission survey would obtain deep Spitzer observations in order to construct extinction maps of Infrared Dark Clouds to measure their structure and determine their masses and thus specify the initial conditions in objects that are likely the future formation sites of rich embedded stellar clusters. Such a survey would require deep IRAC pointed observations of a sample of IRDCs coupled with deep ground-based JHK observations necessary to mitigate strong foreground star contamination (IRDCs are sufficiently distant that foreground star contamination is significant and may be a serious impediment to extinction measurements). There are ~100 objects known to be very dark with mean extinctions of tens of magnitudes. The maximum sizes of these clouds is ~10 arcminutes. Integration times of 240 – 480 seconds per field would be required to penetrate the highest extinction regions within the clouds with a sufficient number of background stars detected to provide reasonable angular resolution. Typical clouds could be covered with a few spatial pointings.

6. SUMMARY

Spitzer's Warm Mission capabilities enable legacy surveys of a substantially larger portion of the Galactic plane than was observed during the cryogenic mission. IRAC Bands 1 & 2 fluxes, with their relative immunity to interstellar extinction, can be combined with near-infrared surveys to exploit knowledge of individual stellar spectral types to construct a three dimensional view of the Milky Way. Given the existing and anticipated coverage of the Galactic plane during the cryogenic portion of the mission (Figure 7), and given knowledge of confusion as a function of Galactic longitude and latitude, the most natural focus of such programs are likely GLIMPSE-like (two 2s integrations per sky position) at latitudes extending up to $-5^\circ < b < 5^\circ$ that extend on the GLIMPSE coverage for longitudes $-90^\circ < l < 90^\circ$ and deeper (e.g. two 12s integrations per sky position) coverage for the outer Milky Way to $-1.5^\circ < b < 1.5^\circ$ with limited extensions to higher latitude to sample vertical structure as was done for GLIMPSE-3D.

ACKNOWLEDGMENTS

The authors would like to acknowledge the hard work and support of the Spitzer Science Center in making this contribution possible.

REFERENCES

1. Whitney, B. A., Wood, K., Bjorkman, J. E., and Cohen, M. 2003, ApJ, 598, 1079
2. Molinari, S. et al. 2007, presentation at Herschel OT KP workshop, 20 Feb 2007.
3. Robitaille, T. P., Whitney, B. A., Indebetouw, R., and Wood, K. 2007, ApJS, 169, 328
4. Indebetouw, R., et al. 2005, ApJ, 619, 931
5. Benjamin, R. A., et al. 2005, ApJ, 630, L149
6. Nishiyama, S., et al. 2005, ApJ, 621, L105
7. Cabrera-Lavers, A., Hammersley, P. L., González-Fernández, C., López-Corredoira, M., Garzón, F., and Mahoney, T. J. 2007, A&A, 465, 825
8. Drimmel, R. 2000, A&A, 358, L13
9. Mercer, E. P., et al. 2004, ApJS, 154, 328
10. Mercer, E. P., et al. 2005, ApJ, 635, 560
11. Churchwell, E., et al. 2006, ApJ, 649, 759
12. Churchwell, E., et al. 2007, ApJ, in prep
13. Churchwell, E., et al. 2004, ApJS, 154, 322
14. Povich, M. S., et al. 2007, ApJ, 660, 346
15. Watson, D. F., et al. 2006, Bulletin of the American Astronomical Society, 38, 1206
16. Kobulnicky, H. A., et al. 2005, AJ, 129, 239
17. Jarrett, T. H., et al. 2007, AJ, 133, 979
18. Reynolds, S. P., Borkowski, K. J., Hwang, U., Harrus, I., Petre, R., and Dubner, G. 2006, ApJ, 652, L45
19. Benjamin, R., and GLIMPSE 2004, Bulletin of the American Astronomical Society, 36, 1569
20. Flaherty, K. M., Pipher, J. L., Megeath, S. T., Winston, E. M., Gutermuth, R. A., Muzerolle, J., and Fazio, G. G. 2007, ApJ, 663, 1069
21. Taylor, J. H., and Cordes, J. M. 1993, ApJ, 411, 674
22. Gerhard, O. 2002, The Dynamics, Structure & History of Galaxies: A Workshop in Honour of Professor Ken Freeman, 273, 73
23. Cole, A. A., and Weinberg, M. D. 2002, ApJ, 574, L43
24. Hammersley, P. L., Garzón, F., Mahoney, T. J., López-Corredoira, M., and Torres, M. A. P. 2000, MNRAS, 317, L45
25. Binney, J., Gerhard, O., and Spergel, D. 1997, MNRAS, 288, 365
26. Cabrera-Lavers, A., Garzón, F., and Hammersley, P. L. 2005, A&A, 433, 173
27. Debattista, V. P., Mayer, L., Carollo, C. M., Moore, B., Wadsley, J., and Quinn, T. 2006, ApJ, 645, 209
28. van der Kruit, P. C., and Searle, L. 1981, A&A, 95, 105
29. Drimmel, R., and Spergel, D. N. 2001, ApJ, 556, 181
30. Majewski, S. R., Skrutskie, M. F., Weinberg, M. D., and Ostheimer, J. C. 2003, ApJ, 599, 1082
31. Belokurov, V., et al. 2006, ApJ, 642, L137
32. Grillmair, C. J. 2006, ApJ, 651, L29
33. Newberg, H. J., et al. 2002, ApJ, 569, 245
34. Yanny, B., et al. 2003, ApJ, 588, 824
35. Rocha-Pinto, H. J., Majewski, S. R., Skrutskie, M. F., and Crane, J. D. 2003, ApJ, 594, L115
36. Rocha-Pinto, H. J., Majewski, S. R., Skrutskie, M. F., Crane, J. D., and Patterson, R. J. 2004, ApJ, 615, 732
37. Majewski, S. R., Ostheimer, J. C., Rocha-Pinto, H. J., Patterson, R. J., Guhathakurta, P., and Reitzel, D. 2004, ApJ, 615, 738
38. Belokurov, V., et al. 2007, ApJ, 657, L89
39. Ibata, R., Chapman, S., Ferguson, A. M. N., Lewis, G., Irwin, M., and Tanvir, N. 2005, ApJ, 634, 287

40. Martin, N. F., Ibata, R. A., Bellazzini, M., Irwin, M. J., Lewis, G. F., and Dehnen, W. 2004, MNRAS, 348, 12
41. Rocha-Pinto, H. J., Majewski, S. R., Skrutskie, M. F., Patterson, R. J., Nakanishi, H., Muñoz, R. R., and Sofue, Y. 2006, ApJ, 640, L147
42. Momany, Y., Zaggia, S. R., Bonifacio, P., Piotto, G., De Angeli, F., Bedin, L. R., and Carraro, G. 2004, A&A, 421, L29
43. Bellazzini, M., Ibata, R., Martin, N., Lewis, G. F., Conn, B., and Irwin, M. J. 2006, MNRAS, 366, 865
44. Martin, N. F., Ibata, R. A., Conn, B. C., Lewis, G. F., Bellazzini, M., Irwin, M. J., and McConnachie, A. W. 2004, MNRAS, 355, L33
45. López-Corredoira, M. 2006, MNRAS, 369, 1911
46. Momany, Y., Zaggia, S., Gilmore, G., Piotto, G., Carraro, G., Bedin, L. R., and de Angeli, F. 2006, A&A, 451, 515
47. Quinn, P. J., Hernquist, L., and Fullagar, D. P. 1993, ApJ, 403, 74
48. Law, D. R., Johnston, K. V., and Majewski, S. R. 2005, ApJ, 619, 807
49. Fellhauer, M., et al. 2006, ApJ, 651, 167
50. Willman, B., Governato, F., Dalcanton, J. J., Reed, D., and Quinn, T. 2004, MNRAS, 353, 639
51. Afflerbach, A., Churchwell, E., and Werner, M. W. 1997, ApJ, 478, 190
52. Bloemen, J. B. G. M., et al. 1984, A&A, 135, 12
53. Cox, P., and Mezger, P. G. 1989, A&ARv, 1, 49
54. Paladini, R., Burigana, C., Davies, R. D., Maino, D., Bersanelli, M., Cappellini, B., Platania, P., and Smoot, G. 2003, A&A, 397, 213
55. Kroupa, P., and Weidner, C. 2003, ApJ, 598, 1076
56. Bessell, M. S., and Brett, J. M. 1988, PASP, 100, 1134
57. Kerton, C. R., and Brunt, C. M. 2003, A&A, 399, 1083
58. Peeters, E., Mattioda, A. L., Hudgins, D. M., and Allamandola, L. J. 2004, ApJ, 617, L65
59. Draine, B. T. 2004, Astrophysics of Dust, 309, 691
60. Draine, B. T. 2006, Pre-Solar Grains as Astrophysical Tools, 26th meeting of the IAU, Joint Discussion 11, 21 August 2006, Prague, Czech Republic, JD11, #8, 11,
61. Noriega-Crespo, A., et al. 2004, ApJS, 154, 352
62. Noriega-Crespo, A., Moro-Martin, A., Carey, S., Morris, P. W., Padgett, D. L., Latter, W. B., and Muzerolle, J. 2004, ApJS, 154, 402
63. Flagey, N., Boulanger, F., Verstraete, L., Miville Deschênes, M. A., Noriega Crespo, A., and Reach, W. T. 2006, A&A, 453, 969

The Spitzer Warm Mission: Prospects for Studies of the Distant Universe

Pieter van Dokkum*, Asantha Cooray†, Ivo Labbé**, Casey Papovich‡ and Daniel Stern§

*Department of Astronomy, Yale University, New Haven, CT 06520-8101, USA
†Center for Cosmology, Dept of Physics and Astronomy, University of California, Irvine, CA 92697, USA
**Carnegie Observatories, 813 Santa Barbara St, Pasadena, CA 91101, USA
‡Steward Observatory, University of Arizona, 933 North Cherry Ave, Tucson, AZ 85721, USA
§Jet Propulsion Laboratory, California Institute of Technology, Pasadena, CA 91109, USA

Abstract.
IRAC excels at detecting distant objects. Due to a combination of the shapes of the spectral energy distributions of galaxies and the low background achieved from space, IRAC reaches greater depth in comparable exposure time at 3.6 and 4.5 μm than any ground- or space-based facility currently can at 2.2 μm. Furthermore, the longer wavelengths probed by IRAC enable studies of the rest-frame optical and near-infrared light of galaxies and AGN to much higher redshift than is possible from the ground. This white paper explores the merits of different survey strategies for studying the distant universe during the warm mission. A three-tiered approach serves a wide range of science goals and uses the spacecraft effectively: 1) an ultra-deep survey of ≈ 0.04 square degrees to a depth of 250 hrs (in conjunction with an HST/WFC3 program), to study the Universe at $7 < z < 14$; 2) a survey of ≈ 2 square degrees to the GOODS depth of 20 hrs, to identify luminous galaxies at $z > 6$ and characterize the relation between the build-up of dark matter halos and their constituent galaxies at $2 < z < 6$, and 3) a 500 square degree survey to the SWIRE depth of 120 s, to systematically study large scale structure at $1 < z < 2$ and characterize high redshift AGN. One or more of these programs could conceivably be implemented by the SSC, following the example of the Hubble Deep Field campaigns. As priorities in this field continuously shift it is also crucial that a fraction of the exposure time remains unassigned, thus enabling science that will reflect the frontiers of 2010 and beyond rather than those of 2007.

Keywords: Spitzer Space Telescope, infrared astronomical observations, external galaxies, quasars, distances, redshifts
PACS: 95.85.Hp, 98.62.Py, 98.80.Bp, 98.54.Aj

1. INTRODUCTION

Infrared observations are crucial for the study of distant galaxies. While blue star forming galaxies can be routinely identified to $z \sim 6$ and beyond using optical selection techniques and follow-up spectroscopy (e.g., Steidel *et al.* [1, 2, 3], Kodaira *et al.* [4], Ouchi *et al.* [5], Stark *et al.* [6], Dow-Hygelund *et al.* [7]), measuring their masses and star formation histories requires access to their rest-frame optical light (see, e.g., Shapley *et al.* [8], Papovich *et al.* [9]). Furthermore, it has become clear that optical samples miss a substantial fraction of the high redshift galaxy population. Near-infrared surveys have discovered substantial numbers of UV-faint red galaxies (Daddi *et al.* [10], McCarthy *et al.* [11], Labbé *et al.* [12], Franx *et al.* [13]) and it appears that these objects domi-

CP943, *The Science Opportunities for the Warm Spitzer Mission Workshop*,
edited by L. J. Storrie-Lombardi and N. A. Silbermann

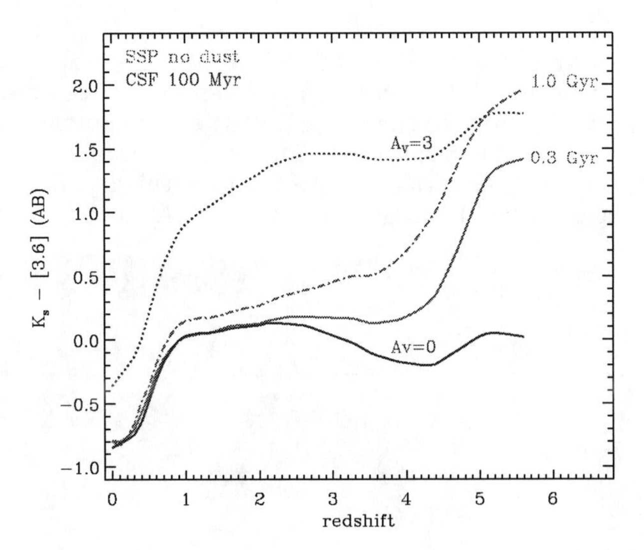

FIGURE 1. Expected $K-[3.6]$ color of galaxies versus redshift from Bruzual and Charlot [16] stellar population synthesis models. The bluest dust-free galaxies have observed $K-[3.6]$ colors > 0 at most redshifts > 1 (blue solid line). Dusty galaxies at $z > 1$ (blue dotted line) and galaxies with old stellar populations at $z > 4.5$ (red solid and dash-dotted line) have much redder colors, reaching $K-[3.6] \sim 2$, which implies that they are ***much*** easier to detect with IRAC than with ground-based near-infrared cameras.

nate the $z = 2-3$ cosmic stellar mass density at the high-mass end (van Dokkum *et al.* [14], Marchesini *et al.* [15]). In addition, surveys at mid-infrared, sub-mm, and radio wavelengths have found highly obscured galaxies, which emit the bulk of their luminosity at IR wavelengths (e.g., Barger *et al.* [17], Blain *et al.* [18]) and may contribute substantially to the global cosmic star formation rate.

The infrared capabilities of the Spitzer Space Telescope have greatly enhanced our understanding of the high redshift Universe. MIPS and IRS are rapidly advancing our knowledge of IR luminous galaxies, such as obscured Active Galactic Nuclei (AGN) and starburst galaxies harboring large amounts of dust (see, e.g., Marleau *et al.* [19], Dole *et al.* [20], Yan *et al.* [21], Houck *et al.* [22], Le Floc'h *et al.* [23], Frayer *et al.* [24], Papovich *et al.* [25]). However, MIPS is not able to study "normal" galaxies out to very high redshift: at redshifts as low as $z \sim 3$ a galaxy has to have a star formation rate exceeding $\sim 200 M_{\odot}\,\mathrm{yr}^{-1}$ to be detectable at 24 μm even in the deepest (10 hr) images, and many times higher to be detected at higher redshift or longer wavelengths.

IRAC, by contrast, excels at detecting distant galaxies of any kind, due to a combination of the shape of their spectral energy distributions (SEDs) and the low background achieved from space. As illustrated in Fig. 1 the bluest galaxies at $z > 1$ have a $K-3.6$ color of ~ 0 in AB units. As IRAC can reach the same AB depth as a ground-based 4m telescope about 20 times faster, Fig. 1 implies that any $z > 1$ object detected with a 4m telescope in the K band can be detected with IRAC in 5 % of the exposure time. For intrinsically red sources this difference is, of course, even larger: dusty galaxies at any redshift and old galaxies beyond $z \sim 4.5$ typically have $K - [4.5] \sim 2$, and for these objects IRAC is a factor of 800 faster than a ground-based 4m telescope! The difference in depth achievable from the ground and from space is illustrated in Fig. 2, which compares a region of the CDF-South field in K to the corresponding $3.6\,\mu$m image. The (VLT) near-IR data in CDF-South are among the best available anywhere in the sky, and yet they are obviously not well matched in depth to the IRAC data.

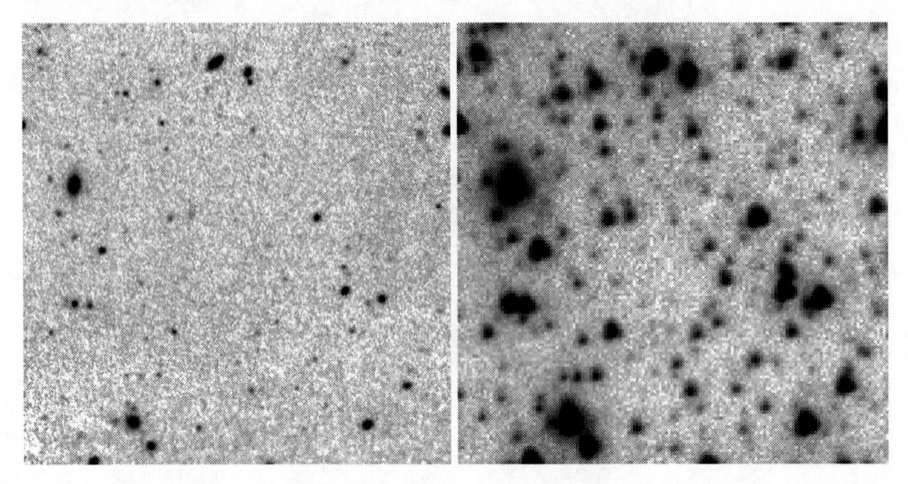

FIGURE 2. Comparison of ground-based K (left) with Spitzer $3.6\,\mu$m (right), for a $1.5' \times 1.5'$ patch in the GOODS CDF-South field. The K band data were taken with ISAAC on the VLT and are of very high quality. Per-pixel exposure times were 7 hrs in K and 20 hrs in the $3.6\,\mu$m band. Despite a very large investment of VLT time in this field (double that of IRAC due to the smaller field of view, for a total of 288 hrs as of writing) the *JHK* depths in GOODS South are poorly matched to the IRAC depth.

For studies of the distant Universe, the key advance allowed by IRAC is not simply survey speed, but the abilitiy to study the rest-frame optical and near-infrared light of galaxies and AGN to much higher redshift than is possible from the ground. As an example, at $z = 7$ the K band samples the rest-frame UV light of galaxies, which is dominated by short-lived O and B stars, whereas the IRAC $4.5\,\mu$m band samples the rest-frame V band, which provides information on Solar type stars and constrains the age and mass of the bulk of the stellar population.

Major achievements with IRAC include: measurements of the abundance of obscured QSOs (Lacy *et al.* [26], Treister *et al.* [27], Stern *et al.* [28], Cool *et al.* [29]); identification of galaxy clusters and groups in the redshift range $1 < z < 2$ (Brodwin *et al.* [30]); identification of massive galaxies with very low star formation rates at $z = 2-3$ (Yan *et al.* [31], Labbé *et al.* [32]); determination of stellar ages and masses of galaxies out to $z \sim 6$ (Eyles *et al.* [33], Yan *et al.* [34], Stark *et al.* [6]); confirmation and characterization of galaxies at $z \sim 7.5$ (Egami *et al.* [35], Labbé *et al.* [36]); and possibly the detection of fluctuations induced by first-light galaxies containing a large fraction of population III stars (Kashlinsky *et al.* [37]).

Nearly all these results were driven by the short wavelength channels of IRAC, as they are the most sensitive. In the warm mission, it will be possible to extend these initial studies to wider areas and larger samples, as well as to fainter luminosities and higher redshifts. Furthermore, very large programs will enable entirely new science, in particular when combined with planned extensive public near-infrared imaging surveys in the next five years.

Here we describe a three-tiered survey program which could be conducted over the course of the warm mission. The surveys comprise ultra-deep observations in a relatively small area, a deep (20 hr per pixel) program over a 2 square degree area, and a shallow (120 s per pixel) program over a 500 square degree area. These programs serve as examples of science that can be done during the warm mission; some other options are briefly discussed in a separate Section.

2. AN ULTRA-DEEP SURVEY

At redshifts above $z \sim 5$ the Balmer break shifts beyond the observed K band, and IRAC is the only instrument until JWST which can provide reliable ages and masses of very high redshift galaxies. As an illustration of the power of IRAC at very high redshifts, Fig. 3 shows 3.6 and 4.5 μm imaging of $z = 7-8$ objects identified in the Hubble Ultra Deep Field. The integration time of ≈ 46 hrs per pixel was sufficient to robustly detect two of the four objects, providing first estimates of the masses, stellar ages, star formation rates, and dust content of these early objects (Labbé *et al.* [36]). Similarly, Egami *et al.* [35] used IRAC to constrain the stellar population of a lensed $z \sim 7$ galaxy.

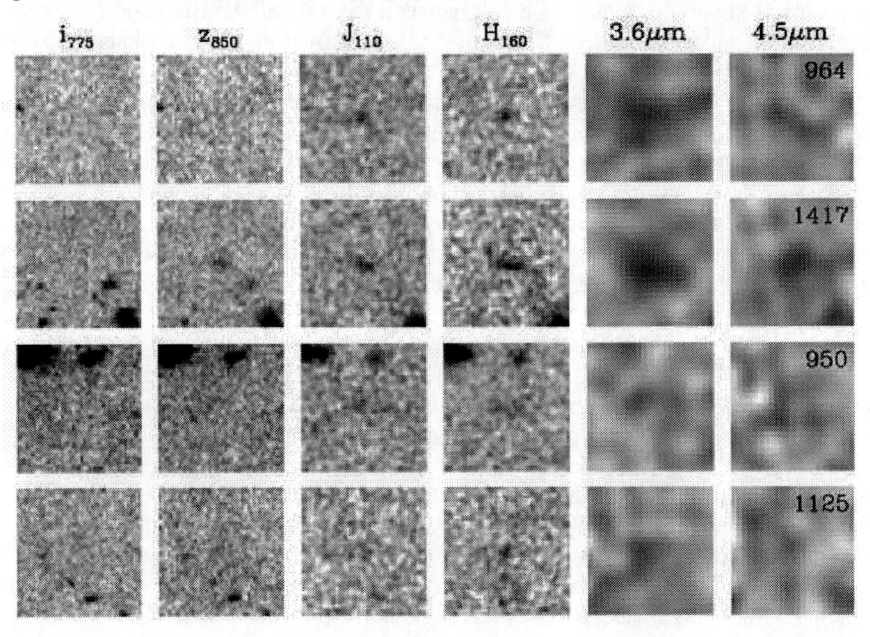

FIGURE 3. IRAC imaging of z-dropouts in the Hubble Ultra Deep Field, from Labbé *et al.* [36]. In ≈ 46 hrs two of these faint z-dropouts are detected with IRAC, and two are marginally detected. To characterize the $z \sim 7+$ galaxy population with IRAC longer integration times and surveys over larger fields are needed.

The galaxy population at $z \sim 7$ may be responsible for reionizing the universe and is of vital importance for understanding feedback and metal production in the earliest stages of galaxy formation. Much deeper IRAC observations over a much wider area than the Hubble Ultra Deep Field are needed to systematically survey the Universe at this important juncture in its history. Extrapolating from the Bouwens *et al.* [38], Bouwens and Illingworth [39] and Labbé *et al.* [36] results, and taking the reduced area due to source confusion into account (see below), a survey over $\sim 150\,\mathrm{arcmin}^2$ with a per-pixel integration time of ~ 250 hrs is needed to obtain a sample of ~ 100 galaxies at $6.5 < z < 7.5$.with $> 5\sigma$ IRAC photometry. The total time required for this survey is ~ 2500 hrs.[1]

An ultra-deep survey would also offer the exciting prospect of a first exploration of the $z \sim 10$ Universe, well in advance of JWST. The depth achieved is $\sim 0.04\,\mu$Jy at $3.6\,\mu$m (~ 27.4 AB), or $M_B \sim -22.6$ at $z = 10$. The expected number of $z \sim 10$ objects is obviously very uncertain, but based on results to $z \sim 7$ one may conservatively expect to detect a handful of galaxies at $9 < z < 11$ (J-dropouts), and ~ 1 object at $12 < z < 14$ (H-dropouts, selected on the basis of their blue $[3.6]-[4.5]$ color and non-detection in HST/WFC3 H).[2] IRAC photometry of galaxies in this redshift range provides very strong constraints on the formation of the first stars. If $z \sim 10$ galaxies experienced their first star formation at this redshift, their K through $4.5\,\mu$m SEDs would be power laws (with the power law index an indication of dust and metal content); if, on the other hand, these objects show a pronounced break between the 3.6 and $4.5\,\mu$m band their spectra have a significant contribution of A stars and star formation must have started several 100 Myr earlier, at $z \sim 20$.

An ultra-deep campaign also offers the possibility of placing limits on the frequency and nature of pair-creation SNe. These SNe are thought to be the end states of very massive ($150-200\,M_\odot$), metal poor stars which may have existed in the early Universe (Abel *et al.* [40], Bromm *et al.* [41]). The peak brightnesses of such SNe are very uncertain, and could range from $0.01-1\,\mu$Jy at $z = 10$ (Scannapieco [42]). The rates are also very uncertain, with estimates ranging from $1-100\,\mathrm{deg}^{-2}\,\mathrm{yr}^{-1}$. With an optimized observing cadence the proposed survey probes the low peak brightness, high rate regime, whereas a wide, shallower survey probes the high peak brightness, low rate regime.

As illustrated in Fig. 3 it is crucial to have supporting near-infrared data that is well matched to the Spitzer depth – in fact, the IRAC data in isolation have very limited value. The near-IR data are needed to identify the high redshift galaxies (by pinpointing their redshifted Lyman break at 1216 Å) and to obtain accurate photometry in the IRAC images (by iteratively modeling the source distribution). Obtaining sufficiently deep K band data is extremely difficult, even if one focuses on the bluest galaxies only. Using the "factor of 20" rule of thumb a 250 hr IRAC depth implies a per-pixel integration time of 5000 hrs on a 4m, or 1000 hrs on a 8-10m class telescope. Fortunately it will be possible to reach the required depth in J and H with HST/WFC3. Based on existing NICMOS data in the Hubble Ultra Deep Field and the expected sensitivity of WFC3, ~ 9 orbits are needed to match the depth of a 250 hr IRAC observation. To cover an area of $150\,\mathrm{arcmin}^2$

[1] A 3×3 pointing mosaic; total survey times in this document include overheads.

[2] These numbers are somewhat conservative as they assume very blue SEDs redward of $\lambda_{\mathrm{rest}} \sim 1400$ Å.

in J and H would require about 700 orbits. Given the importance of supporting near-IR data an ultra-deep IRAC program should probably only be undertaken in coordination with an investment of HST time of this order.

A drawback to an ultra-deep field is the limited efficiency of IRAC at faint flux levels due to crowding. At the GOODS depth only $\sim 30\,\%$ of pixels are uncontaminated background, that is, not affected by the wings of the PSFs of identified sources. Source confusion is not a hard limit, and can be greatly reduced with the use of a prior image with better resolution (typically a K-band image). However, confusion reduces the efficiency of IRAC observations in two ways: the fraction of the field in which good photometry can be done steadily diminishes when going deeper, and the S/N increases slower than $\sqrt{t}$ due to the steadily rising "background" of PSF wings.[3] At the time of writing, no results are yet available from the deepest – 100 hr per pixel – region that has been obtained with IRAC so far (GOODS HDF-N); when these results are in it will be easier to quantify the effects of crowding with integration times > 50 hrs per pixel.

Another drawback is that this type of science in particular can be done with much greater efficiency with JWST. There is little doubt that JWST will image GOODS-sized fields (and larger), and that the depth of IRAC data can be surpassed very rapidly indeed: quite apart from its vastly superior PSF ($0.1'' - 0.2''$ at $3 - 5\,\mu$m) the required exposure time to reach a given point-source depth is about three orders of magnitude shorter. Assuming a typical high redshift galaxy size of $0.5''$ ($1.0''$) FWHM and factoring in the respective detector sizes, JWST/NIRCam can cover small areas about 200 (40) times faster than Spitzer/IRAC. Although this may limit the legacy value of an ultra-deep Spitzer survey, such considerations have to be weighed against the long lead time for JWST and the uncertainties associated with any space mission.

3. A DEEP SURVEY OVER 2 DEG^2

Although much larger than the original Hubble Deep Fields, the $10' \times 15'$ GOODS fields (Dickinson *et al.* [43]) are too small to provide a fully representative sample of the distant Universe: the correlation length r_0 of massive galaxies is $\sim 8h_{100}^{-1}$ Mpc (roughly independent of redshift), which is $\sim 8'$ at $z = 2$ (e.g., Daddi *et al.* [10], Somerville *et al.* [44], Adelberger *et al.* [45], Quadri *et al.* [46]). The GOODS fields are also too small for clustering studies (except for populations with small r_0), and for studies of the relation between galaxy properties and density. The importance of sampling large volumes at high redshift is dramatically illustrated by the identification of structures of several tens of Mpc up to $z \sim 6$ (e.g., Ouchi *et al.* [5]).

Furthermore, the relatively small size of GOODS does not sample the bright end of the luminosity function well, which means that the brightest galaxies at high redshift are missed even if the depth is sufficient to detect them. As an example, the Bouwens *et al.* [38] $z = 6$ luminosity function implies that only ≈ 5 $L > 3L_*$ galaxies at $5.5 < z < 6.5$

[3] For example, data in the Hubble Ultra Deep Field (≈ 46 hrs) suggests that the depth increase compared to 1 hr is only 1.7 mag instead of the 2.1 mag expected from $\sqrt{t}$, even after reducing the source confusion using available NICMOS near-IR data (Labbé *et al.* [36]).

are expected in a 150 arcmin^2 area. Although these bright examples may not contribute greatly to the total luminosity density at these early epochs (see, e.g., Bouwens *et al.* [38]), they may be accessible for morphological studies with WFC3 and spectroscopic follow-up with 20m – 30m telescopes and JWST.

Motivated by similar concerns, several programs are underway to extend the area covered by deep ground- and space-based observations. Examples are the $30' \times 30'$ Extended CDF-South (E-CDFS, aka the GEMS field); the $50' \times 50'$ UKIDSS Ultra Deep Survey (aka the Subaru/XMM deep field); the $10' \times 60'$ Extended Groth Strip; and the $1.4° \times 1.4°$ COSMOS field. All these fields have excellent supporting data, although different fields have different strengths. Current IRAC coverage of these fields varies. The E-CDFS and the Groth Strip have both been covered with IRAC to ~ 3 hr depth. The UDS will be done with IRAC to ~ 0.7 hr depth in Cycle 4, and the COSMOS field has relatively shallow (~ 0.3 hr) IRAC coverage over the entire field.

Given the large investments of ground- and space-based observatories in these fields it seems likely that they will continue to play important roles in studies of the distant Universe. New instrumentation on existing telescopes (e.g., multi-object near-IR spectrographs on 10m class telescopes and WFC3 on HST) will likely be utilized in these fields, as well as future telescopes (Herschel, ALMA, 20-30m telescopes, JWST). There is therefore a strong legacy argument to be made for covering several or all of these fields with substantially deeper 3.6 and 4.5 μm imaging than is currently available.

The availability of near-IR imaging that is well matched to the IRAC depth is crucial for correctly measuring the IRAC fluxes and for determining photometric redshifts. Interestingly, *none* of the fields mentioned currently has near-IR coverage approaching the depth achieved in a few hours (per pixel) with IRAC. However, this situation will change in the near future thanks to ambitious public surveys with new large field near-IR imagers on 4m class telescopes. WFCAM on UKIRT will cover the Subaru/XMM deep field to a 5σ AB depth of $K = 25$ (with additional J and H) in the context of the UKIDSS Ultra Deep Survey (Dye *et al.* [47]). UltraVISTA (an approved public survey on the soon to be commissioned VISTA telescope) aims to cover 1/3 of the COSMOS field to a depth of $K = 24.5$ and 1/3 to a depth of $K = 25.6$ (with additional Y, J, and H). An IRAC depth of 20 hrs per pixel is well matched to the K band depths of UKIDSS/UDS and UltraVISTA, in the sense that every K-detected source will have a $3.6\,\mu\text{m} > 5\sigma$ counterpart.

Covering the other two fields should also be a high priority. Their areas are small compared to the UKIDSS/UDS and COSMOS UltraVISTA fields — which implies that the investment with Spitzer would be relatively modest — and they offer qualitatively different legacy value. Covering only the 0.7 deg^2 UDS field and the 0.8 deg^2 COSMOS/UltraVISTA field to the GOODS depth would cost $\sim 6,000$ hrs, whereas covering all four fields would require $\sim 7,500$ hrs.[4]

The area and depth of such a ≈ 2 deg^2 survey should be sufficient to detect 1000s of galaxies at $5 < z < 8$. At these redshifts IRAC uniquely samples the rest-frame optical emission beyond the Balmer break, allowing measurements of star formation histories and stellar masses (see, e.g., Labbé *et al.* [36], Stark *et al.* [6]). The intrinsic brightness

[4] In practice, it may be beneficial to vary the exposure time within a field or between fields somewhat.

of these objects implies that they can be observed spectroscopically, either with existing telescopes or with future 20m-30m telescopes and/or JWST. In combination with the ultra-deep survey discussed above, which samples the luminosity function at $L < L_*$, the evolution of the rest-frame optical luminosity function and the stellar mass function can be accurately measured at $5 < z < 8$.

Furthermore, the survey will characterize the relation between galaxies and the emerging large scale structure over the redshift range $2 < z < 6$. GOODS-depth IRAC observations over 2 deg^2 would allow characterization of the stellar populations of several tens of thousands of red and blue galaxies in this redshift range to low stellar mass limits (e.g., $\sim 10^{10} M_\odot$ at $z = 3$) and accurately determine their density and evolution in relation to their environment. The combination of clustering and stellar population measurements is an extremely powerful tool to determine the properties and evolution of galaxies as a function of halo mass (e.g., Adelberger *et al.* [45], Lee *et al.* [48], Quadri *et al.* [46]), thus linking the hierarchical build-up of dark matter halos to the formation and evolution of their constituent galaxies.

Deep IRAC observations over such a large area also offer intriguing possibilities for studies of faint galaxies below the detection threshold. If first-light galaxies during reionization were to contain a substantial fraction of massive population III stars then their redshifted rest-frame UV emission will be present at IR wavelengths. While none of these sources will be individually detected even in the ultra-deep survey described earlier, the unresolved emission will be clustered (as these sources are expected to trace the large-scale structure at $z > 7$) and this clustering component can be extracted to the extent that any correlated systematics and noise sources are understood. A first detection of such a clustered component in the unresolved IRAC pixels in the first-look survey was interpreted as evidence for massive population III stars (Kashlinsky *et al.* [37]), although this result is somewhat controversial (Cooray *et al.* [49]). A deep survey over 2 deg^2 will make it possible to accurately measure the clustering strength of the undetected sources, allowing a direct comparison to model predictions for the clustering of first-light objects.

4. A SHALLOW SURVEY OVER 500 DEG^2

Areas of several square degrees are sufficient to obtain representative samples of the Universe at $z \leq 1$, but they are not large enough for studies of extreme objects such as luminous quasars or high-redshift galaxy clusters. Although the instantaneous field-of-view of IRAC is small, it can do very efficient mapping over large areas of sky; as an example, the SWIRE survey covered 49 deg^2 to a depth of 120 s. An order of magnitude larger survey than SWIRE would take $\sim$ 4000 hrs and serve a wide range of science goals.

High redshift quasars can be efficiently identified by their relatively flat mid-IR SEDs and their extremely red optical – mid-IR colors (Cool *et al.* [29], Stern *et al.* [50]). The clustering strength of these objects will constrain the masses of their dark matter halos, and spectroscopic follow-up will provide information on the build-up of supermassive black holes and the interplay of star formation and nuclear activity in the earliest phases of galaxy formation. Quasars also provide useful probes of the intervening universe; indeed, the most distant quasars provide some of our most powerful tools for probing the epoch of reionization (Becker *et al.* [51]).

Galaxy clusters can easily be identified out to $z \sim 2$ with IRAC in integration times as short as a few minutes (see Eisenhardt *et al.* [52] and Fig. 4). Based on the WMAP3 cosmology, a 500 deg^2 survey to the SWIRE depth would provide $\sim$ 1500 clusters at $1 < z < 2$ with masses $> 10^{14} M_{\odot}$, and a handful of extremely massive objects with masses $> 5 \times 10^{14} M_{\odot}$. The evolution of galaxies in these clusters provides information on the fate of the earliest objects that formed in the Universe, and the observed mass-dependent evolution of the abundance of clusters over $1 < z < 2$ provides strong constraints on cosmological parameters (particularly w).

FIGURE 4. Color composite of B, I, and IRAC 4.6 μm images of a galaxy cluster at $z = 1.41$, from Stanford *et al.* [53]. The ground-based B and I images required several hours of exposure time on a 4m telescope, but the integration time for the IRAC 4.5 μm image was only 90 seconds!

IRAC 3.6 and 4.5 μm data alone can provide a crude redshift estimate, as the $[3.6] - [4.5]$ color fairly cleanly separates galaxies with redshifts below or above 1 (see Fig. 5). However, the returns from this survey will be greatly enhanced when it is performed in an area, or areas, of sky with existing or planned ancillary data. Examples of such areas are the South Pole Telescope's SZ survey, and the fields imaged by the near-IR VISTA Kilo Degree Survey (KIDS). The combination of these data will not just allow detection of the clusters, but also enable redshift and mass estimates.

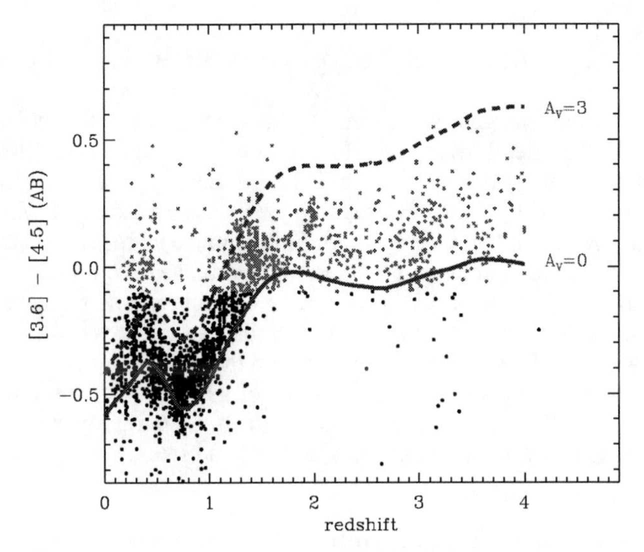

FIGURE 5. The observed $[3.6]-[4.5]$ colors of galaxies versus redshift in the GOODS-south field. The points show a sample of K-selected galaxies (Wuyts *et al.* in prep). The tracks show the envelope of colors spanned by stellar population models with a range of dust attenuations. A simple cut only in $[3.6]-[4.5]$ can efficiently isolate galaxies at $z > 1$ regardless of SED shape.

We note that wide-area, shallow surveys in high latitude fields could also prove useful for Galactic programs, most notably for detecting and characterizing the coldest brown dwarfs (e.g., Stern *et al.* [50]). Objects cooler than about 700 K, so-called "Y dwarfs", must exist. Objects with inferred masses down to $\approx 5M_{\rm Jup}$ have been identified in star-forming regions and, according to theoretical models, dwarfs less massive than $30M_{\rm Jup}$ with ages > 4 Gyr should have $T < 600$ K. However, none have been found to date. This is primarily because their SEDs peak at $\approx 4.5\,\mu$m (e.g., Burrows *et al.* [54]), making them very faint at ground-based optical through near-IR wavelengths. For instance, a 600 K brown dwarf could only be detected in the 2MASS PSC to about 1 pc. In contrast, it would be detectable in a 120 sec IRAC $\approx 4.5\,\mu$m image out to about 50 pc.

5. OTHER PROGRAMS

We consider the three-tiered approach outlined above an excellent starting point for designing observational programs for studies of the distant Universe in the warm era. Many other survey programs could, of course, be considered, and we briefly discuss several alternatives here.

5.1. A Medium Deep Survey Over Several 10s of deg^2

There is a conspicuous gap in the three surveys discussed in this document, as we jumped from a 20 hr depth over several deg^2 to a 120 s depth over 100s of deg^2. A survey over several 10s of deg^2 to a $\sim$ 1 hr depth would require a similar investment as each of the surveys discussed in more detail in the preceeding Sections. This territory is out of reach of JWST and a unique niche for Spitzer in the warm era.

This committee failed to come up with a broadly defined, high impact science case for this type of survey, but that may simply reflect the biases and preconceptions of its members. A survey of this type would map large scale structure at $z = 2-4$ over a very wide area, which could lead to new constraints on the growth of dark matter halos and perhaps cosmology. Thanks to the large number of galaxies that would be observed it would also be possible to split the sample into many bins, and study galaxy evolution as a function of luminosity, mass, color, AGN-activity, and size.

5.2. An Extremely Wide Survey of 1000s of deg^2

It may seem odd to consider using a $5' \times 5'$ imager to cover areas requiring hundreds of thousands of pointings. Nevertheless, the unique wavelength regime and sensitivity of IRAC, combined with the large amount of time that is potentially available in the warm era, warrant a discussion of this question.

An ultra-wide survey will identify the most extreme objects in the Universe, such as very luminous quasars and galaxy clusters with masses $\sim 10^{15} M_{\odot}$. However, the high overheads associated with very short integrations make such a program either very inefficient or extremely costly. The spacecraft overheads are such that they start to dominate over the on-sky time for integration times per pointing significantly shorter than $\sim$ 100 s. As an example, a survey of 125 square degrees with a 120 s exposure time (comprising 4 dithered 30 s exposures) takes about 1000 hrs. A survey of 2500 square degrees with a 6 s exposure time (comprising 3 dithered 2 s exposures) would have the same total on-sky integration time, but cost more than 6000 hrs due to greatly increased overheads.

Taking 120 s exposure time as a minimum, covering 2000 square degrees would be extremely costly as it would require 16,000 hours. Such a large expenditure may be difficult to justify given the somewhat limited additional science that can be accomplished above and beyond the 500 deg^2 scenario discussed earlier.

5.3. Gravitationally Lensed Galaxies

Gravitational lensing by foreground clusters allows the study of high redshift galaxies fainter than the limits achievable in unmagnified fields (e.g., Ellis *et al.* [55], Stark *et al.* [56]), and detailed analysis of intrinsically more luminous galaxies (e.g., Pettini *et al.* [57]). The gain in S/N is substantial: the exposure time needed to reach a given lensing-corrected limiting magnitude decreases as A^{-2} (for point sources), with the lensing amplification A reaching values of 20 in extreme cases.

This technique has great potential, although there are some drawbacks: the small volume that is sampled at high redshift (as the relevant region is limited to a $\sim 1'$ diameter annulus whose lensing-corrected area decreases with A), the requirement that the mass distribution in the inner parts of the cluster can be adequately modeled (to correct the measured properties for the effects of lensing), and crowding. The latter aspect is particularly problematic for IRAC, due to its large PSF compared to the distances between galaxies in the central parts of clusters.

IRAC has already yielded interesting results in this area: Egami *et al.* [35] report the detection of a significant Balmer break in a previously identified lensed $z \sim 6.7$ galaxy, based on 3.6 and 4.5 μm IRAC data of the well-studied cluster Abell 2218. A program is currently underway to systematically image 30 lensing clusters with IRAC, and it may be very interesting to extend this type of work in the warm era.

5.4. The Stellar Populations of $z = 10$ Galaxies

Although these things are difficult to predict, it seems likely that WFC3 on HST, VISTA, HAWKEYE on the VLT, or some other new capability will identify a robust sample of J band dropouts in the near future (see Bouwens *et al.* [58]). IRAC imaging of these objects will both confirm them (by establishing whether they have a blue continuum redward of Lyα) and constrain their stellar populations by measuring the strength of the redshifted Balmer break (which falls between the 3.6 and 4.5 μm bands at this redshift). We note that extremely deep IRAC imaging may already be available if the objects are found in a combined WFC3/IRAC survey, as advocated above.

5.5. Future Priorities

The program described in the preceeding paragraph is an example of science that cannot currently be planned (although anticipated), and it is almost certain that many other exciting possibilities will emerge during the remaining lifetime of Spitzer. Such future programs can be large surveys, but could also be small, very high impact observations of special objects, special sky areas, or time-variable objects (e.g., a $z = 10$ gamma-ray burst).

It is crucial that a fraction of the time available in the warm period will remain unassigned, to accomodate the shifting frontiers in the field. However, there will be limitations imposed by the anticipated reduction in user support. It may be possible to have a TAC process twice during the 5 year warm mission (rather than yearly) to assign remaining survey time and to accomodate a limited number of small, high-impact programs which do not require a large support effort on the part of the SSC.

6. CONCLUSIONS

The end of Spitzer's cryogenic lifetime will leave its most sensitive and versatile capability for studying the distant Universe intact, enabling very ambitious survey programs addressing a wide range of science. Nearly anything that is done in the warm mission will explore unique parameter space, as there is no competitive instrument in this wavelength regime until JWST. Among the various possibilities, we feel that the three-tiered approach outlined in this document would extend currently available samples by at least an order of magnitude, enable qualitatively new science, and serve a wide community. The survey parameters are summarized in Table 1.

TABLE 1
Recommended Surveys

Area	Depth	Total time	Fields	Main science drivers
150 arcm^2	250 hr	~ 2500 hr	TBD	galaxies at $z = 7-14$
2 deg^2	20 hr	~ 7500 hr	COSMOS, UDS, EGS, E-CDFS	bright galaxies at $z > 6$ AGN at $z = 1-7+$ clustering at $z = 2-6$
500 deg^2	120 s	~ 4000 hr	TBD	quasars to $z \sim 7$ galaxy clusters at $1 < z < 2$

Chosing survey fields is a charged subject, as several large groups in the high redshift community have invested significant effort and resources in particular areas of the sky. This document leaves this issue open for the ultra-deep and shallow surveys, as there are no fields that can be easily identified as superior to all others. Distributing these surveys may also be an option, e.g., covering two widely separated 75 arcmin^2 fields in the ultra-deep survey rather than a single 150 arcmin^2 area.

However, we are explicit about the fields that can be covered in the deep 2 deg^2 survey. Despite the large investment of IRAC time that would be required, we believe a case can be made for covering all four well-studied > 0.25 deg^2 fields. Each of these fields offers qualitatively different legacy value: the UDS and COSMOS fields will have the best near-IR coverage, the EGS has the best spectroscopy, a lower mid-IR background than the equatorial fields, and is well placed for Northern telescopes, and the E-CDFS has very low mid-IR background and is ideally placed for Chilean telescopes (including ALMA). These four fields have been vetted for their legacy value by many time allocation committees for ground- and space-based facilities, and one may question whether it is sensible to do that yet again.

An important consideration in this context is not just the quality of the supporting data, but their access. The survey programs that are considered in this document require such a large investment of Spitzer time that a level playing field is absolutely crucial. A field should therefore only be observed if access to crucial supporting data (e.g., near-IR imaging) is completely unrestricted. This would be an extension of the usual process, where proposers use their (often partially proprietary) data to argue for a certain survey strategy or sky area, and then promise to make the space-based data publicly available in reduced form.

The TAC process is also unusual, in the sense that the size of the envisioned proposals will exceed even the largest programs that have been executed on space observatories to date. It is unlikely that proposers will have a chance to revise their proposals for a future round, as a large fraction of the available time over the entire warm mission may be reserved in a single proposal round. TACs inevitably vary in their composition, priorities, and expertise, and special care needs to be taken to ensure that the best science is selected for this unique opportunity.

In practice it may be desirable to have the SSC implement one or more of the TAC-approved surveys, following the example of the various Hubble deep field campaigns. This will capitalize on the experience and expertise of the SSC staff, ensure a timely distribution of reduced data, and allow the community to focus their efforts on the science enabled by these surveys rather than their execution.

ACKNOWLEDGMENTS

We thank Rychard Bouwens, Ranga-Ram Chary, Alastair Edge, Eiichi Egami, Jonathan Gardner, Subha Majumdar, and Adam Stanford for their input, and Lisa Storrie-Lombardi for her expert coordination of the Warm Mission preparatory efforts.

REFERENCES

1. Steidel, C. C., Giavalisco, M., Pettini, M., Dickinson, M., and Adelberger, K. L. 1996, ApJ, 462, L17
2. Steidel, C. C., Giavalisco, M., Dickinson, M., and Adelberger, K. L. 1996, AJ, 112, 352
3. Steidel, C. C., Shapley, A. E., Pettini, M., Adelberger, K. L., Erb, D. K., Reddy, N. A., and Hunt, M. P. 2004, ApJ, 604, 534
4. Kodaira, K., *et al.* 2003, PASJ, 55, L17
5. Ouchi, M., *et al.* 2005, ApJ, 620, L1
6. Stark, D., Bunker, A., Ellis, R., Eyles, L., and Lacy, M. 2007, ApJ, 659, 84
7. Dow-Hygelund, C., *et al.* 2007, ApJ, 660, 47
8. Shapley, A., Steidel, C., Adelberger, K., Dickinson, M., Giavalisco, M., and Pettini, M. 2001, ApJ, 562, 95
9. Papovich, C., Dickinson, M., and Ferguson, H. 2001, ApJ, 559, 620
10. Daddi, E., Cimatti, A., and Renzini, A. 2000, A&A, 362, L45
11. McCarthy, P. J., *et al.* 2001, ApJ, 560, L131
12. Labbé, I., *et al.* 2003, AJ, 125, 1107
13. Franx, M., *et al.* 2003, ApJ, 587, L79
14. van Dokkum, P. G., *et al.* 2006, ApJ, 638, L59
15. Marchesini, D., *et al.* 2007, ApJ, 656, 42
16. Bruzual, G., and Charlot, S. 2003, MNRAS, 344, 1000
17. Barger, A. J., Cowie, L. L., and Richards, E. A. 2000, AJ, 119, 2092
18. Blain, A. W., Smail, I., Ivison, R. J., Kneib, J.-P., and Frayer, D. T. 2002, Phys. Rep., 369, 111
19. Marleau, F. R., *et al.* 2004, ApJS, 154, 66
20. Dole, H., *et al.* 2004, ApJS, 154, 93
21. Yan, L., *et al.* 2005, ApJ, 628, 604
22. Houck, J. R., *et al.* 2005, ApJ, 622, L105
23. Le Floc'h, E., *et al.* 2005, ApJ, 632, 169
24. Frayer, D., *et al.* 2006, AJ, 131, 250
25. Papovich, C., *et al.* 2006, ApJ, 640, 92
26. Lacy, M., *et al.* 2004, ApJS, 154, 166
27. Treister, E., *et al.* 2004, ApJ, 616, 123

28. Stern, D., *et al.* 2005, ApJ, 631, 163
29. Cool, R., *et al.* 2006, AJ, 132, 823
30. Brodwin, M., *et al.* 2006, ApJ, 651, 791
31. Yan, H., *et al.* 2004, ApJ, 616, 63
32. Labbé, I., *et al.* 2005, ApJ, 624, L81
33. Eyles, L., Bunker, A., Stanway, E., Lacy, M., Ellis, R., and Doherty, M. 2005, MNRAS, 364, 443
34. Yan, H., *et al.* 2005, ApJ, 634, 109
35. Egami, E., *et al.* 2005, ApJ, 618, L5
36. Labbé, I., Bouwens, R., Illingworth, G., and Franx, M. 2006, ApJ, 649, L67
37. Kashlinsky, A., Arendt, R., Mather, J., and Moseley, S. 2005, Nature, 438, 45
38. Bouwens, R., Illingworth, G., Blakeslee, J., and Franx, M. 2006, ApJ, 653, 53
39. Bouwens, R., and Illingworth, G. 2006, Nature, 443, 189
40. Abel, T., Bryan, G., and Norman, M. 2002, Science, 295, 93
41. Bromm, V., Coppi, P., and Larson, R. 2002, ApJ, 564, 23
42. Scannapieco, E. 2006, astro-ph/0609208
43. Dickinson, M., *et al.* 2004, BAAS, 36, 701
44. Somerville, R. S., Lee, K., Ferguson, H. C., Gardner, J. P., Moustakas, L. A., and Giavalisco, M. 2004, ApJ, 600, L171
45. Adelberger, K. L., Erb, D. K., Steidel, C. C., Reddy, N. A., Pettini, M., and Shapley, A. E. 2005, ApJ, 620, L75
46. Quadri, R., *et al.* 2007, ApJ, 654, 138
47. Dye, S., *et al.* 2006, MNRAS, 372, 1227
48. Lee, K., Giavalisco, M., Gnedin, O., Somerville, R., Ferguson, H., Dickinson, M., and Ouchi, M. 2006, ApJ, 642, 63
49. Cooray, A., *et al.* 2007, ApJ, 659, L91
50. Stern, D., *et al.* 2007, ApJ, 663, 677
51. Becker, R., *et al.* 2001, AJ, 122, 2850
52. Eisenhardt, P., *et al.* 2006, BAAS, 209, 161.05
53. Stanford, A., *et al.* 2005, ApJ, 634, L129
54. Burrows, A., Sudarsky, D., and Lunine, J. 2003, ApJ, 596, 587
55. Ellis, R., Santos, M., Kneib, J-P, and Kuijken, K. 2001, ApJ, 560, L119
56. Stark, D., Ellis, R., Richard, J., Kneib, J-P., Smith, G., and Santos, M. 2007, ApJ, 663, 10
57. Pettini, M., Steidel, C., Adelberger, K., Dickinson, M., and Giavalisco, M. 2000, ApJ, 528, 96
58. Bouwens, R., Illingworth, G., Thompson, R., and Franx, M. 2005, ApJ, 624, L5

Star Formation: Answering Fundamental Questions During the Spitzer Warm Mission Phase

Steve Strom*, Lori Allen†, John Carpenter**, Lee Hartmann‡, S. Thomas Megeath§, Luisa Rebull¶, John R. Stauffer¶ and Michael Liu||

*NOAO, P.O. Box 26732, Tucson, AZ, 85726-6732, USA
†Harvard-Smithsonian Astrophysical Observatory, 60 Garden St., Cambridge, MA, 02138, USA
**California Institute of Technology, MC 105-24, Pasadena, CA, 91125, USA
‡University of Michigan, 500 Church St., 830 Dennison Bldg., Ann Arbor, MI, 48109, USA
§University of Toledo, Ritter Observatory, M-113, Toledo, OH, 43606, USA
¶Spitzer Science Center, Caltech MC 314-6, Pasadena, CA, 91125, USA
||Institute of Astronomy, University of Hawaii, 2680 Woodlawn Dr., Honolulu, HI, 96822, USA

Abstract.
Through existing studies of star-forming regions, Spitzer has created rich databases which have already profoundly influenced our ability to understand the star and planet formation process on micro and macro scales. However, it is essential to note that Spitzer observations to date have focused largely on deep observations of regions of recent star formation associated directly with well-known molecular clouds located within 500 pc. What has **not** been done is to explore to sufficient depth or breadth a representative sample of the much larger regions surrounding the more massive of these molecular clouds. Also, while there have been targeted studies of specific distant star forming regions, in general, there has been little attention devoted to mapping and characterizing the stellar populations and star-forming histories of the surrounding giant molecular clouds (GMCs). As a result, we have yet to develop an understanding of the major physical processes that control star formation on the scale or spiral arms. Doing so will allow much better comparison of star-formation in our galaxy to the star-forming complexes that dominate the spiral arms of external galaxies.

The power of Spitzer in the Warm Mission for studies of star formation is its ability to carry out large-scale surveys unbiased by prior knowledge of ongoing star formation or the presence of molecular clouds. The Spitzer Warm Mission will provide two uniquely powerful capabilities that promise equally profound advances : high sensitivity and efficient coverage of many hundreds of square degrees, and angular resolution sufficient to resolve dense groups and clusters of YSOs and to identify contaminating background galaxies whose colors mimic those of young stars. In this contribution, we describe two major programs: a survey of the outer regions of selected nearby OB associations, and a study of distant GMCs and star formation on the scale of a spiral arm.

Keywords: Spitzer Space Telescope, infrared astronomical observations, star formation, circumstellar shells, clouds, stellar activity
PACS: 95.85.Hp, 97.10.Bt, 97.10.Fy, 97.10.Jb

1. INTRODUCTION

The sensitivity, angular resolution and wavelength coverage of the Spitzer Space Telescope have enabled fundamental contributions to the study of star formation during its first three years of operation. Among the most noteworthy are:

1. Providing a thorough census of forming stars in nearby (d < 500 pc) molecular

CP943, *The Science Opportunities for the Warm Spitzer Mission Workshop*,
edited by L. J. Storrie-Lombardi and N. A. Silbermann

cloud complexes – from optically-obscured protostars, still enshrouded by their natal, collapsing cores, to optically revealed stars surrounded by circumstellar accretion disks, the birthplaces of planetary systems. As a result of these surveys, astronomers are now engaged in a panoply of research programs using both Spitzer and ground-based optical and radio telescopes to answer questions key to understanding the star and planet formation process:

   (a) How are initial conditions in protostellar cores (mass, density, temperature, turbulent speeds, rotation, infall rates) surrounding individual forming star-disk systems linked to outcome stellar properties?

   (b) Is there evidence for ongoing planet formation in circumstellar disks? If so, what is the relationship between the properties of forming planetary systems and the physical conditions of their parent disks?

2. Providing clear evidence that multiple star formation modes are manifest in nearby star-forming regions, ranging from young stars formed in isolation or in small aggregates of a few tens of stars, to clusters with 100 to 1000 stars. As a result, we are now armed with a database that promises over the next few years to uncover:
   (a) What conditions in parent molecular clouds lead to the observed outcome stellar populations (isolated/aggregate; cluster)?
   (b) How does environment influence star formation and disk evolution?
3. Providing a census of stars surrounded by circumstellar disks and spanning a wide range of masses (from brown dwarfs to stars as massive as 10 $M_\odot$) and ages from less than 1 Myr to 5-10 Myr. As a result, we now have in hand
   (a) An initial understanding of the range of timescales for the survival of accretion disks around stars of differing mass, and thus constraints on the timescales for completion of the initial phases of planet-building;
   (b) Evidence of disk evolution: (i) grain settling to the mid-plane – the first step in the planet-building process; (ii) evidence of grain growth and processing from analysis of mineralogical signatures; and (iii) evidence of major changes in the distribution of solid material within the disk, including evidence of large (multiple AU-sized) 'holes', which combined with other measurements suggest strongly that giant planet formation is underway in some systems.
4. Providing the first large-scale maps of the young stellar populations associated with a few selected giant molecular cloud complexes located both in the inner (Carina arm) and outer (Perseus arm) galaxy. As a result, we are beginning to see clearly the morphological relationships between multiple episodes of star formation in these complexes, and from these relationships to address questions essential to understanding large-scale star-formation processes in other galaxies:
   (a) What is the role of feedback from young stars in triggering, regulating and terminating star formation? Does formation of massive stars result in 'triggering' of star-forming episodes in nearby molecular material as a result of propagating HII regions?
   (b) How do the properties of stellar populations formed 'initially' in molecular clouds differ from YSOs formed as a result of 'triggering'?

Spitzer has clearly created rich databases which have already profoundly influenced our ability to understand the star and planet formation process on micro and macro scales, and have as well provided grist for a decade or more of follow-up analysis using complementary tools – ground-based radio and O/IR telescopes and next generation facilities in space (e.g. JWST and Herschel).

However, it is essential to note that Spitzer observations to date have focused largely on deep observations of regions of recent star formation associated directly with well-known molecular clouds located within 500 pc (e.g., Taurus-Auriga; Ophiuchus; NGC 1333 and IC 348; the Orion A and Orion B molecular clouds, as well as the Orion Nebula Cluster itself). In most cases, the IRAC integrations were chosen in order to detect and characterize young stars with masses at or in some cases well below the hydrogen-burning limit. What has not been done is to explore to sufficient depth or breadth is a representative sample of the much larger regions surrounding the more massive of these molecular clouds: larger T- and OB- associations comprised of multiple older generations of stars (some as old as 15 Myr) which have long since dispersed their natal material. As a result, we have little information regarding the likely sequence of star formation and the timescales associated with key events such as the dissipation of planet-building circumstellar disks.

Furthermore, while there have been studies of a small selection of more distant star-forming regions located at d $\sim$1-2 kpc from the Sun, these have tended to focus on small regions centered on known centers of active star formation. In general, there has been little attention devoted to mapping and characterizing the stellar populations and star-forming histories of the surrounding giant molecular clouds (GMCs). As a result, we have yet to develop an understanding of the major physical processes that control star formation on the scale of the large complexes that dominate the spiral arms of galaxies, or more globally on the scale of a spiral arm.

2. THE POWER OF SPITZER DURING ITS WARM MISSION PHASE

Despite restriction to the 3.6 and 4.5 μm IRAC channels, the Spitzer Warm Mission Phase will provide two uniquely powerful capabilities that promise equally profound advances – (i) sensitivity sufficient to detect infrared emission from dust around young stars and protostars, and hence to locate representatives of young stellar populations spanning areas of many hundreds of square degrees; and (ii) angular resolution sufficient to resolve dense groups and clusters of YSOs and to identify contaminating background galaxies whose colors mimic those of young stars. Together, these capabilities will enable us (i) to survey young stars in nearby OB associations in which episodes of star formation have continued for timescales approaching 15 Myr; and (ii) to conduct large-scale surveys of rich star formation complexes unbiased by the presence of already known 'signposts' of star formation for a representative sample of GMC complexes located at 1-2.5 kpc.

Carrying out these surveys will allow us to understand:

1. The number and spatial distribution of star-forming episodes in OB associations,

and how these episodes may be linked to the initial conditions in their parent molecular cloud complex;

2. The timescales for circumstellar disk evolution for stars spanning a wide range of masses and ages, including in particular, long-lived (t > 10 Myr) disks;
3. The properties of emerging stellar populations in rich, dense clusters and associations whose characteristics are far more similar to the star-forming complexes we observe in other galaxies than those of well-studied regions located within 0.5 kpc of the Sun;
4. The processes that initiate and sustain star-formation on the scale of a galactic spiral arm – an essential complement to extant Spitzer studies of targeted small regions or well-known GMCs;

We describe below two major programs aimed at addressing the above issues, outlining notional target regions, nominal sensitivity limits, corresponding survey times, along with a third program which offers the possibility of serendipitous discovery of variability patterns that may provide fundamental insight into the star and disk assembly processes. Following a brief description of each of the proposed programs, we outline the potential 'legacy' value of the databases for follow-on ground- and space- based optical and radio observations, as well as the potential combined value of the Spitzer and ground-based programs. We also summarize the relative capabilities of the Spitzer Warm Mission and other planned space facilities (e.g. WISE; JWST) for carrying out aspects of the proposed science programs. Finally, we discuss the value of an ongoing archival research program, both during the Spitzer Warm Mission and beyond.

3. PROGRAM I. SURVEY OF SELECTED OB ASSOCIATIONS

3.1. Outline of a Possible Warm Mission Program

In large star-forming regions, molecular clouds and their associated newly-formed stellar populations are often found embedded within much larger OB associations which harbor evidence of multiple star-forming events which in some cases span up to 15 Myr. Exploiting the Spitzer Warm Mission's ability to carry out deep IRAC 3.6 and 4.5 μm observations offers the potential of identifying young stars surrounded by circumstellar disks and/or envelopes and located within large, nearby OB associations. Candidate regions and their characteristics are listed in Table 1.

TABLE 1. Candidate OB Associations for Spitzer Warm Mission Mapping

Associations	Distance	S-F age range (Myr)	Size (sq deg)	Time to Map (hrs)
Orion OB	0.5	0.1 - 12	100	500
Cyg OB1	0.8	1 - 10	50	250
Per OB1	2	1 - 20	225	1125

Extant studies of the proposed regions suggest each has harbored multiple episodes of star formation, in some cases spanning ages between 1 and 20 Myr.

The areas mapped range from 50 to 200 square degrees depending on the distance and extent of the association. We propose a survey depth of 12 μJy at 3.6μm and 17.8 μJy at 4.5μm. At a distance of 500 pc, these correspond approximately to a young stellar object of mass $\sim$ 0.08 $M_\odot$ at 5 Myr assuming an excess above photospheric emission of 1 mag at 3.6 μm and extinction $A_V \sim$10 mag. Assuming that we map a total of $\sim$2000 square degrees (sufficient to study a representative sample of OB associations) to the indicated sensitivity levels, an integration time of 2000 hours will be required.

The motivation for such large-scale surveys is well illustrated in Fig. 1, in which we present maps of the Orion Ia-d association. The green areas denote the largest portions of this region that have been mapped with Spitzer to date, the blue dots represent the known OB stars in the association, while the red areas represent the distribution of molecular gas to a limiting column density corresponding to $A_V \sim$5 mag. For reference (top), the mapped regions (left hand panel) are shown adjacent to the molecular map and an image of the well-known outline of the Orion constellation (which fairly well coincides with the extent of the OB association.) As the figure illustrates dramatically, current surveys with Spitzer have targeted the regions showing the highest column density of molecular gas and hence the youngest, most active star-forming regions. Regions with lower gas column density, small globules external to the cloud, and the older regions of the association which have dispersed their molecular gas have largely been ignored, even though they contain a plethora of B stars formed within the past 12-15 Myr, and presumably a large associated population of low mass young stars and substellar objects. As a result, we have no information regarding the distribution of the low mass population with ages 1-15 Myr, nor do we know what fraction of that population might still be surrounded by circumstellar accretion disks.

Figure 2 (a map of W5) further vivifies the importance of surveying well beyond the present boundaries of molecular cloud complexes and signposts of the most recent episodes of star formation. Here we see a significant population of young, low mass stars with excess emission indicative of circumstellar disks and/or envelopes extending well beyond the recently formed OB stars and associated dense molecular gas.

By carrying out the proposed survey of nearby OB associations we will provide

1. A complete survey, typically down to masses at least as low as 0.2 $M_\odot$, of the distribution of protostellar sources (forming star-disk systems still surrounded by infalling envelopes; Class I) and sources in which the envelopes have dissipated, but which are still surrounded by circumstellar accretion disks. This will provide a clear indication of the sequence of star formation (youngest dominated by Class I sources; older dominated by Class II sources) in these regions;
2. A complete catalog of stars still surrounded by circumstellar disks, including those located well away from the boundary of actively star forming molecular clouds, some in the vicinity of 10-15 Myr old B stars and presumably formed during earlier star-forming episodes within the larger association. Because studies of nearby moving groups (e.g. TW Hya) demonstrate some disks surrounding low mass stars can survive for ages as least as great as 10-15 Myr (see, e.g., Chen *et al.* [1]), it is almost certain that the proposed survey will locate a significant number of long-lived accretion disks. Understanding the range of disk lifetimes and the properties of older accretion disks can provide important information regarding the range of

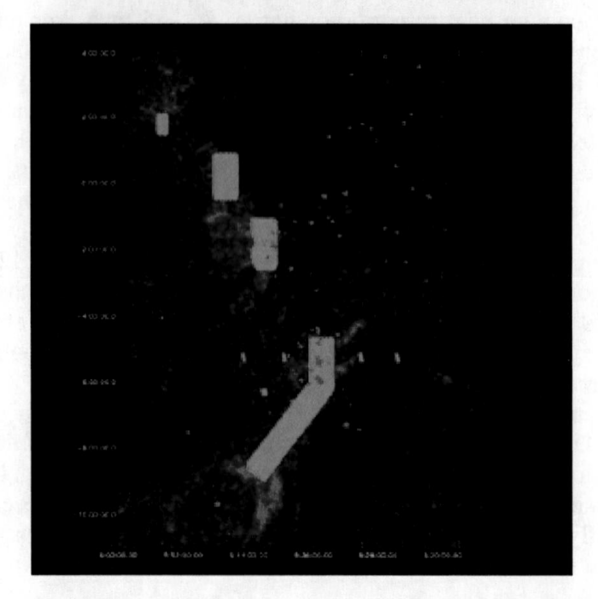

FIGURE 1. The top figure illustrates (LHS, bright green) most of the largest regions mapped with IRAC as part of GTO or GO programs carried out by Spitzer. The top RHS image represents a molecular line contour map (from the CfA CO surveyed) on which sources surrounded by disks and/or envelopes are superposed (green). The outlines of the Orion constellation and its signature bright stars are included for reference. The bottom figure depicts most of the regions mapped with IRAC as part of GTO or GO programs (bright green) overlaid on an extinction mass generated from the 2MASS database. Known OB stars with ages as great as 15 Myr are shown in blue, while the outlines of the molecular contour map shown in red (as in the top map). Note that the current surveys have been targeted largely at the main molecular clouds and signposts of recent ($t < 3$ Myr) or ongoing star-formation. The much more extensive regions of the Orion association in which star formation has proceeded for almost 15 Myr have yet to be surveyed.

times available for building planetary systems.

The proposed survey comprises three representative nearby OB associations to the limits needed to survey down to the hydrogen burning limit will require approximately 2000 hours of Spitzer time during the Warm Mission phase.

The Spitzer Warm Mission represents an ideal facility for carrying out this survey because:

1. It can provide a complete survey of protostars and stars surrounded by disks to below the hydrogen burning limit (at ages t $\sim$ 5 Myr) provided reddening is A_V<10 mag.
2. Its ability to survey large regions (100 sq deg or more) will for the first time enable study of the star-forming history of complex OB associations in which star-forming events have been spread out over time and space. Previous surveys have largely targeted well-known 'signposts' of recent star formation.

3.2. Complementary Ground-Based Observations Motivated by this Spitzer Warm Mission Program

The Spitzer Warm Mission survey of nearby OB associations will provide strong motivation for complementary ground-based observations, which, in combination with the Spitzer data, will add significantly to our understanding of star formation and disk evolution. For example:

1. Multi-epoch ground-based JHK imaging surveys will provide (i) via variability studies, a complete census of young, low mass stars formed throughout the star-forming history of the association, including those stars presently lacking circumstellar accretion disks (e.g., Carpenter *et al.* [2]); and (ii) via near-IR photometric measurements, a tool to weed-out heavily reddened background stars (A_V > 20 mag), and/or extragalactic interlopers of sizes near the limit of Spitzer's angular resolution – in those cases, the IRAC colors mimic those of star-disk systems, but the JHK colors differ significantly from those of young stars surrounded by circumstellar disks. In combination with the Spitzer observations, these measurements will yield a complete census of the low mass stellar population spanning ages 1-15 Myr, and provide thereby the statistical basis for quantifying disk evolutionary timescales as a function of stellar mass – essential input for constraining the timescales available for planet formation.
2. High resolution optical spectroscopic studies will provide (i) a map of the kinematic structure of the young stellar populations comprising the association; and (ii) quantitative measurements of disk accretion rates as a function of time and mass via measurements of Hα and Ca II triplet emission line profiles. The accretion rate measurements will help to quantify the rate of disk spreading as a function of age and mass, and provide the basis for identifying objects which lack excess infrared emission in the inner disk (0.1 to 1 AU) probed by IRAC and ground-based measurements, but are nevertheless still accreting gas through the inner disk regions. If so, such objects have the characteristics expected of disks in the process of building either terrestrial or Jovian mass planets. Choosing among these possibilities will require follow-up high resolution near- and mid-IR spectroscopy , as well as ground-based photometric studies with 8-10m class telescopes and facilities such as the SMT or ALMA.

FIGURE 2. IRAC color composite (rgb = 8, 4.5, 3.6 μm) of W5, with the distribution of Spitzer-identified (IRAC only) young stars overlaid (Koenig *et al.* 2007, in prep). Green symbols are Class II, red are Class I, and blue are probably background galaxies. There is a rich population of young stars inside the cavity, not coincident with the molecular cloud material in the rim.

3.3. The Interplay of Spitzer Warm Mission Surveys and Theoretical Studies of Molecular Cloud Collapse

In combination with molecular line maps and numerical models of cloud collapse, these data will provide essential insight into the relationship between the outcome spatial distribution and kinematics of young stars comprising the association, and the initial conditions in their natal GMC.

Recent developments have suggested a new approach to understanding star formation that builds on the idea that molecular clouds are formed by swept-up, shocked gas. Instabilities in the post-shock gas can produce dense concentrations that can serve as the sites of star formation. To illustrate the idea, in Fig. 3 we show a recent simulation of colliding flows, demonstrating the formation of dense protostellar cores as a result of focusing instabilities along with rapid cooling (Heitsch et al. 2007 in preparation).

The simulations suggest that there are three qualitatively different "modes" of star formation: (1) "distributed" star formation, where initial instabilities are so strong that runaway growth of protostellar cores occurs before global cloud collapse; (2) filamentary formation, where perturbations are not strong enough to form stars before collapse into filaments; and (3) clustered star formation, which typically happens when gravity focuses material into dense regions, often at the ends of clouds.

As an example of how global collapse can explain cloud structure, in Fig. 4 we show the results of a (gas-only) simulation (Hartmann and Burkert [3]) which can

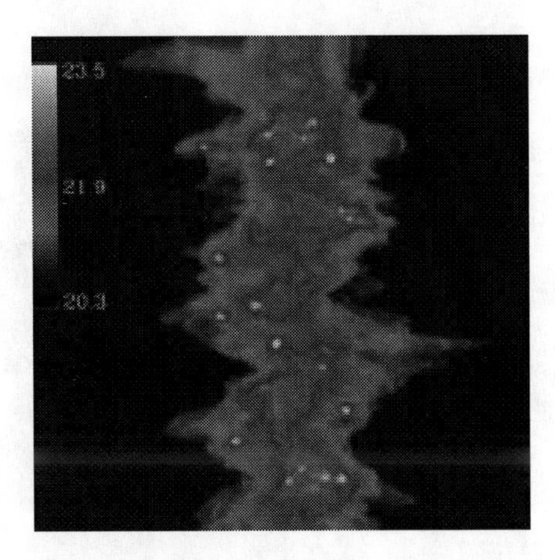

FIGURE 3. Edge-on view of instabilities developing from opposing uniform flows in the horizontal direction. Small, long-wavelength perturbations produce initial seeds which can grow initially through cooling and the ultimately through gravity. From Heitsch et al. 2007, in preparation.

reproduce the peculiar morphology of the Orion A cloud, including the integral-shaped filament, by global gravitational collapse of an initially elongated, rotating cloud. A dense concentration of material forms at the upper end, near the location where the present Orion Nebula Cluster of thousands of stars is situated.

These theoretical calculations make specific predictions about the relationship between cloud and stellar population morphology and kinematics. A warm Spitzer mission can help test these ideas and make advances in developing a more predictive theory of star formation in concert with kinematic studies.

4. PROGRAM II: DISTANT GMCS AND STAR FORMATION ON THE SCALE OF A SPIRAL ARM

4.1. Outline of a Possible Warm Mission Program

This program exploits the power of IRAC 3.6 and 4.5 μm colors to (a) select young stellar objects surrounded by disks or envelopes; and (b) map the extent of the UV radiation field within molecular clouds, and the relationship between forming stars and the boundaries of propagating HII regions and superbubbles. Potential targets would include:

1. Selected GMCs known to contain rich stellar populations, including dense clusters, comparable to those found in actively star-forming galaxies outside the Milky Way. These GMCs are located at distances between 1 and 2.5 kpc from the sun in the Perseus and Carina arms. The goal here is to understand how star formation in regions more typical of those that dominate external galaxies differs from that in well-observed molecular clouds located within 0.5 kpc of the Sun.

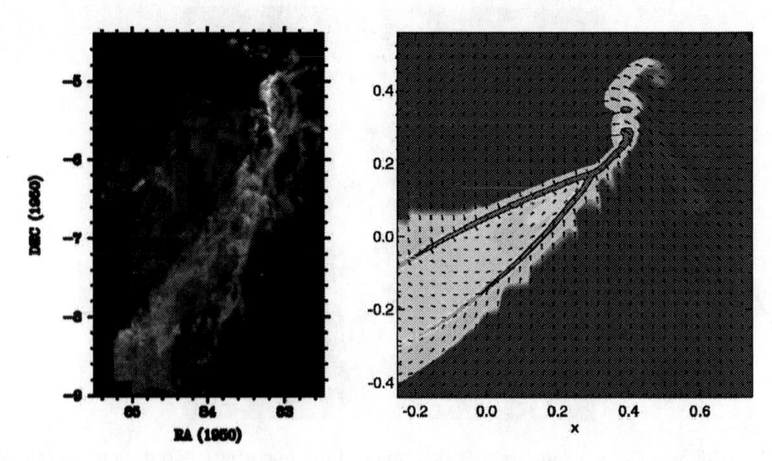

FIGURE 4. ^{13}CO map of Orion A (left) compared with the simple collapse model of Hartmann and Burkert [3]. Global collapse with rotation focuses material at the upper end into the integral-shaped filament, with a dense core which might fragment into a Orion Nebula-like cluster with higher resolution. Dense ridges are present at the lower (southernmost) end, which the Spitzer survey shows are locations of filamentary star formation.

2. A broad region (approximately 10 degrees in latitude and 30 degrees in longitude) of the Perseus arm (d∼2-3 kpc). The target region will be selected solely from examination of the contours of molecular line maps of the outer Galaxy suggesting extinctions $A_V > 5$ mag. The proposed observations will provide the first detailed and objective survey of star formation on the scale of a spiral arm. The study is motivated not only by this overriding goal, but also by a surprising discovery from Spitzer surveys of star forming regions: that a large population of young stars (< 5 Myr) are found outside the confines of molecular clouds (Megeath et al. [5]). Depending on the number and spatial extent of this population, the proposed observations could alter significantly our understanding both of the sequence of star formation and the lifetime of molecular clouds.

Our observing strategy is predicated on the ability of IRAC 3.6 and 4.5 μm colors to locate protostars and young stars still surrounded by circumstellar disks based on the unique colors produced by heated dust located in the inner regions of circumstellar envelopes and disks. In Fig. 5, we present a histogram (from Hartmann *et al.* [4]) depicting the frequency distribution of 3.6 - 4.5 μm colors for well-studied stars in the nearby Taurus-Auriga complex for which (a) there is no evidence from IR or optical data for the presence of accretion disks (open histogram); and (b) those for which there is clear evidence of a disk (shaded histogram). The figure clearly demonstrates the ability of 3.6 - 4.5 μm colors to select stars surrounded by circumstellar disks from stellar objects with bare photospheres. This strategy is robust for objects obscured by $A_V < 30$ mag; typical extinctions in GMC complexes range from 5-20 mag, except in the densest, cluster-forming clumps. For more heavily reddened objects, deep ground-based (H and K-band) observations will be required to complement the Spitzer survey; several

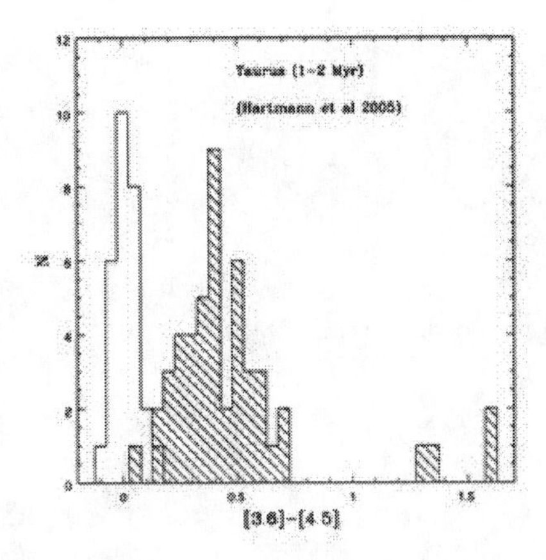

FIGURE 5. Histogram depicting stars surrounded by accretion disks (hatched region) and those known to lack accretion disks. After Hartmann *et al.* [4].

methods have been developed for identifying and classifying young stellar objects using combined near-IR and Spitzer 3.6 and 4.5 μm photometry (see, e.g., Hartmann et al. [6]). We note that although many galaxies, particularly AGN, have 3.6 - 4.5 μm colors similar to young stars surrounded by disks, they can typically be distinguished from young stars by their faintness, and/or because they will appear extended in 3.6 μm or even 2MASS (or deeper JHK) images.

Our strategy is to obtain four 12 second integrations for each of the $5'$x $5'$ frames for each position; with 10% overlap between frames; with these specifications the mapping speed is approximately 0.18 sq degrees per hour. Our integration times have been chosen to achieve 5-σ sensitivity limits of 12 μJy and 17.8 μJy at 3.6 μm and 4.5 μm respectively. At a distance of 2 kpc, these correspond to a 10σ detection of a young stellar object of mass $\sim$ 0.3 $M_\odot$ and age $\sim$5 Myr, assuming an excess above photospheric emission levels of 1 mag at 3.6μm, and cloud extinction A_V = 10 mag. In addition to locating stellar sources, the proposed survey will enable detection of the extent and boundaries of photodissociation regions diagnosed via strong PAH emission at 3.6 μm (see Fig. 6) and illuminate the possible role of propagating HII regions in triggering star-forming events.

By carrying out the proposed surveys, we will be able to detect recently formed stars in a wide range of environments as well as the outer boundaries of propagating HII regions. Together, these observations will allow us to

1. infer the distribution and sequence of star formation and study the role of triggering mechanisms such as colliding flows, or expanding superbubbles and HII regions;
2. relate the patterns and nature (isolated or cluster) of star forming events to the parent molecular cloud morphology and kinematics;
3. develop an understanding of star formation in a wide range of environments and an understanding of how star formation differs in these environments (e.g. molecular cloud density and turbulence; radiation fields from nearby star forming episodes);

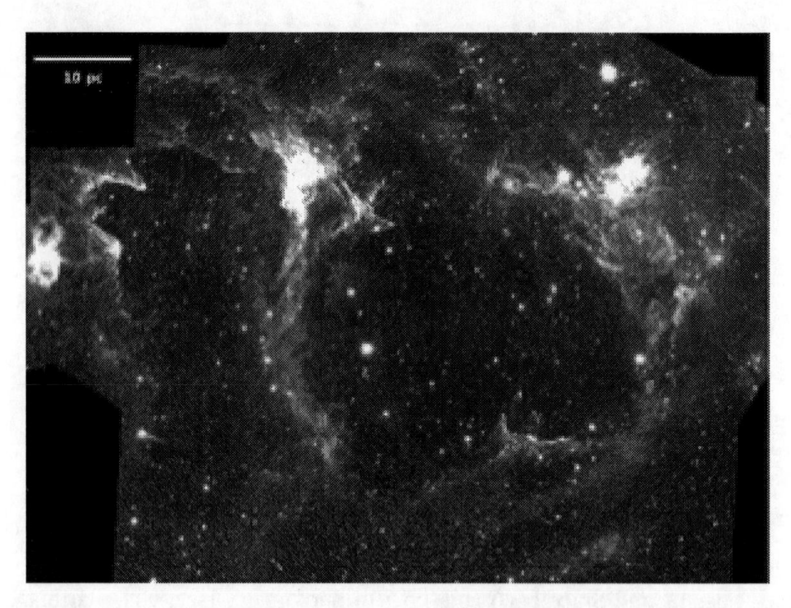

FIGURE 6. W5 mosaic at 3.6 μm (Koenig *et al.* 2007, in prep). The PAH emission lines in this bandpass clearly delineate the ionized/molecular gas boundary, or photodissociation region, powered by multiple O stars in this large HII region cavity.

4. infer the initial and environmental conditions that control the star formation process, from analysis of both Spitzer stellar and UV-radiation field tracers, and from extant molecular line maps.
5. understand the relative contributions to the Galactic disk from differing star-forming environments (small molecular clouds; GMCs; molecular clouds dominated by isolated star-formation; clouds dominated by cluster formation) from an unbiased survey of star-formation on the scale of a spiral arm.

We anticipate that a survey of selected GMCs in the Perseus and Carina arms will cover 300-500 square degrees and will require between 1600 and 3000 hours to cover to the desired depth.

An unbiased survey of the outer Galaxy would ideally cover $\sim$300 sq degrees (see Fig. 7) selected using various molecular tracers and are characterized by relatively unconfused lines of sight. The total time to survey such a region to the indicated depths is $\sim$1600 hours.

The Spitzer Warm Mission represents an ideal facility for carrying out this survey because:

1. Even with its restricted wavelength coverage, it has the ability to provide a complete census of protostars and young stars surrounded by accretion disks, in contrast to JHK ground-based surveys which miss a significant fraction of such sources (see Fig. 8).
2. Large areas can be probed at sufficiently fine ($1.7''$) angular resolution in times of

FIGURE 7. A map of the outer galaxy between 102.5 and 141.5 deg (longitude) and -3 to +5.4 latitude: representative of a region we propose for an unbiased 3.6 – 4.5 μm survey. Top panel is ^{12}CO J=1-0 from the FCRAO outer galaxy survey, middle panel is average J-K stellar colors from from 2MASS (a measure of extinction), bottom panel is three color JHK 2MASS surface density map.

a few hundred hours to sensitivities sufficient to detect 0.3 $M_{\odot}$, young ($t < 5$ Myr) stars surround by disks provided they are reddened by 10 magnitudes of extinction or less.

3. The angular resolution is sufficient to probe all but the densest regions of active star formation without suffering from significant source confusion.

4.2. Complementary Ground-Based Observations Motivated by this Spitzer Warm Mission Program

The Spitzer Warm Mission survey of distant GMCs will provide strong motivation for complementary ground- and space-based observations, which, in combination with the Spitzer data will add significantly to our understanding of star formation and disk evolution over the next decade. For example:

1. The proposed Spitzer survey will likely identify many hundreds of star-disk systems still surrounded by their natal protostellar cores and in various stages of assembly. Because we are probing rich star-forming regions known to be forming large numbers of intermediate and high mass stars, we expect that our sample will include protostars in the process of assembling not only large numbers of low mass stars, but a significant number of high mass stars as well. The Spitzer sample will

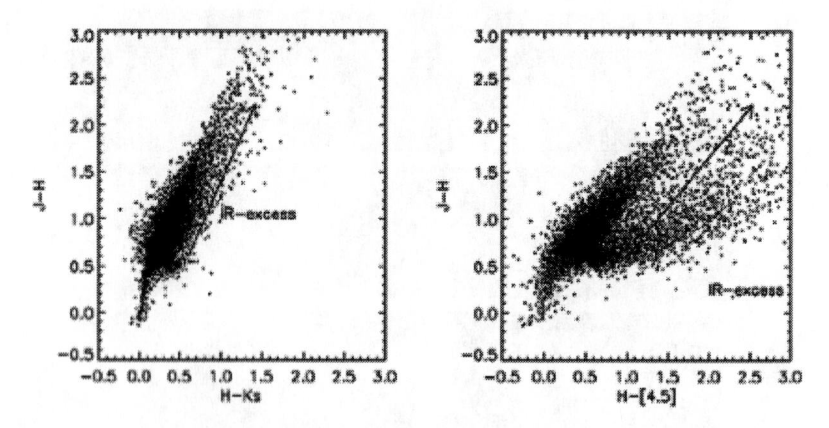

FIGURE 8. . Left panel: a J-H, H-K diagram depicting the location of stars survey in Orion A. The vector depicts the locus of reddened late-type stars. Objects that lie to the right of this vector have excess emission arising from warm dust in the inner regions of circumstellar accretion disks. Those to the left have colors consistent with pure photospheric emission. Right panel: a J-H, IRAC 3.6-4.5 μm color-color plot for the same region. Again, stars with disks lie to the right of the vector have colors indicative of the presence of circumstellar disks. Note the much larger fraction of detected disks manifest when 3.6 μm and 4.5 μm colors are used.

thus enable detailed study of the relationship between the physical properties of natal cores (size, mass, rotation, turbulent speeds, infall rates) and the mass of the forming star – thus providing a fundamental link between initial conditions and outcome stellar properties.

While Spitzer will provide the candidate target lists, follow-up observations with the CARMA, SMA, and ALMA interferometers will map the distribution of molecular gas and dust in the cloud core that feed the circumstellar environment. As an example, CARMA and ALMA will have the power to diagnose gas kinematics, temperature, density and chemistry in cores on scales extending from their outer radii (0.05 – 0.1 pc) to within 250 AU of the forming star-disk system for protostars located at a distance of 2.5 kpc. Ground-based O/IR telescopes will be able to probe infalling gas within 5-10 AU of the star-disk system using high resolution near- and mid- IR spectrographs to measure line profiles measured for a variety of gas tracers, yielding thereby temperature, density, and velocity along lines of sight through the core to the embedded star-disk system. These observations can yield complementary estimates of the infall rates characterizing cores surrounding stars of differing mass (and luminosity), thereby providing another essential link between initial conditions and outcome stellar properties.

2. While Spitzer will identify candidate protostars and young stars surrounded by accretion disks, a full picture of the young stellar population associated with these clouds will require tools to detect stars that no longer are surrounded by accretion disks. Spitzer will serve as a pathfinder to identify candidate regions for followup Chandra and XMM observations. Their ability to detect coronal X-ray emission

associated with young stars will enable identification of the cohort of stars that lack disks, i.e., those that cannot be distinguished from Spitzer IRAC observations. Chandra and XMM's ability to penetrate optically opaque GMC clouds will be particularly valuable in detecting heavily embedded young stars lacking disks. Optical and near-IR surveys can also provide an important tool for detecting diskless stars via surveys aimed at locating stars that exhibit variability (typically driven by starspot-modulated light variations). Having a complete census of all recently formed stars will be essential to establishing both the full extent of star formation in the GMCs, as well as understanding the sequence and timescales for multiple star-forming events (see below).

3. Ground-based spectroscopic surveys (optical and near-IR) to provide spectral classification essential for placing young stars in HR diagrams and determining masses and ages. These measurements, combined with Spitzer-detected star-disk systems and space- and ground-based detections of diskless young stars will provide the basic information needed to quantify accretion disk lifetimes as a function of stellar mass. Moreover, the derived ages will provide the quantitative complement to the morphological information regarding the timing and sequencing of star-forming events.
4. Follow-up observations of targeted regions identified in Spitzer imaging surveys with JWST imaging and spectroscopy will enable deep studies of the stellar content, particularly of dense, source-confused forming clusters uncovered with the Spitzer survey. Such studies will enable studies of the initial mass function as well as the disk population in regions which find no analog in nearby star-forming complexes.

5. PROGRAM III: A SEARCH FOR VARIABILITY PATTERNS AMONG YSOS

5.1. Outline of a Possible Warm Mission Program

It would be of significant potential discovery value to select one of the nearby OB Associations (Orion I c-d is our preferred region) as a target for a systematic survey of variability patterns among YSOs. In particular, a 1 x 3 degree region centered on the Trapezium contains a thousand young stars that could be monitored photometrically. Monitoring observations, e.g., with 2MASS, have already identified variability time scales of days, months, and years (see, e.g., Carpenter *et al.* [2] or Grankin *et al.* [5]). The potential sources of variability include: rotational modulation by star spots, obscuration by remnant material in protostellar envelopes, and variability of accretion rates (and associated disk heating) in the inner disks. We propose to observe the Orion the inner regions of the Orion I c-d association to identify variability events, and to probe all time scales accessible with Spitzer to completely characterize the relevant time scales for disk variability. Since most stars in Orion have K$<$ 14, high precision photometry is required on 0.5 mJy sources at 4.5 μm. In a 12 second integration, we can reach a signal to noise of 50 on the faintest cluster members.

The 1 x 3 degree region can be covered in 432 non-overlapping IRAC frames. Assuming 12 second frame times (an HDR mode), it would take about 3.5 hours to survey the entire region (based on Spot). The visibility window for Orion is about 44 days in duration, with windows in the early spring and fall each year. We would propose to survey the 1 x 3 degree region for one entire visibility window (44 days, or about 300 individual images at each point in the region), and then once per day for the next two visibility windows. The total survey would require of order 1300 hours.

The scientific return of the proposed survey is more difficult to quantify than the survey programs discussed above, simply because we currently know relatively little about the physical causes of variability among protostars and young stars surrounded by circumstellar disks. The proposed program offers the possibility of providing deeper insight into (i) the factors that control variability in the inner (0.1 to 1 AU) regions of circumstellar accretion disks, including stochastic processes associated with accretion and events driven by the interaction of orbiting giant planets and accreting material; (ii) variability during the protostellar infall/accretion phase and insight into the interactions between disk and envelope; and (iii) the relation between stellar rotation, age, and the presence/absence of accretion disks. We note that while ground-based surveys can contribute significantly to (iii), (i) and (ii) above require the sensitivity and rapid surveying power of Spitzer.

Spitzer can undertake this program based on its unique combination of sensitivity, area coverage, and photometric precision. Ground-based L-band surveys are generally limited to areas smaller than $\lesssim 0.03$ deg and typically achieve 10% photometry at the sensitivity limit required to detect stars with masses near the hydrogen burning limit in nearby young clusters (e.g. Haisch, Lada, and Lada [6]). Therefore, the challenges of ground-based 3.5 μm observations prohibit for all practical purposes an extensive variability survey over large regions. While near-infrared (JHK) observations are more feasible in this context, mid-infrared observations are essential to probe variability from the disk. Spitzer on the other hand has demonstrated it can achieve the combination of coverage, sensitivity, and precision to probe variability in hundreds of young stars. These combined attributes are unlikely to be replicated in the foreseeable future.

6. COMPARISON WITH WISE AND JWST

The IR survey satellite WISE will be launched in late 2009, while JWST is expected to be launched in late 2013. Both include bands in the wavelength range that will be covered by Spitzer during its Warm Mission Phase. Hence, it is of some importance to examine the strengths of each of these missions for carrying out the programs proposed here.

WISE's 40 cm telescope will be able to image a $47'$ FOV at wavelengths of 3.3, 4.7, 12, and 23 μm with typical image size of $6''$ (FWHM). WISE will be an all-sky survey; as a polar orbiter, its sensitivity is a strong function of ecliptic latitude (as well as background). By contrast, JWST will be a pointed mission. JWST NIRCAM will have filters covering the 0.7 to 5 μm region and a detector/camera combination that will cover a $2.3'$ x $4.5'$ FOV and deliver images with FWHM $\sim 0.1''$. JWST will also provide a mid-IR camera (MIRI) enabling imaging over the wavelength region between

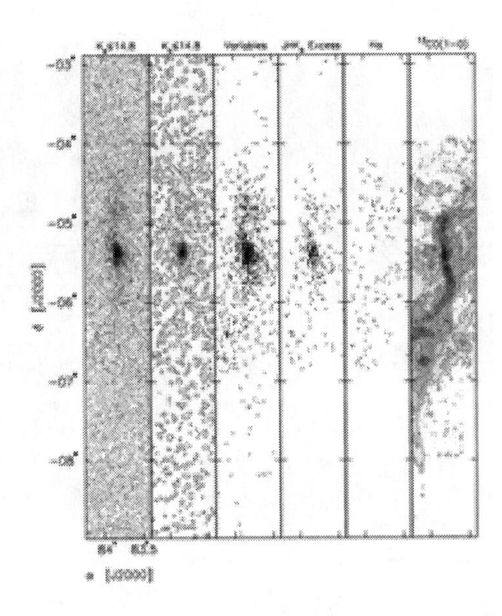

FIGURE 9. A plot depicting the variable stars found in the 2MASS survey of a region 1x3 degrees centered on the Orion Nebula Cluster. 1st panel: individual stars, K < 14.8 mag); 2nd panel: contour map of previous panel; 3rd panel: all variables found during the month-long survey period; 4th panel: stars showing excess emission at K-band; 5th panel: stars with Hα emission; and 6th panel 13CO contour map. Reprinted from Carpenter *et al.* [2].

5 and 27 μm and a FOV of $1.9'$. For comparison, the Spitzer Warm Mission will provide imaging at 3.6 and 4.5 μm with image sizes (FWHM) of $\sim 1.7''$ over a FOV of $5.5'$. Of relevance to the surveys proposed in this are the times required by WISE, JWST, Keck and Spitzer to survey 1 square degree to fixed sensitivity in 1 hour, along with the angular resolution provided by each of the facilities. Figure 10 presents such a plot for a wavelength of 3.6 microns.

The plot reveals the enormous power of WISE to map 1 square degree to modest sensitivity in very short times (0.1), but at reduced angular resolution. By comparison, Spitzer can cover the same area to 10 times the sensitivity in only 1 hour, and to nearly 300 times the sensitivity of WISE in a few hundred hours. JWST can achieve much greater sensitivity that of Spitzer IRAC, but design constraints (primarily downlink data volume and slew/acquisition times) severely limit its ability to map large areas of sky. Also plotted on this graph is the sensitivity required by our proposed programs.

We conclude from this plot that

1. Surveys of the nature proposed here are impractical with JWST. Rather, JWST

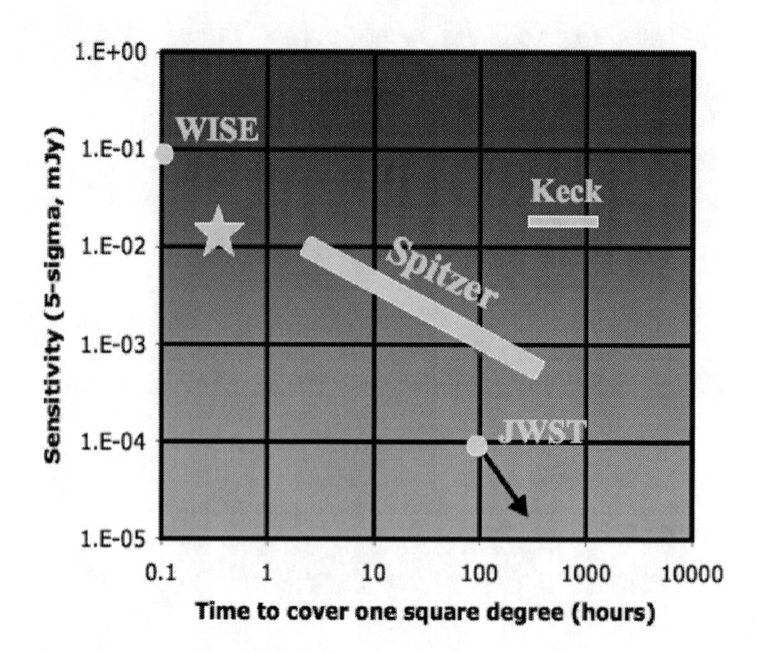

FIGURE 10. A comparison of the time to cover a one square degree field with WISE, Spitzer, JWST and Keck. The star (upper left) indicates the sensitivity limit required by the programs discussed here. Adapted from the Spitzer Science Center WISE memo, 3 April 2007.

pointings should be reserved for crowded regions, where its sensitivity provides overwhelming advantages, and for targets selected both from the Spitzer Warm Mission and the Spitzer archives that require higher sensitivity and/or higher angular resolution measurements at MIRI wavelengths

2. Surveys with Keck cannot reach the desired sensitivities in practical times – a natural consequence of working in the thermal regime with a warm telescope. Keck may be of value in making adaptive optics images of selected sources (e.g. protostars) where its angular resolution advantage (20 times Spitzer) can be brought to bear.
3. WISE lacks the sensitivity needed to survey significantly below the hydrogen-burning mass limit in nearby OB associations at ages up to >10 Myr needed to fully characterize the distributed population. For the more distant associations where source confusion is an issue, the lower angular resolution of WISE would compromise the ability to survey young clusters and embedded proto-clusters.
4. Spitzer (warm) hits a 'sweet spot' in both sensitivity and angular resolution that make it an ideal facility for carrying out the three science programs outlined here.

7. THE VALUE OF ARCHIVAL RESEARCH DURING THE SPITZER WARM MISSION ERA

Over the past three years, Legacy teams (C2D; FEPS; GLIMPSE), GTOs and GO teams have assembled databases of enormous value to addressing key problems in star formation. These programs include:

1. IRAC and MIPS surveys of nearby star-forming regions (e.g., Taurus-Auriga; NGC 1333 and IC 348) to sensitivities well below the hydrogen burning limit;
2. Pathfinder IRAC and MIPS surveys of the large scale star-forming processes operative in the inner galaxy;
3. IRS spectroscopic studies of large numbers of forming stars.

As a result, the Spitzer database contains IRAC, MIPS and in some cases, IRS data for hundreds of protostars unknown before the launch of Spitzer, thousands of young stellar objects surrounded by disks in various stages of evolution, and maps of large-scale star-forming events in the inner galaxy and near the galactic (clusters and associations) heretofore too obscured by intervening dust to enable detection or detailed study).

The database is so rich that neither the proposing teams or individuals nor archival researchers have begun to mine the data and carry out the ancillary observations at other wavelengths needed to enable full interpretation of key physical and chemical processes. To cite a few examples:

1. The compilation of protostars will provide a catalog for follow-up CARMA, ALMA, SMA, Herschel and in several years, ELT observations capable of probing the characteristics of protostellar cores and linking initial core conditions to outcome stellar properties;
2. The discovery of perhaps hundreds of 'transition disks' – disks which have developed inner opacity holes possibly associated with planet formation – demands follow-up high spectral resolution mid-IR and mm-wave observations aimed at diagnosing the gas content and kinematics of the circumstellar material orbiting parent suns – sine qua non for understanding whether planet formation is ongoing in these environments, what kinds of planets may be present, and where they were born.
3. The discovery of circumstellar debris disks surrounding relatively young stars is already providing the basis for adaptive optics observations aimed at understanding the distribution of material in these disks and how that distribution is linked to the presence of orbiting planets.

Together, further mining of the Spitzer databases combined with supporting complementary ground- and space-based data will provide the basis for a scientific legacy far exceeding the extraordinary results which have already emerged from the Spitzer mission. A good analog is the IRAS mission, which provided for nearly two decades a rich source both for discoveries emerging from the database itself as well as from complementary observations at other wavelengths.

8. CONCLUSIONS

Our proposed programs – and we suspect those of others – will no doubt aim to exploit the extraordinary ability of Spitzer during its warm mission phase to survey large areas at unprecedented sensitivity. In our major programs, we try to suggest the power of Spitzer to carry out large-scale surveys unbiased by prior knowledge of ongoing star formation or the presence of molecular clouds, and regions containing star forming complexes far more similar to those that dominate the appearance of external galaxies.

As we discuss above, Spitzer data alone will provide insight into (i) modes of star formation (isolated; aggregate; cluster); (ii) the sequence of star-formation over large regions of molecular clouds, as well as the role of 'triggers'; and (iii) the evolution of protostellar envelopes and circumstellar accretion disks. Moreover, we suggest a time domain study which is aimed at characterizing the variability characteristic of both protostars and the inner regions of accretion disks on timescales of hours, weeks, months and years – a program which may provide important insight into the accretion process.

The resulting databases will be incredibly rich – containing 1000s of protostars (including, we believe, those which will form very massive stars), and 10s of thousands of stars surrounded by circumstellar accretion disks in various stages of evolution.

These data not only promise new insights into the physical processes that initiate and propagate star formation, star formation in complexes analogous to those in other galaxies, and star-formation on the scale of a spiral arm, but as well represent critical pathfinder observations for follow-up with a variety of tools: complementary ground-based, Chandra and XMM imaging surveys; JWST imaging of targeted regions; AO-imaging of selected objects with large telescopes on the ground; high resolutions mid-IR and mm-wave spectroscopy of protostellar cores and clouds.

We believe that support of an archival research program aimed at mining both the extant database from the cold mission phase and the database that should emerge from the warm mission will yield enormous scientific return. The ideal program would be one that supports analysis not only of the Spitzer per se, but for the complementary data and theoretical studies that will enable deeper understanding of the fundamental physical processes at work in forming stars and planets.

REFERENCES

1. Chen, C. *et al.*, 2005, ApJ, 643, 1372
2. Carpenter, J. *et al.*, 2001, AJ, 121, 3160
3. Hartmann, L., and Burkert, A., 2007, ApJ, 654, 988
4. Hartmann, L. *et al.*, 2005, ApJ, 629, 881
5. Grankin, K.N., *et al.*, 2007, A&A, 461, 183
6. Haisch, K.E., Lada, E.A., and Lada, C.J., 2001, AJ, 121, 2065

Stellar Astronomy in the Warm Spitzer Era

Gillian R. Knapp*, Peter R. Allen†, Željko Ivezić**, Massimo Marengo‡, Fergal Mullally§ and Paula Szkody**

*Department of Astrophysical Science, Princeton University, Princeton NJ 08544, USA
†Department of Astronomy and Astrophysics, Pennsylvania State University, University Park, PA 16802, USA
**Department of Astronomy, University of Washington, Seattle, WA 98195, USA
‡Harvard-Smithsonian Center for Astrophysics, 60 Garden St., Cambridge, MA 02138, USA
§Department of Astronomy, University of Texas, Austin, TX 78712, and Department of Astrophysical Science, Princeton University, Princeton NJ 08544, USA

Abstract. We consider the impact on the study of normal stars of large-scale pointed and mapping observations at 3.6μm and 4.5 μm with the Spitzer IRAC imager. Deep observations at these wavelengths are particularly sensitive to very cool stellar and substellar objects, both as companions to other stars and in the field. A wide-angle survey can be expected to detect 50 – 100 cool T dwarfs and up to 5 "Y" dwarfs in the field, and AGB stars throughout the Galactic halo. Pointed observations of white dwarfs at these wavelengths will be sensitive to unresolved cool companions and to circumstellar dust disk remnants of planetary system objects. The cumulative photometry of normal stars in the imaging fields will be invaluable for understanding stellar colors and atmospheres.

Keywords: Spitzer Space Telescope, infrared astronomical observations, stellar atmospheres, brown dwarfs, mass loss
PACS: 95.85.Hp, 97.10.Yp, 97.20.V, 98.35.Ln

1. INTRODUCTION AND OVERVIEW

This paper discusses science from IRAC 3.6μm and 4.5μm photometry of stars in the Galaxy and Local Group made possible by large surveys using IRAC on Spitzer in the post-cryogen era. As a strawman we consider that the "Warm Spitzer" observations, and the post-cryogen mission, may fall into three broad categories: (1) One or two very large surveys, perhaps consisting of a shallow survey over a wide area accompanied by embedded deeper surveys over smaller areas, to acquire data to support a very wide range of science investigations of interest to the entire astronomical community. What strategies would be optimal for stellar science, i.e. depth, cadence, region(s) of sky covered, existence of data sets at other wavelengths? (2) Moderate-sized (PI team) pointed surveys of individual classes of rare objects. (3) Reduction of the imaging data from the accumulated Spitzer mission, to provide photometric catalogues of serendipitous objects (stars in particular of course) found in the fields of the observed targets.

We consider four main science topics: (1) searches for brown dwarfs, in particular those cooler than any found to date, (2) a survey of white dwarfs to detect low luminosity companions and circumstellar dust and to provide infrared data to extend the spectral energy distribution coverage and investigate atmospheric models, (3) studies of AGB stars in the Galactic halo, and (4) stellar photometry from IRAC imaging to extend the available broad-band photometry for normal stars and contribute to measures of gravity,

CP943, *The Science Opportunities for the Warm Spitzer Mission Workshop*, edited by L. J. Storrie-Lombardi and N. A. Silbermann

metallicity and effective temperature.

2. WIDE AREA SURVEYS: CONSIDERATIONS FOR STELLAR ASTRONOMY

With the exception of studies of the stellar content of clusters, streams etc., we could think of no particular rationale for a large *ab initio* filled area as far as stellar science goes - in principle a wide-area survey designed for stellar astronomy could consist of a large number of individual $5' \times 5'$ patches of sky. Of greater importance is a uniform sensitivity limit, very accurate photometric calibration, a large total area, and wide coverage in Galactic latitude. Thus in principle the areas to be observed could be driven by ease of scheduling. However, there are two strong arguments for a contiguous area: calibration, and the existence of data at other wavelengths.

2.1. Calibration

Accurate photometric calibration is probably the single hardest problem facing observational astronomy. As Worthey and Lee [1] put it: "Perhaps the reason no-one has done this exercise (stellar effective temperatures from multiple broad-band colors) before is that photometric systems are such a mess". The problem can be separated into two parts: accurate *internal* calibration, in which the relative measurements of all sources with a given instrument through a given filter are accurate with respect to each other across the entire survey area, and accurate *external* calibration, in which the instrumental quantities are accurately converted to physical units, so that they can be used with observations at other wavelengths to construct broad-band spectral energy distributions (SEDs). There has been enormous progress in this area of late, with both 2MASS and SDSS employing multiple dithered or overlap observations to ensure end-to-end internal calibration of their surveys to better than 2% (Skrutskie *et al.* [2], Padmanabhan *et al.* [3]). Therefore, the Spitzer 3.6/4.5 μm wide area survey needs to be done in such a way as to ensure consistent calibration across its area. One way or the other, this will be done with multiple exposures, thereby enabling a second type of science, that of time variability. The appropriate observing pattern is to spread the observations over the entire six years of the warm mission, with a consistent flux limit across the whole survey area at all times (in case the warm mission were to terminate early), sensitivity built up by multiple passes of the entire survey area, and an observing cadence analogous to that proposed for LSST (Ivezić [4]), i.e. with a roughly logarithmic distribution of Δt, the time interval between any pair of observations. This would optimize the ability to find photometric variability on a wide range of time scales, and to find secular position variability (i.e. proper motions).

2.2. Observations at Other Wavelengths

Since Spitzer will not observe the whole sky during the warm mission, all science will be optimized if the Spitzer surveys cover areas with data at as many other wavelengths as possible. USNO-B, 2MASS, IRAS, Akari, GALEX and ROSAT cover the whole sky, and large-area surveys covering part of the sky include UKIDSS, FIRST and SDSS. We highlight two surveys here: the SDSS southern equatorial survey and a planned Subaru very deep optical and near-infrared survey over several hundred square degrees.

The SDSS southern stripe is a 2.5° wide stripe along the celestial equator between right ascensions 20^h to 4^h, about 300 square degrees in all, which has been observed multiple times by the SDSS mosaic telescope in the u, g, r, i, and z bands over almost ten years. Co-added, the survey depth is about 24^m (AB), and there is a wealth of variability and proper motion information. There are about 500 spectra per square degree and deep surveys available in the near-infrared (UKIDSS), the ultraviolet (GALEX), and the radio (FIRST). The Atacama Cosmology Telescope (ACT) will image this region at 1 mm wavelength in 2007-2009 at a resolution of about 1 arcminute. While Spitzer is not affected by limits on sky coverage, regions near the celestial equator can be observed by most ground-based telescopes and naturally cover a range of Galactic and Ecliptic latitudes.

The second complementary survey is only at present in the planning stages. A group including the National Astronomical Observatory of Japan and Princeton University is planning a multiband optical survey with Subaru over several hundred square degrees in the g(27.3), r(26.8), i(26.4), z(25.8) and y(25.3) filters - the numbers in parentheses are the 5σ limiting (point source) magnitudes. The region to be surveyed is not yet decided, but since its primary driver is extragalactic science it will be at high latitudes. This region will also be surveyed by ACT, and is the only current survey whose sensitivity to very cool substellar objects will approach that of Spitzer.

2.3. Catalogues

We consider a most important aspect of the warm mission to be the production of catalogues from all Spitzer imaging, including that acquired during the cryogenic mission, exploiting the measurement of objects in the large IRAC field of view. This archive should be searchable both for objects of particular properties and for objects matched with those discovered at other wavelengths. A position query of this data base would then return the Spitzer flux densities and uncertainties plus the epoch of observation, or upper limits and the epoch of observation, or a notation that this position had not been observed by Spitzer. This facility would have particular use in several of the stellar programs discussed below. As examples: (1) it would provide a large archive of the colors of normal stars. These would be useful for identifying infrared excesses, for defining colors which could help identify chemically peculiar stars, and for input to determinations of temperature, metallicity, and gravity based on multiband observations and model atmospheres; (2) it could provide earlier epoch measurements to help identify proper motion pairs, including very low temperature wide pairs and companions; (3) it could help verify the existence of single-band sources, for example possible very cool dwarfs found only at 4.5μm by WISE (Mainzer *et al.* [5]), as pointed out by Stauffer *et al.* [6]; and (4) there may be rare objects in these fields.

The general usefulness of the Spitzer "Point Source Catalogue" and "Extended Source Catalogue" would be greatly enhanced if the data base were to return not only the IRAC and MIPS photometry and Spitzer spectroscopy, but also matches, positions and photometric data from other catalogues. This effort will integrate into the National Virtual Observatory to provide a comprehensive UV to mid-infrared photometric catalogue for all of the sky observed by Spitzer during the cryogenic and warm missions.

3. T AND Y DWARFS

3.1. Expected Flux Densities, Colors, Numbers

The Warm Spitzer Mission has the potential of finding several tens of brown dwarfs with temperatures less than 1000 K, i.e. cool T dwarfs and the as-yet-undiscovered "Y" dwarfs.

Recent deep large-area surveys at optical and near-infrared wavelengths, primarily 2MASS and SDSS, have led to the discovery of many field L and T dwarfs, allowing the definition of these two spectral types and the measurement of their effective temperatures (Kirkpatrick *et al.* [7, 8], Burgasser *et al.* [9], Geballe *et al.* [10], Golimowski *et al.* [11]). The stellar- substellar boundary appears to occur at spectral type about L5, $T_{eff} \sim 1700$ K), and the spectral transition between L and T at $T_{eff} \sim 1300$ K, although exactly what happens at this transition is still uncertain because of the complicating effects of close binaries on the observations (Liu, Leggett, and Chiu [12]). Spectroscopically, the L/T transition is defined by the onset of CH_4 absorption in the *H* and *K* bands, although absorption in the methane fundamental at 3.3μm is seen as early as spectral type L5 (Noll *et al.* [13]).

T dwarfs as cool as 700 K have been observed (Golimowski *et al.* [11], Warren *et al.* [14]). The next spectral type cooler than T9, dubbed "Y", remains to be discovered (unless you count Jupiter and recently directly detected extrasolar planets). These objects will have effective temperatures less than 600 K and NH_3 absorption bands. Their discovery in the field, should this prove to happen, will be invaluable for their study uncontaminated by the light of a nearby/primary star.

Finding the very faint T dwarfs has proved challenging both for SDSS (where the objects are often detected only in the *z* band) and in 2MASS, where the *JHK* colors are degenerate with those of main sequence stars over much of the T dwarf temperature range. The discovery of very cool T and Y dwarfs in the field will require a sensitive, wide area survey in at least two bands where they can be expected to be detected. Further, the two bands should be selected to produce a T/Y dwarf color which is very different from that of any other astronomical objects. In addition to the upcoming WISE mission (see below) there are two surveys which will meet these criteria: deep mapping in the near-infrared, and a Spitzer survey in IRAC bands 1 and 2.

UKIDSS is a set of surveys in *JHK* which is reaching several magnitudes deeper than 2MASS (Lawrence *et al.* [15]) and has already led to the discovery of very cool T dwarfs

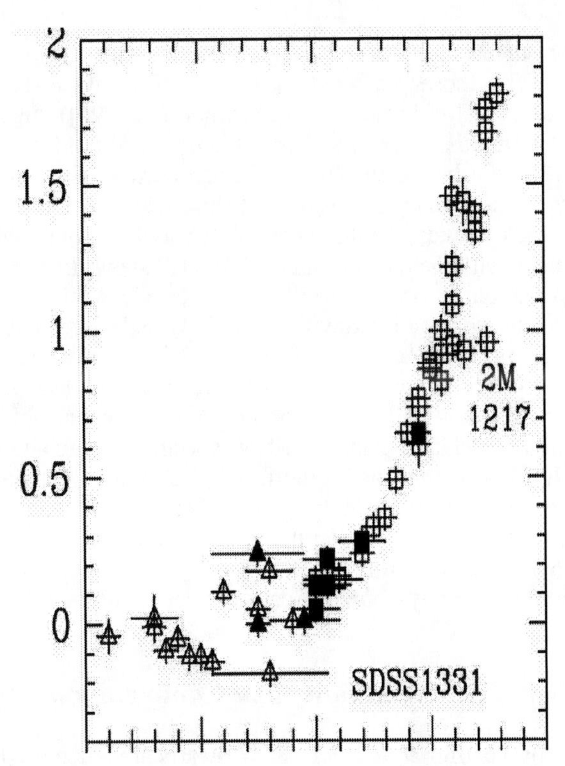

FIGURE 1. [3.6] - [4.5] color vs spectral type for L and T dwarfs, from Leggett *et al.* [16]. The horizontal axis is spectral type between L0 and T9.

(Kendall *et al.* [17], Warren *et al.* [14], Lodieu *et al.* [18]). The proposed Subaru survey described in the previous section will reach 25.8 in *z* and 25.3 in *y* (AB magnitudes). A medium sensitivity Spitzer survey would be orders of magnitude more sensitive than UKIDSS for very cool T dwarfs.

The Spitzer instrumentation is almost ideal for this work. Bands 1 and 2 cover the sky at different times, so that asteroids can be eliminated by proper motion. The [3.6] - [4.5] colors of M and L dwarfs are close to zero, while towards later spectral types the colors rapidly become redder, with values of about 2 for the latest T dwarfs observed (Patten *et al.* [19], Leggett *et al.* [16], Stern *et al.* [20]); see Fig. 1. The reasons for this behavior are twofold. First, the emission peak moves through these bands for objects cooler than about 3000 K. Second, the 3.5μm band contains the CH_4 fundamental at 3.3μm, which greatly suppresses the 3.6μm flux towards lower temperatures. Although the CO fundamental is at 4.6μm and therefore affects the 4.5μm band, it becomes increasingly weak with decreasing temperature and does not have a strong effect on the colors. (CO is weakly present in very cool objects with methane-dominated atmospheres due to vertical mixing and non-equilibrium chemistry, which can be probed with larger samples (Mainzer *et al.* [21]).

The colors measured by IRAC (Patten *et al.* [19], Leggett *et al.* [16]) are in qualitative agreement with those predicted by models (Burrows *et al.* [22], Jones *et al.* [23], Burrows *et al.* [24], Marley *et al.* [25], Saumon *et al.* [26]). Figure 2 shows a plot of the predicted 3.5μm flux density at 10 pc vs. [3.6] - [4.5] for dwarfs of effective temperature below 1200 K (the 3.6μm flux is selected because the objects are far fainter at 3.6μm, so it is the sensitivity of this band, fortunately the more sensitive of the two because of lower backgrounds, that limits our ability to detect and characterize T and Y dwarfs). Comparison with the data of Patten *et al.* [19] shows that present observations have probed only the bluest objects, to [3.6] - [4.5] $\sim$ 2. While the quantities plotted in Fig. 2 are somewhat schematic (gravity, metallicity and weather all affect the colors, (Leggett *et al.* [27]) the important point from Fig. 2 is that the dwarfs continue to get redder in the [3.5] - [4.5] color with decreasing temperature and therefore if they are detected at both bands, they will not be confused with faint stars. Neither will they be confused with highly redshifted galaxies, whose spectra, containing roughly equal contributions from the almost single age turnoff, red giant and AGB stars are basically flat at the Spitzer wavelengths over a wide range of redshift.

Also plotted in Fig. 2 are the 5σ depths for the "shallow" and "moderate" surveys (Stauffer *et al.* [6]), showing that the shallow survey can detect almost all T dwarfs in the survey area in the presently-known effective temperature range.

3.2. Serendipity and Confirmation

As pointed out by Stauffer *et al.* [6] T and Y dwarfs are much brighter at 4.5μm and will therefore produce a lot of 4.5μm single-band detections by both Spitzer and WISE. Since WISE is an all sky survey, any such object found by WISE can be compared with Spitzer 4.5 μm data. For this and many other reasons, it will be very useful to have an archive of all fully-reduced Spitzer data, i.e. catalogues derived from all images ever observed. Spitzer will also be able to confirm faint single-band 4.5μm detections by ISE.

Of the data sets at other wavelengths; the existing 2MASS, UKIDSS and deep SDSS surveys, and the upcoming or proposed VISTA and Subaru surveys: only the last will rival Spitzer in sensitivity. Its magnitude limits of 25.8 and 25.3 in z and y will allow the detection of dwarfs to essentially the same depth as will the moderate Spitzer survey, as illustrated in Fig. 2, i.e. down to objects as cool as 400 K at a distance of 10 pc. Thus inasmuch as possible, the Spitzer survey should be carried out in the same part of the sky as the proposed Subaru survey.

How many T and Y dwarfs can Spitzer be expected to detect? Since these objects cool with time, the answer depends on both the initial mass function and its possible variations with time, and on the star formation history. Models by Burrows *et al.* [28, 22, 24], together with mass-function fits to the existing counts of cool dwarfs (Allen [29]) indicates that the "wide" survey (Table 1 of Stauffer *et al.* [6]) will find up to 100 T dwarfs, many of them very cool, and 1-5 Y dwarfs. If they are detected in both IRAC bands, these objects will have unique colors, but the selection efficiency for follow-up can be greatly enhanced by the inclusion of J and K photometry using

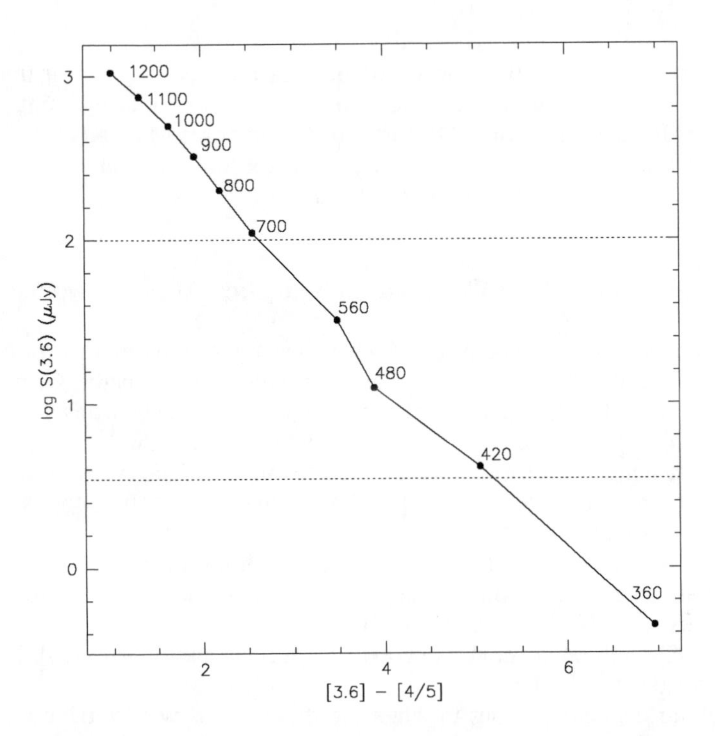

FIGURE 2. Model flux density in IRAC band 1 at a distance of 10 pc versus [3.6] - [4.5], from the models of Burrows *et al.* [22, 24]. The points are labelled by the effective temperature of the model. The horizontal lines show the 5σ sensitivities of the "shallow" and "moderate" surveys from Stauffer *et al.* [6].

methods developed by Marengo *et al.* [30] and, as mentioned above, by very deep optical imaging.

3.3. Resolved Companion Searches

Spitzer has already led to the discovery of resolved T dwarf companions to nearby stars, most notably of the T7.5 companion to HD 3651, which has a planetary system (Luhman *et al.* [31]). Spitzer is unsurpassed by any current project for this sort of discovery because of its great sensitivity to very cool objects, its good spatial resolution, and the fact that the parent stars are so much fainter at the Spitzer wavelengths than at shorter wavelengths (Marengo *et al.* [32]). A survey of the 1000 nearest stars would sample stars within about 20 pc to separations of about 600 AU and greater for the most distant stars (this estimate is based on an assumed ability to reliably detect a cool companion as close as $20''$ to the primary star) and could detect companions as cool as about 400 K in modest amounts of observing time (about one hour total observing time per star). Ideally, the exposure would be built up over the six-year timescale of the warm mission, to allow the identification of faint companions by common proper motion. The above estimate of 1000 stars mostly comprises M dwarfs, so the star list could be modified to more uniformly sample the spectral type range, the metallicity range, the present or absence of massive planetary systems, etc.

3.4. Unresolved Companion Searches: White Dwarfs

Cool companions can also be detected when they are too close to their primary to be spatially resolved by searching for color excesses at long wavelengths. Given the huge luminosity contrast between main sequence stars (even late M stars) and T dwarfs, plus the fact that the long-wavelength colors of normal stars are not well characterized (see below) such searches are too difficult for current technology. The exception is the search for cool companions to white dwarfs, whose low luminosities and blue colors ensure that cool companions are brighter then the star in the infrared and can be found as measurable infrared excesses. Indeed, the first known object of spectral type L was discovered as a companion to the nearby white dwarf GD 165 (Zuckerman and Becklin [33]), and large numbers of dM/WD pairs are known (Silvestri *et al.* [34]). Currently, only a small handful of L/T dwarf companions is known (Zuckerman and Becklin [33], Farihi *et al.* [35, 36], Maxted *et al.* [37]).

A search for cool companions to white dwarfs offers, as well as the opportunity to discover more ultracool dwarfs, the opportunity to measure the mass function at the bottom of the main sequence and, perhaps, to identify the coolest, lowest mass main sequence star. This transpires because white dwarfs are old, and substellar companions may have faded below detectability (Burrows *et al.* [28]). There are some 20,000 white dwarfs known at present, and the number is rising rapidly, including the discovery of many new nearby white dwarfs (Bergeron *et al.* [38], McCook and Sion [39], Eisenstein *et al.* [40], Subasavage *et al.* [41]). Several searches for cool companions have recently been made by matching known white dwarfs with infrared surveys, in particular 2MASS (Wachter *et al.* [42]), but unfortunately 2MASS is not quite sensitive enough to detect substellar companions around most known white dwarfs. A targeted search with well-controlled exposure times of a sample of 100-200 carefully selected white dwarfs - nearby, with decent photometric or trigonometric distances, with well-determined masses, ages and temperatures, and known from optical and 2MASS photometry not to have companions of spectral type M8 or earlier could answer this question, while a sample twice as large would also allow the investigation of metallicity effects. The search could also possibly find extremely cool brown dwarfs - late T and perhaps even Y, may yield resolved ultracool companions, and would also be of enormous interest for studies of fossil planetary systems around WDs (see below).

3.5. Unresolved Companion Searches: Cataclysmic Variables

Cataclysmic variables (CVs) are interacting binaries whose primary is a white dwarf and secondary a low mass donor star. Mass accretion onto the white dwarf causes sporadic huge increases in luminosity due to a short-lived phase of nuclear burning of material accreted to the white dwarf from the circumstellar disk produced by Roche-lobe overflow from the secondary. Until recently, most known CVs were discovered while in their high state, but recent surveys, especially the SDSS, have discovered hundreds of CVs in their low state, where the spectroscopic signature of the low-mass M stars companion can almost always be discerned (Szkody *et al.* [43]).

Quite apart from their interest as variable stars, as possible nova and supernova precursors, and as laboratories for studying accretion under extreme conditions, including very high magnetic fields, CVs can also be used to study the properties of their low-mass secondaries. Star formation and stellar evolution theory predict that a fair number of CV secondaries should be substellar, so that observations of CVs allow the measurement of masses (via measures of the orbital parameters of these binaries), luminosities (via measurements of the infrared excess), and radii (since the secondary is experiencing Roche lobe overflow) for low-mass stars and substellar objects. Indeed, recent Spitzer-IRAC observations have discovered infrared excesses from several short-period CVs, and substellar masses are inferred for some companions from radial velocity measurements (Harrison *et al.* [44], Howell *et al.* [45, 46], Littlefair *et al.* [47], Brinkworth *et al.* [48]). There is some question (Howell *et al.* [45]) as to whether the infrared excesses are due to dust or to a substellar companion, but this can be answered by deeper near-infrared photometry and spectroscopy, with Spitzer observations providing the candidates.

4. WHITE DWARFS

As well as the search for unresolved companions described above, 3.6μm and 4.5μm observations of large samples of white dwarfs will allow the investigation of two other important areas: a search for circumstellar dust disks (see Fig. 3), and the investigation of model atmospheres.

4.1. Circumstellar Dust

White dwarfs come in two main flavors, DA and DB (the analogues of main sequence A and B stars, with H and He lines respectively). Some DA stars have long been known to have in addition absorption lines of heavy elements, especially Ca. This has been hard to explain, since the heavy-element settling times are much shorter than the evolutionary timescales (Fontaine and Michaud [49]). While metals can be radiatively levitated in the atmospheres of hot white dwarfs (Chayer *et al.* [50]), metal lines are also seen in cooler white dwarfs. Recent accretion is therefore suggested, but the interstellar medium is, over almost all of its volume, of too low density for sufficient accretion. A natural possibility is then the accretion of left-over planetary system objects - comets or asteroids (Alcock *et al.* [51], Jura [52]). This phenomenon can give new and quite

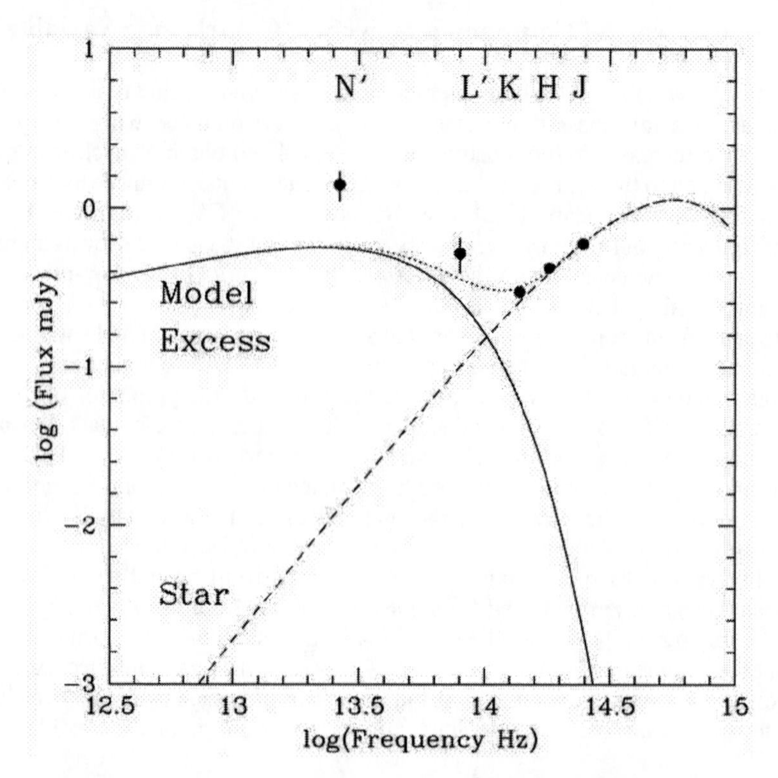

FIGURE 3. Spectral energy distribution of the white dwarf GD 362, showing the excess due to circumstellar dust (Becklin *et al.* [53]). The Spitzer spectrum of this feature shows strong silicate emission (Jura *et al.* [54]).

different insights into planetary systems. These white dwarfs must have an asteroid system not dissimilar to that of the Solar System - well outside the radius of an AGB star and well within the orbits of the giant planets. This suggestion, that these DAZ stars might be temporary phenomena, in that they are accreting comets from a left-over planetary system, has very likely been confirmed by the discovery of dust disks around a small number of cool white dwarfs (Zuckerman and Becklin [55], Becklin *et al.* [53], Jura *et al.* [54], Mullally *et al.* [56], von Hippel *et al.* [57]). At the same time, Gänsicke *et al.* [58] have found several hotter white dwarfs showing double-peaked emission lines of Ca, showing the presence of metal rich gas disks around these stars. These exciting discoveries provide new information on planetary systems, the type of stars they form around, and how they survive the rigors of stellar evolution. The currently known number of white dwarfs with circumstellar disks is small, but judicious observing campaigns will likely discover many more – about 25% of DA white dwarfs have CaII absorption (Zuckerman *et al.* [59]), and about 14% of DAZ white dwarfs have detectable circumstellar disks (Kilic and Redfield [60]).

A large 3.5μm and 4.5 μm search for dusty disks around nearby white dwarfs is clearly called for, focussing for this purpose on DAZ and DZ white dwarfs. The spectroscopic data set for white dwarfs has improved enormously in recent years, with some 10,000 good spectra extant, the discovery of hundreds of WDs with metal lines, the characterization of some 150 DZ white dwarfs (Dufour *et al.* [61, 62]), and the development of white dwarf models to the point of deriving reliable masses and surface temperatures. The two short IRAC bands are especially well suited to this work. Unlike the situation for debris disks around main-sequence stars, where the bulk of the dust is far enough from the stars to be too cool for detection at these relatively short wavelengths (see Marengo *et al.* [32] for a discussion of sensitivity limits), the dust disks around white dwarfs are close enough to the star to have temperatures close to 1000 K and therefore emit strongly in the near-infrared suggesting that the dust is produced by the tidal destruction of asteroids or comets (Jura *et al.* [54], Zuckerman *et al.* [63]).

4.2. White Dwarf Atmospheres

The Spitzer observations which have led to the above discoveries have also shown that the observed WD colors at IRAC wavelengths often do not agree well with the predictions of model atmospheres, with, in particular, flux deficits observed for cool ($<$ 7000 K) white dwarfs (Kilic *et al.* [64], Mullally *et al.* [56]). It is well known that ultracool ($<$ 4000 K) white dwarfs (Gates *et al.* [65], Harris *et al.* [66]) have flux deficits in the optical red and near infrared bands due to collisionally-induced H_2 absorption, and presumably similar molecular processes are producing the longer-wavelength deficits in somewhat warmer stars. Understanding this is important not only for searches for flux enhancements due to cool companions or dusty disks, but for the modeling of white dwarf cooling, age-dating of white dwarfs, and the chronology of star formation in the local Galactic disk and halo (Harris *et al.* [67]).

4.3. A Strawman Observing Project on White Dwarfs

All of the above discoveries (ultracool companions, dust disks and flux deficits) are based on observations of fewer than 200 white dwarfs, including the Spitzer survey of 124 white dwarfs at 4.6μm and 8 μm of Mullally *et al.* [56]. For a sample of white dwarfs and cataclysmic variables within 30 pc, the required sensitivity to detect a 500 K companion is about 1.8 μJy, which can be achieved in both bands for 200 objects in 400 hours, and will also find very low mass disks. A further 50 hours of observing will reach 4μJy for 500 stars. All observations should be repeats, to verify the reality of the detections. The white dwarfs can be selected from the spectroscopic surveys which include ultracool white dwarfs (Gates *et al.* [65], Harris *et al.* [66]), about 150 DZ WDs (Dufour *et al.* [62]) about 200 DQ WDs (Halford *et al.* [68]), as well as large samples of magnetic WDs, normal DA and DB WDs, and cataclysmic variables. Using 2MASS and the SDSS *z* band data, where available, white dwarfs with M dwarf companions can be rejected.

During the discussions at this meeting, it was apparent that there is a lot of interest in this search for infrared excess emission due to circumstellar dust disks around white dwarfs (see e.g. the paper by M. Jura, this volume). The exact moment at which the cryogen will run out is not known, and there needs to be an observational program ready to go when that event happens and the system verification checks have been carried out. A program to begin imaging white dwarfs is ideal – there are large numbers of them all over the sky and therefore targets available at any time, the observing priority and strategy are reasonably easy to work out for each star, the observations are simple, and the science is exciting and of interest to many areas of current astronomical research. We encourage the development of this survey to provide the plan for the initial Warm Spitzer observations.

5. AGB STARS

Spitzer's sensitivity is sufficient to detect AGB stars throughout the Galaxy and the Local Group in all the IRAC bands. In particular, the sensitivity is enough to detect the color excesses (with respect to flux densities at shorter wavelengths from 2MASS, for example) due to mass loss rates as low as a few $\times 10^{-8}$ $M_{\odot}$ yr^{-1}.

Spitzer Galactic plane surveys (GLIMPSE, Benjamin *et al.* [69] and surveys of the Magellanic Clouds (Meixner *et al.* [70], Groenewegen *et al.* [71]) will discover very large numbers of AGB stars. In addition, we believe it to be worth while to carry out searches for mass-losing AGB stars in loose halo structures such as the Sagittarius stream and in the distant halo.

Any star which leaves the red clump/horizontal branch with more than about 0.6 $M_{\odot}$ can become an AGB star, but as the mass decreases the amount of time spent on the AGB, and the amount of fuel burned there, decreases, so that AGB stars are extremely rare in old populations, in Galactic globular clusters for example. Nevertheless, there are some hundred distant carbon AGB stars found at high Galactic latitudes, some with significant infrared excesses indicating mass loss (Liebert *et al.* [72], Ibata *et al.* [73], Mauron, Gigoyan and Kendall [74], Downes et al., in prep).

The halo AGB stars can be found in two ways, via their large infrared excesses, and via variability (typical periods are 1-3 years), in wide-angle surveys. In addition, the chemistry, i.e. whether the star has "normal" abundances with more oxygen than carbon, or is a carbon star, with carbon more abundant than oxygen, can be determined from the infrared colors. Figures 4-6 show a series of color-color diagrams combining JHK and IRAC photometry. The JHK photometry is from 2MASS, and the IRAC colors are derived by convolving ISO SWS spectra with the IRAC filters for a sample of AGB stars and supergiants (Margengo *et al.* [75]). Some scatter will be present because of variability (Smith, Price and Moffett [76]), since these observations were taken at different times. Also shown in Figs. 4-6 are data from the Spitzer First Look Survey, containing both Galactic stars and extragalactic objects.

AGB stars of all types are redder than main sequence stars in the near-IR colors in all diagrams, especially in J-K, even for AGB stars with little circumstellar dust ([3.6] - [4.5] $<$ 0.5). AGB stars with thick circumstellar envelopes are easily separated from galaxies by color. In addition, the oxygen and carbon-rich stars partly separate, in that

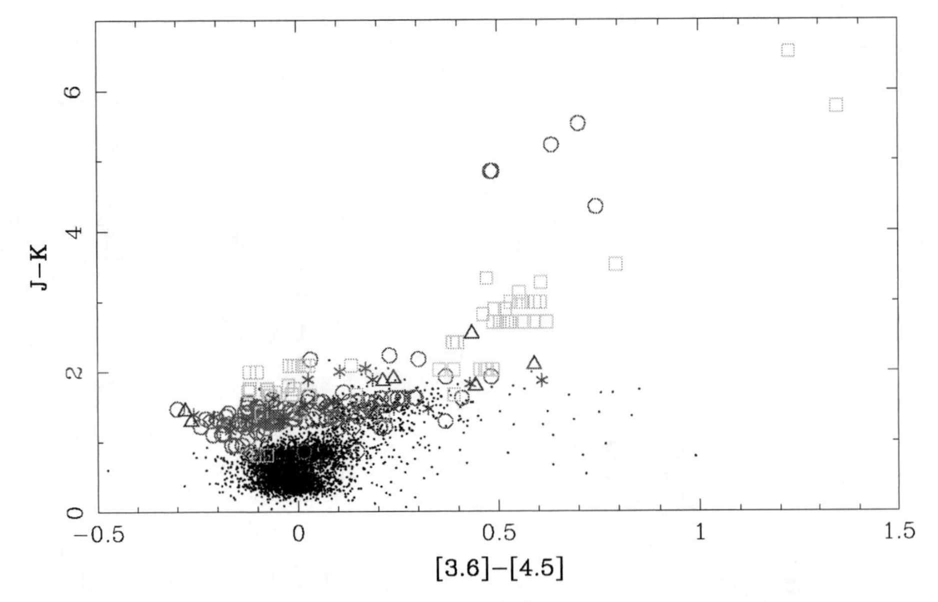

FIGURE 4. J-K versus [3.6] - [4.5] color for a sample of AGB stars. Oxygen rich stars are red circles, Carbon stars are green squares, S stars are blue triangles and supergiants are shown by brown asterisks. The data are from 2MASS and synthesized for the IRAC bands from ISO SWS spectra. The black dots show sources from the IRAC First Look Survey.

dust-poor carbon stars have slightly redder near-infrared colors than do oxygen stars, while for stars with larger infrared excesses the oxygen-rich stars have redder near-IR colors than do the carbon stars.

Any wide-angle survey at high latitudes, such as the shallow survey over 400 square degrees, will detect 100-200 distant halo AGB stars, and more will be found in the fields around targeted sources. These objects are invaluable for probing the structure and formation history of the very distant Milky Way.

6. THE COLORS OF NORMAL STARS

The GALEX, SDSS and 2MASS surveys provide precision photometry of millions of stars. The 2MASS limit of 14^m at K_s together with an assumed Rayleigh-Jeans spectral index gives a desired 5σ depth of 0.61 mJy at 3.6μm and 0.39 Jy at 4.5 μm. This is achievable with the "shallow" survey in Table 1 of Stauffer *et al.* [6]. The expected stellar density to this limit is about 5000 per square degree at high Galactic latitudes, so that data added by Spitzer during its warm mission will produce precision 12-band photometry between 0.14 and 4.5μm for 50,000 - 100,000 stars as a byproduct, and these can reliably be separated from galaxies and quasars by color and image size at optical wavelengths.

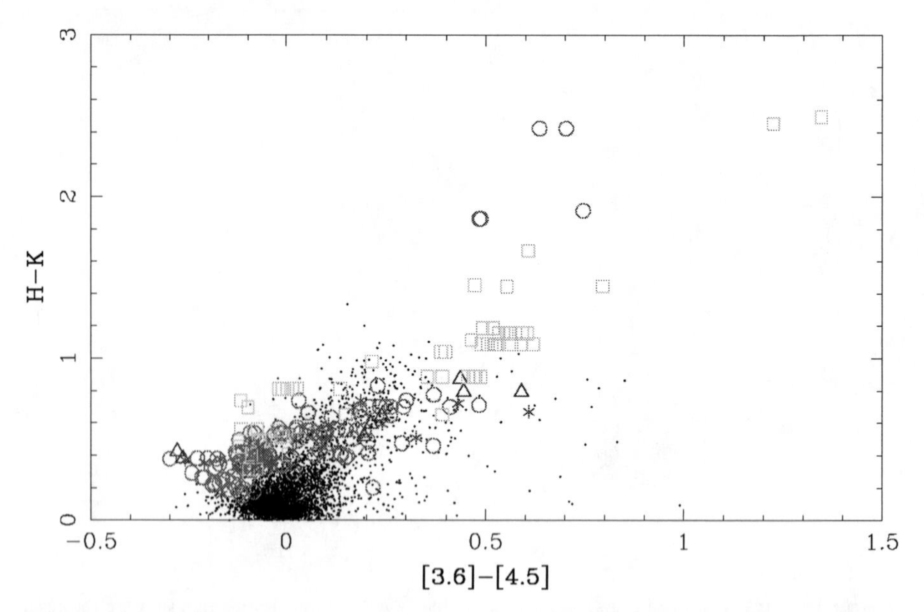

FIGURE 5. H-K versus [3.6] - [4.5] color for a sample of AGB stars. Oxygen rich stars are red circles, Carbon stars are green squares, S stars are blue triangles and supergiants are shown by brown asterisks. The data are from 2MASS and synthesized for the IRAC bands from ISO SWS spectra. The black dots show sources from the IRAC First Look Survey.

While data from each of these four surveys are expected to be extremely well calibrated internally, some of the challenge in exploiting the data will be on consistency between surveys. Here, the stellar data play a vital rôle and will likely provide fundamental broad-band calibration across this wavelength range for all other classes of object as well. A huge effort is underway at present to make multiband stellar photometry internally consistent using synthetic magnitudes calculated from stellar spectral libraries, both observed and synthesized (Cohen *et al.* [77], Coelho *et al.* [78], Martins and Coelho [79], Worthey and Lee [1], Davenport et al. in prep., Lee et al. in prep., and many others), and this work is driven by the need to understand both stellar populations and galaxy colors. These efforts use compilations of data from the literature, and demonstrate that different stellar colors have sensitivities to stellar effective temperature, metallicity, gravity, chemical peculiarity and even α element enhancement. For example, V-I is weakly sensitive to α element abundance, V-K is an excellent indicator of effective temperature, J-K is degenerate with temperature for M dwarfs but depends on metallicity while H-K does not, and so on. These findings are based on small samples of bright stars for which high-dispersion spectroscopy yields accurate element abundances but whose photometry is heterogeneous. Major efforts are now underway to use moderate resolution spectroscopy to derive stellar parameters for these very large numbers of stars with well calibrated photometry (Re Fiorentin *et al.* [80]).

What will Spitzer add to this effort? It extends the longest wavelength by almost a factor of two, into a spectral region sensitive to spectral lines of common molecules,

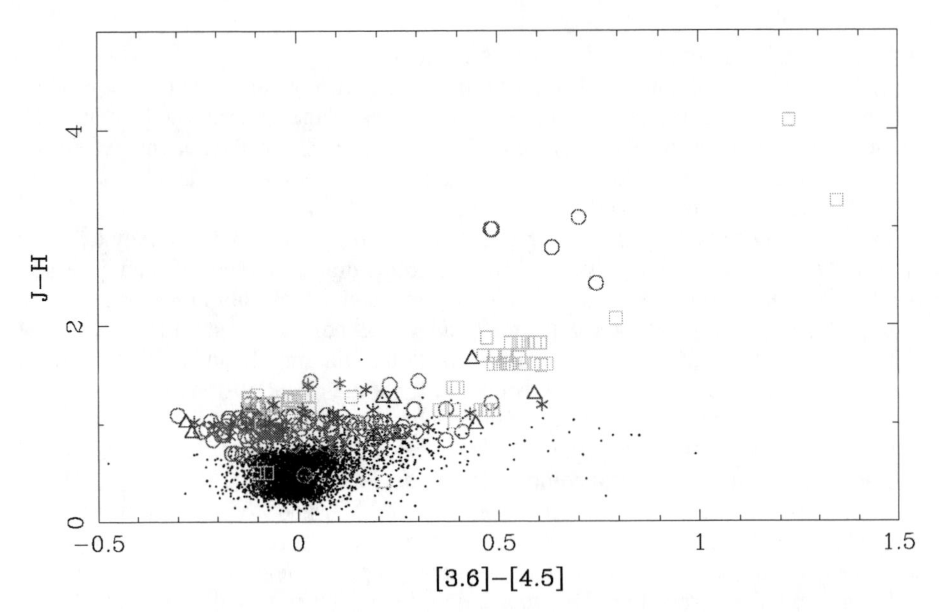

FIGURE 6. J-H versus [3.6] - [4.5] color for a sample of AGB stars. Oxygen rich stars are red circles, Carbon stars are green squares, S stars are blue triangles and supergiants are shown by brown asterisks. The data are from 2MASS and synthesized for the IRAC bands from ISO SWS spectra. The black dots show sources from the IRAC First Look Survey.

especially those containing carbon. At present only very crude metallicity values are available for the cooler stars (spectral types K, M and later). These stars are so cool that their flux densities are negligible at UV and blue wavelengths, and the Spitzer observations will thus provide wavelength coverage equal to that available for bluer stars. This effort can extend reliable measures of basic stellar parameters to much cooler stars than are accessible by optical photometry alone, including the most numerous stars, the M dwarfs. The likely science outcomes include: more reliable effective temperatures for cool stars, photometric identification of chemically peculiar cool stars (giant and especially dwarf carbon stars, cool subdwarfs, extremely low metallicity stars etc.) and the characterization of the broad-band colors of normal stars of all spectral types which will greatly aid in the identification of excess emission at wavelengths from 2- 5 μm due to cool companions and circumstellar dust.

7. SUMMARY

The advances in stellar astronomy that can be achieved by the Warm Spitzer Mission include:

(a) The discovery of 50-100 very cool T dwarfs (< 1000 K) and possibly up to 5 Y dwarfs from large, wide angle surveys. The location in the sky is to first order irrelevant, since these objects are nearby enough to be isotropic, so this project can piggy back on any general purpose survey that is sensitive enough.

(b) The discovery of tens of resolved companions to nearby stars, which will provide important ancillary information: distance, metallicity and age, as well as information on the relative frequencies of planetary and substellar companions, the dependence of the presence of companions on stellar age and metallicity, and, perhaps, indirect information on the formation of brown dwarfs and planets.

(c) Three important areas can be investigated with the same pointed survey of white dwarfs: the detection of unresolved L, T and Y companions via their infrared excesses, and the determination of the luminosity function and lowest luminosity star at the bottom of the main sequence: the detection of dust disks possibly produced by remaining planetary system members: and the characterization of the broad-band SEDs to 4.5μm, useful for white dwarf models atmospheres and for white dwarf cosmochronology.

(d) The combination of Spitzer and near-infrared colors will identify several tens of distant, high-latitude AGB stares in the Galactic halo and allow their mass loss rates and, with some uncertainty, chemical composition, to be determined.

(e) Accurate broad-band photometry of tens of thousands of normal stars of all spectral types between 0.15μm and 4.5μm and the relation of these colors to metallicity, effective temperature, gravity and other chemical properties (carbon excess, α-element variation, etc.) These spectra will form a calibration set for use in all multi-wavelength astronomy surveys.

ACKNOWLEDGMENTS

We thank the Spitzer Science Center for the invitation to this meeting and for all their help and support in preparing this manuscript - thanks especially to Lisa Storrie-Lombardi, Tom Soifer and Luisa Rebull. We're also grateful to Michael Liu, Ted von Hippel, Davy Kirkpatrick, Mike Jura and Dieter Engels for very useful and helpful discussions and communications.

REFERENCES

1. G. Worthey and H.-C Lee, *ApJS* (in press: astro-ph/0604590) (2007).
2. M.F. Skrutskie et al., *AJ* **131**, 1163 (2006).
3. N. Padmanabhan et al., *ApJ* (in press: astro-ph/0703454) (2007).
4. Ž Ivezić, "Selected Science Topics for LSST", LSST Science Working Group White Paper (2003).
5. A.K. Mainzer et al., *SPIE* **6265**, 626521 (2006).
6. J.R. Stauffer et al., "The Spitzer Warm Mission Science Prospects" white paper draft (this volume) (2007).
7. J.D. Kirkpatrick et al., *ApJ* **519**, 802 (1999).
8. J.D. Kirkpatrick et al., *AJ* **120**, 447 (2000).
9. A.J. Burgasser et al., *ApJ* **564**, 421 (2002).
10. T.R. Geballe et al., *ApJ* **564**, 466 (2002).
11. D.A. Golimowski et al., *AJ* **127**, 3516
12. M.C. Liu, S.K. Leggett, and K. Chiu, *ApJ* **660**, 1507 (2007).
13. K.S. Noll, T.R. Geballe, S.K. Leggett, and M.S. Marley, *ApJ* **541**, L75 (2000).
14. S.J. Warren et al., *MNRAS* (in press: arXiv0708.0655) (2007).
15. A. Lawrence et al., *MNRAS* **379**, 1599 (2007).

16. S.K. Leggett et al., *ApJ* **655** 1079 (2007a).
17. T.R. Kendall et al., *A&A* **466**, 1059 (2007).
18. N. Lodieu et al., *MNRAS* **379**, 1423 (2007).
19. B. Patten et al., *ApJ* **651**, 502 (2006).
20. D. Stern et al. *ApJ* **663**, 677 (2007).
21. A.K. Mainzer et al. 2007, *ApJ* **662**, 1245 (2007).
22. A. Burrows, D. Sudarsky, and J.I. Lunine, *ApJ* **596**, 587 (2003).
23. H.R.A. Jones et al., *AN* **326**, 920 (2005).
24. A. Burrows, D. Sudarsky, and I. Hubeny, *ApJ* **640**, 1063 (2006).
25. M. Marley et al. 2007, *ApJ* **655**, 541 (2007).
26. D. Saumon et al., *ApJ* **656**, 1136 (2007).
27. S.K. Leggett, et al., *ApJ* (in press: arXiv:0705.2602) (2007b).
28. A. Burrows, W.B. Hubbard, J.I. Lunine, and J. Liebert, *Rev. Mod. Phys.* **73**, 719 (2001).
29. P.R. Allen, Ph.D. Thesis, University of Pennsylvania (2005).
30. M. Marengo et al., *AAS* **209**, 161.07 (2006a).
31. K.L. Luhman et al., *ApJ* **654**, 570 (2007).
32. M. Marengo et al., *ApJ* **647**, 1437 (2006b).
33. B. Zuckerman and E.E. Becklin, *Nature* **336**, 656 (1988).
34. N. Silvestri, N. et al., *AJ* **131**, 1674 (2006).
35. J. Farihi and M. Christopher, *AJ* **128**, 1868 (2004).
36. J. Farihi, E.E. Becklin, and B. Zuckerman, *ApJS* **161**, 394 (2005).
37. P.F. Maxted et al., *Nature* **442**, 543 (2006).
38. P. Bergeron, P., F. Wesemael, and A. Beauchamp, *PASP* **107**, 1047 (1995).
39. G.P. McCook, and E.M. Sion, *ApJS* **121**, 1 (1999).
40. D.J. Eisenstein et al., *ApJS* **167**, 4 (2006).
41. J.P. Subasavage et al., *AJ* **134**, 252 (2007).
42. S. Wachter, D.W. Hoard, K.H. Hansen, R.E. Wilcox, H.M. Taylor, and S.L. Finkelstein, *ApJ* **586**, 1356 (2003).
43. P. Szkody et al., *AJ* **131**, 973 (2006).
44. T.E. Harrison et al., *AJ* **125**, 2609 (2003).
45. S.B. Howell et al., *ApJ* **646**, L65 (2006a).
46. S.B. Howell et al., *ApJ* **652**, 709 (2006b).
47. S.P. Littlefair et al., *Science* **314**, 1578 (2006).
48. C.S. Brinkworth et al., *ApJ* **659**, 1541 (2007).
49. G. Fontaine and G. Michaud, *ApJ* **231**, 826 (1979).
50. P. Chayer et al., *ApJ* **454**, 429 (1995),
51. C. Alcock, C.C. Fristom, and R. Siegelman, *ApJ* **302**, 462 (1986).
52. M. Jura, *ApJ* **653**, 613 (2006).
53. E.E. Becklin et al., *ApJ* **632**, L119 (2005).
54. M. Jura, J. Farihi, B. Zuckerman, and E.E. Becklin, *AJ*, **133**, 1927 (2007).
55. B. Zuckerman and E.E. Becklin, *Nature* **330**, 138 (1987).
56. F. Mullally et al., *ApJS* **171**, 206 (2007).
57. T. von Hippel et al., *ApJ* **662**, 544 (2007).
58. B. Gänsicke et al., *Science* **314**, 1908 (2006).
59. B. Zuckerman, D. Köster, I.N. Reid, and M. Hünsch, *ApJ* **596**, 477 (2003).
60. M. Kilic and S. Redfield, *ApJ* **660**, 641 (2007).
61. P. Dufour, P. et al., *ApJ* **651**, 1112 (2006).
62. P. Dufour et al., *ApJ* **663**, 1291 (2007).
63. B. Zuckerman, D. Köster, C. Melis, B. Hansen, and M. Jura, *ApJ* (in press: arXiv0708.1098) (2007).
64. M. Kilic et al., *ApJ* **642**, 1051 (2006).
65. E. Gates et al., *ApJ* **612**, L129 (2004).
66. H.C. Harris et al., submitted to *AJ* (2007).
67. H.C. Harris, H.C. et al., *AJ* **131**, 571 (2006).
68. K.B. Halford et al., *BAAS* **37**, 1377 (2005).
69. R.A. Benjamin et al., *ApJ* **630**, L149 (2005).
70. M. Meixner, M. et al., *AJ* **132**, 2268 (2006).

71. M.A.T. Groenewegen et al. *MNRAS* **376**, 313 (2007).
72. J. Liebert et al., *PASP* **112** 1315 (2000).
73. R. Ibata et al., *ApJ* **551**, 294 (2001).
74. N. Mauron, K.S. Gigoyan, and T.R. Kendall, *A&A* **463**, 969 (2007).
75. M. Marengo et al., *ASP. Conf. Proc.* (in press: astro-ph/0611346) (2007).
76. B.J. Smith, S.D. Price, and A.J. Moffett, *AJ* **131**, 612 (2006).
77. M. Cohen, W.A. Wheaton, and S.T. Megeath, *AJ* **126**, 1090 (2003).
78. P. Coelho, B. Barbuy, J. Meléndez, R.P. Schiavon, and B.V. Castilho, *A&A* **443**, 735 (2005).
79. L.P. Martins and P. Coelho, *MNRAS* (in press: astro-ph/0705.1994) (2007).
80. P. Re Fiorentin et al., *A&A* **467**, 1373 (2007).

The Warm Spitzer Mission For The Investigation Of Nearby Galaxies

D. Calzetti*, M. Regan†, L. van Zee**, L. Armus‡, R. Chandar§, K.D. Gordon¶, K. Sheth‡ and M. Yun*

*Department of Astronomy, University of Massachusetts, 710 N. Pleasant St., Amherst, MA 01003, USA
†Space Telescope Science Institute, 3700 San Martin Dr., Baltimore, MD 21218, USA
**Department of Astronomy, Indiana University, 727 E. 3rd St., Bloomington, IN 47405, USA
‡Spitzer Science Center, Caltech MC 220-6, Pasadena, CA 91125, USA
§Carnegie Observatories, 813 Santa Barbara St. Pasadena, CA 91101, USA
¶Steward Observatory, University of Arizona, 933 N. Cherry Ave., Tucson, AZ 85721, USA

Abstract. Spitzer in the warm phase will provide a unique opportunity for large-field and deep observations of nearby galaxies to address a number of fundamental science questions that cannot be adequately answered within the confines of current sample sizes/depths as produced by regular general observer, or even Legacy, proposals. Potential science goals include the morphological classification of galaxies, the investigation of the edges of galaxy disks and of intragroup/intracluster environments, the use of hot dust emission to trace star formation and AGNs, the monitoring of supernovae and of other variable sources, and the calibration of stellar population models in the mid-IR. The local Universe is the only available benchmark against which distant galaxies can be compared, and every effort should be made to ensure that the properties of nearby galaxies are fully investigated by the 'warm' Spitzer.

Keywords: Spitzer Space Telescope, infrared astronomical observations, external galaxies, galaxy formation, stellar populations
PACS: 95.85.Hp, 98.62.Ai, 98.62.Ck, 98.62.Lv

1. INTRODUCTION

Spitzer in the warm phase will provide a unique opportunity for large-field and deep observations of nearby galaxies, to address a number of fundamental science questions that cannot be adequately answered within the confines of current sample sizes/depths as produced by regular general observer, or even Legacy, proposals.

The science goals, which will be detailed in the next few sections, include the morphological classification of galaxies, the investigation of the edges of galaxy disks and of intragroup/intracluster environments, the use of hot dust emission to trace star formation and AGNs, the monitoring of supernovae (for the study of dust production) and of other variable sources, and the calibration of stellar population models in the mid-IR.

The exploration of the nearby Universe is not only important *per se*, but it also provides the ultimate benchmark for unraveling the distant Universe. Hence, the utilization of the 'Warm' Spitzer for the investigation of the properties of nearby galaxies will have the double benefit of providing for the understanding of both the local Universe and the distant galaxies populations.

These goals can be accomplished with a combination of both Super-Legacy Surveys

CP943, *The Science Opportunities for the Warm Spitzer Mission Workshop*,
edited by L. J. Storrie-Lombardi and N. A. Silbermann

and targeted observations. Independent of the nature of the observing project, the science goals described below have the potential to generate data that both provide an enduring archival research value and are pathfinders or foundation science for existing or upcoming facilities, both at infrared/submillimeter wavelengths (e.g., Herschel, JWST, etc.), and mm/radio wavelengths (ALMA, CARMA, LMT, EVLA, etc.). As an example, by the time ALMA is operational at full capacity in 2012, JWST will not yet be in space, and other NASA Great Observatories need to provide the critical complementary, multiwavelength data while they are still operating.

IRAC on the Warm Spitzer will be the most sensitive camera available at 3.6 and 4.5 μm for such applications until the advent of JWST. IRAC still has one advantage over JWST, in that its field of view is 2.6 times larger, a crucial feature for observations of extended galaxies. IRAC on the Warm Spitzer will also be about twice as sensitive than WISE, and with a much higher angular resolution ($\sim$2 arcsec for IRAC versus $\sim$6 arcsec for WISE). These capabilities are used in what follows as guidance to identify the unique science contributions that the Warm Spitzer mission can provide in the field of nearby galaxies studies.

2. SCIENCE DRIVERS

2.1. Stellar Masses And The Morphological Classification Of Galaxies

The stellar mass and the morphology of a galaxy are two of the primary tools to investigate galaxy evolution. Galaxies are routinely decomposed into their primary stellar components: bulge, disk, bars (and/or into other objective elements, such as concentration parameters, second order moments, asymmetry parameters, and Gini coefficients (Bershady et al. [1], Conselice [2], Lotz et al. [3]), to map their history across cosmological times.

At optical wavelengths, stellar masses and morphological classifications have been historically hampered or complicated by the presence of dust, and by the sensitivity of some of the indices to presence of active regions of star formation (which tend to make some features more prominent than others, without necessarily tracing the underlying stellar mass). At infrared wavelengths, the impact of dust extinction and of localized star formation decreases by at least one order of magnitude. At these wavelengths, spiral structures become less evident, and bulges and bars become more prominent.

Tracing the underlying mass and structure of galaxies requires statistically large samples of nearby galaxies observed at infrared wavelengths. Because of their proximity, nearby galaxies are the natural test-beds for extant and new classification schemes, thanks to the higher spatial resolution they provide relative to their more distant counterparts.

2MASS (Skrutskie et al. [4]) has provided a first 'cut' to a sample like the one proposed here, but far from optimal in terms of depth. The typical 2MASS galaxy has been observed with a depth that is about 1-2 orders of magnitude shallower than the surface brightness at the R25 isophotal radius. This isophotal radius is the canonical value used in most classical morphological surveys. Deep infrared observations from the ground are far less efficient than from space, due to the many hundred-fold increase in the background. Furthermore, observations in the K band suffer from larger contamination from red supergiants (which trace recent star formation) than longer wavelengths.

A 'Spitzer RC3' would be the ultimate survey of this kind, but a unbiased sample selected from existing catalogs of about $1,000 - 1,500$ targets (about $150 - 200$ galaxies per morphological bin) would still accomplish much of the science goals above. This is the type of project, both in terms of required observing and data products uniformity and of sheer amount of observing time, that would easily lend itself to be SSC-led as a service to the community.

2.2. The Edges Of Disks: Pathfinders Of Galaxy Evolution

A number of studies, mostly based on HST optical and GALEX UV imaging, have revealed the presence of low surface brightness features at very extended radii associated with most galaxies. GALEX has shown that in a number of nearby galaxies the UV emission extends well beyond the 'edge' of the H-alpha disk, and beyond 2-3 times the Holmberg radius (Thilker et al. [5], Gil de Paz et al. [6]). This UV emission is due to stellar populations, for the most part 'evolved' (i.e., no longer producing ionizing photons, but still young enough to produce significant UV, i.e. 10 Myr < ages < a few hundred Myr; Dong et al. 2007, in prep.). Longer wavelength observations have further shown the presence of stellar populations older than a few hundred Myr beyond the standard edge of the disks. Clearly, there is and/or has been star formation in these extreme environments, possibly with multiple episodes or complex histories.

Presence of extended disks, and the presence or absence of an edge to the stellar population, have profound implications for our understanding of the formation and evolution of disks. The extreme questions one may be tempted to ask are: Where does one galaxy end and another begin? Are all galaxies just one humongous galaxy? Optimal targets for this type of investigation are inclined galaxies, which maximize the contrast against the background.

Not less important are the implications of current or recent-past star formation for our understanding of the physics of the scaling laws of star formation and of local versus global star formation thresholds (in a potentially low-gas-density environment like the outer disk regions). Outer-disk star formation has also the potential to play a crucial role for the pollution of the intergalactic medium, which would be much more easily done from the 'edges' of the potential wells. Examples of targets include the galaxies for which GALEX has detected extended (beyond the disks) UV emission.

Along the same lines, the intragroup and intracluster environments provide excellent laboratories for observing galaxy evolution in action, including harrassment, ram pressure stripping, and tidal encounters. Tidal tails or accretion driven events may produce gaseous features that last a significant period of time ($\sim$ 1 Gyr). These tidal features are attractive for the investigation of the local star formation threshold, as they provide a high-density environment far away from the inner disks of galaxies. Blind surveys of nearby groups and clusters will enable identification of so-far unknown faint galaxies (red and dead) and the investigation of tidal streams and intragroup/intracluster old stellar populations. An example is a blind survey of all of the Virgo Cluster, or at least the inner 20deg x 20deg portion. Virgo is both the nearest large cluster and is non-virialized, implying that galaxies in a range of environments are present.

Deep Spitzer observations at 3.6 and 4.5 μm would target the evolved stellar populations at the edges of galaxies and/or in the intragroup/cluster medium, thus providing a more direct link than shorter wavelength observations to the mass of the stellar populations formed in these environments. As already remarked in the previous section, the 3.6 and 4.5 μm bands have lower contamination from red supergiants than the ground-based K band, and are therefore much more efficient tracers of the old, evolved stellar populations. The full census of these populations provided by Spitzer, together with existing supporting observations (e.g., HI deep maps, UV images, etc.), will shed light on their nature, and address questions on the formation and evolution of their parent galaxies.

2.3. Hot Dust And Star Formation

The investigation of the SINGS galaxies sample (Kennicutt et al. [7]) shows that the ratio 4.5/3.6 varies significantly across galaxy disks, and it correlates strongly with tracers of star formation, like the IRAC 8 μm emission, in the sense that higher 8 μm emission corresponds to large values of the 4.5/3.6 ratio. The interpretation is that the recently formed stars heat the dust to temperatures high enough to contribute to the emission in the 4.5 IRAC band.

Further investigations are needed to provide a full characterization of the 4.5/3.6 ratio as a SFR indicator, and if a reliable calibration can be derived, this ratio would provide a potentially interesting tracer of star formation because of its insensitivity to extinction. However, existing 3.6 and 4.5 images are generally not deep enough to probe the full dynamical range of the ratio between the two bands.

Deep 3.6 and 4.5 IRAC images of nearby galaxies would provide the needed characterization (e.g., by comparison with archival MIPS 24 μm images, which are more closely related to current SFR, Calzetti et al. [8]), and, subsequently, a potentially powerful application for the mapping of SFRs in nearby galaxies. This would complement efforts to map the evolved stellar populations (see below) and stellar masses.

2.4. Hot Dust And AGNs

A common infrared spectral signature of Active Galactic Nuclei (AGN) is an excess of very hot ($T > 500$K) dust emission over a stellar continuum. This dust is thought to exist at the inner edge of a torus, or on the nuclear-facing surfaces of individual clouds, at small radii ($r < 50$ pc), directly illuminated by the AGN accretion disk. The dust produces strong blackbody emission throughout the mid–infrared, and in some cases, a 'bump' in the spectrum at 10μm from silicate emission. In sources with powerful AGN, or a lot of obscuring dust near the nucleus, the hot dust continuum can be detected over a very large range in wavelengths, extending into the near-infrared and far-infrared parts of the spectrum. In fact, the 25/60 μm flux density ratio was used very effectively over 20 years ago to isolate active galaxies in the IRAS database (De Grijp et al. [9], Miley et al. [10]).

The IRAC 3.6 and 4.5 μm bands are very sensitive to this hot dust emission, and AGN can be separated from star-forming galaxies in large Spitzer surveys (e.g. FLS, SWIRE, GOODS) using IRAC alone (Lacy et al. [11], Stern et al. [12]). While all four IRAC bands are typically used to constrain the mid-infrared spectral energy distribution, the 3.6 and 4.5 μm bands provide most of the power in detecting the excess emission from the hottest dust, over that from stars, since the stellar light is dropping while the hot dust is climbing.

A warm mission science program combining near-infrared (J, H, K) data with 3.6 and 4.5 μm IRAC band data would effectively isolate AGN for redshifts $z < 1$. At these redshifts the ground-based K and L bands move through the 3.6 and 4.5 IRAC bands. For $z > 1$, stellar emission dominates those bands , making detection of hot dust extremely difficult, except for the most luminous AGN (whose near-infrared light shortward of the K-band can still be dominated by dust emission). Starburst galaxies tend to get redder in their IRAC colors for $z > 1$. AGN typically have $m_{3.6}-m_{4.5} = 0.5-1.0$ mag (Vega), while starburst galaxies tend to be 0.5–1.0 mag bluer for $z < 0.3$ and $0.5 < z < 1.0$ (Stern et al. 2005).

A program that targets AGN at $z < 1$, in order to quantify the amount of hot dust, and compares this to the optical emission line (e.g. [OIII]) luminosity or the radio power, would be useful for modeling the properties of the torus or obscuring clouds. Samples could be built from SDSS, FIRST, or other surveys that focus on properties not directly associated with near-nuclear emitting dust. A careful accounting of the detection limits (as a function of AGN-to-host luminosity) would be required in order to properly assess the statistics, since the IRAC observations would have the largest contribution from host (stellar) light. However, the IRAC data would be extremely sensitive to classes of (moderately) obscured AGN, such as those uncovered in the 2MASS survey (Cutri et al. [13]), which are not readily found via more traditional means. The sensitivity of IRAC coupled with the mapping speed of Spitzer makes it possible to build up complete samples of AGN, extremely rapidly. Very large, near-infrared based surveys (e.g., UKIDSS) could take full advantage of pointed or mapped IRAC follow-up, even with only the 3.6 and 4.5 bands. The planned availability of extremely large or all sky surveys in the optical and mid-infrared (e.g., LSST, WISE) when combined with IRAC 3.6 and 4.5 μm data, would provide an extremely valuable tool for both finding hidden AGN, and understanding accretion process, by tying optical or UV variability to the response of the (reprocessed) dust emission. Even if a focused survey for buried AGN is not done, a complete assessment of the number density of galaxies hosting obscured AGN, as revealed through their excess hot dust emission, can arguably be a key science driver for other extragalactic surveys being planned for the warm mission.

2.5. A Census Of Evolved Stellar Populations: The Local Group

The proximity of Local Group Galaxies offers the unique opportunity to map their evolved stellar populations with a spatial resolution that is unmatched by more distant galaxies. This can provide a complete census of such populations within a well defined volume of the Universe, and complement similar observations at other wavelengths, to investigate the evolution of Local Group Galaxies.

During the Warm Spitzer mission, an investigation of the archive should be performed to establish which Local Group galaxies have been observed (down to R25), which need supplementary observations, and which are missing altogether and thus need observing. The complete local volume observed at 3.6 and 4.5 μm will provide an enduring legacy for future observations, and lends itself to also be SSC-led, and be provided as a service to the community.

As an example of scientific application for such observations, we mention the investigation of the dust versus stellar emission in Local Group Galaxies. Although the 3.6 and 4.5 μm IRAC bands are contributed mainly by photospheric emission from stars (at least when observing external galaxies), both bands receive contribution from dust as well: aromatic emission at 3.6 μm and small grain continuum emission at 4.5 μm. The proximity of the Local Group Galaxies will provide the opportunity to discriminate the contribution from dust from that of stars, and the two dust contributions as well. The spatial decomposition between stars and dust will help understand how much of the integrated emission in more distant galaxies in these two bands is contributed by dust emission versus stellar emission.

2.6. Variability Monitoring In Local Group Galaxies

Spitzer observations have been instrumental in establishing that supernovae may be more efficient at destroying dust than producing it, both for Type II and Type 1a (Borkowski et al. [14], Williams et al. [15]). However, the issue is not entirely settled yet, and some evidence for dust production has been reported, also using Spitzer data (Sugerman et al. [16], Smith et al. [17]). Whether massive-star supernovae are efficient dust factories or not has profound implications for our understanding of the primordial Universe, in particular for our understanding of the dust detected in quasars and GRBs at epochs when the Universe was less than 1 Gyr old (Bianchi et al. [18]). During the warm phase, Spitzer could periodically monitor nearby galaxies for which data already exist, to observe supernovae explosions. Observations at 3.6 and 4.5 μm of such events within a short period of their happening will provide insights into the evolution of the warm dust emission (warm enough to emit in the short IRAC bands) from Supernovae.

Recent investigations (e.g., with the SAGE Spitzer data of the LMC) are beginning to explore the connection between stellar variability and mass loss in Local Group galaxies. One idea that is being pursued is that mass loss in stars is driven by variability. Mass loss from stars is traced via IRAC 8 and MIPS 24 μm, which are good indicators of dust in the winds of stars which are injecting dust+gas into the ISM. Variability could be traced by multiple-epochs observations of those stars at the short IRAC wavelengths, over multiple years during the warm phase. This could be done not only for the LMC, but also for other Local Group galaxies for which suitable data for tracing the dust emitted from stars exist. Indeed, IRAC and MIPS data already exist for most Local Group galaxies, which can be used to trace the mass loss. Even at the distances of M31 and M33, IRAC 3.6 and 4.5 μm observations would probe deep enough that we should see a significant fraction of the total mass injecting stars. Thus, Spitzer in the warm phase would provide an excellent tool to monitor the variability of the stars for which mass loss has been detected.

2.7. The Infrared Spectral Energy Distributions Of Stellar Populations

Some of the science that Spitzer will be able to accomplish during the warm phase could be labeled under the term ‘foundation science’ for other facilities. Within this category is a Spitzer survey of star clusters covering ages from a few tens of millions of years to 12 billion years and a range of metallicity in nearby galaxies. These observations will provide empirical measurements of simple stellar populations which will be used to improve population synthesis model predictions in the mid-IR (as is currently being done in the near-IR). Such a survey would also enable a direct calibration between metallicity and color in the IRAC bands. In addition, the Spitzer data can be combined with 2MASS observations to create color-magnitude and color-color diagrams of the infrared bright RGB and AGB stars, in order to quantify the relative contributions from the stars and dust as a function of metallicity. These observations will provide benchmarks for interpreting the colors of galaxies and star forming regions within galaxies in the mid-IR (at both high and low redshift) during the JWST era.

3. CONTRIBUTION OF WARM SPITZER DATA

The premise for the science goals described in the previous section is to leverage on the unique capabilities that the Warm Spitzer mission can offer in comparison with other facilities available roughly within the same timeframe and operating at similar wavelengths. Relative to WISE, Spitzer has higher angular resolution (2 arcsec versus 6 arcsec for WISE at similar wavelength) and a 30 second IRAC exposure is about twice as sensitive as 8 passes with WISE. The higher angular resolution is a critical requirement for discriminating as much as possible various components in the crowded fields that characterize galaxy observations.

Relative to ground-based near infrared facilities, Spitzer offers a few hundred times lower background and a time-stable PSF (no seeing). These two characteristics are required to probe deeply within and outside galaxies; the lower background from space is a requirement, for instance, for investigating the outer fields of galaxies or the intra-cluster/intergalactic medium. The superior sensitivity of Spitzer in the 3-5 μm window relative to ground-based instrumentation is also instrumental for reaching the sensitivity limits required for identifying and discriminating AGNs in galaxies via their hot dust emission.

Relative to telescopes with adaptive optics, Spitzer has larger Field-of-View (required for mapping extended objects, like nearby galaxies), stable, repeatable, and uniform PSF, and smaller and stable PSF wings (which enable accurate photometry). Furthermore, the efficient mapping capability is essential for obtaining large maps of nearby objects.

4. SCOPE OF DATA

Each of the science goals described above requires its own datasets, although some observations may address more than one goal.

For goal #1 (§2.1), a SINGS-like approach can be adopted, which will require between 1 and 3 hours per galaxy (depending on the size of the galaxy, as the total exposure time per pointing will be fixed at 240 seconds to reach R25). For about 1,500 targets, this translates into approximately $1,500-3,000$ hours of observing time.

About 10 times deeper images would be required for characterizing and calibrating the 4.5/3.6 ratio as a SFR tracer, but the number of galaxies required for such investigation would be much smaller, of order of a few tens.

For the outer regions of galaxies and the intragroup/intracluster medium, the goal will be to observe at least small, evolved stellar clusters (masses around 5,000 solar masses and ages about 100 – 300 Myr). This would correspond to the detection of 1 microJy at 3.6 μm, which can be obtained with 2,000 seconds of exposure to achieve S/N=5. For a 20 deg x 20 deg map (i.e., the inner region of the Virgo Cluster), this would correspond to a total integration time of about 80,000 hours (or about the entire observing time available during the Warm Spitzer mission). Shallower depths (and shorter times) will still yield considerable information on the Virgo structure and the intracluster medium. The outer regions of galaxies will require considerably less total time, because smaller areas will be required to be mapped (although the exposure time per pointing will be roughly the same). In the local Unverse, about 50 – 100 targets will have the characteristics required to address science goal #2 (§2.2).

Observing time requirements for Local Group Galaxies will be dependent on how many of the galaxies will already be available in the archive by the end of the Cold Spitzer Mission. Exposure times will be also dependent on the extent of the galaxy; as example, a galaxy as extended as the LMC will require a few hundred hours, but a galaxy of similar size as NGC6822 will only require 2 hours.

Variability studies can be used to augment the observations of Local Group Galaxies, by progressively accumulating maps that are observed with a cadence of about 30 – 60 days over 5 – 10 epochs.

Finally, observations of stellar clusters in external galaxies (goal #7, §2.7) will have exposure time requirements per pointing that are similar to those of goal #2 (§2.2). Optimal targets will be galaxies within 5-10 Mpc that host stellar clusters with a large range of ages and masses within relatively small regions, which can help reduce requirements for mapping. Technically, one could use the same galaxies of goal #3 (§2.3) for goal # 7 (§2.7).

ACKNOWLEDGMENTS

The authors would like to heartfully thank Nancy Silbermann (Spitzer Science Center) for her extensive help with this manuscript.

REFERENCES

1. Bershady, M.A., Jangren, A., and Conselice, C.J. 2000, AJ, 119, 2645
2. Conselice, C.J. 2003, ApJS, 147, 1
3. Lotz, J.M., Primack, J., and Madau, P. 2004, AJ, 128, 163
4. Skrutskie, M.F. 2006, AJ, 131, 1163
5. Thilker, D.A., Bianchi, L., Boissier, S., Gil de Paz, A., Madore, B.F. et al., 2005, ApJ, 619, L79
6. Gil de Paz, A., Madore, B., Boissier, S., Thilker, D., Bianchi, L., et al.,. 2007, ApJ, 661, 115
7. Kennicutt, R.C., Armus, L., Bendo, G., Calzetti, D., Dale, D.A., Draine, B.T., Engelbracht, C.W., Gordon, K.D., Grauer, A.D., Helou, G., et al., 2003, PASP, 115, 928
8. Calzetti, D., Kennicutt, R.C., Engelbracht, C.W., Leitherer, C., Draine, B.T. et al., 2007, ApJ, 666, 870
9. De Grijp, M.H.K., Miley, G.K., Lub, J., and de Jong, T. 1985, Nature, 314, 240
10. Miley, G.K., Neugebauer, G., and Soifer, B.T. 1985, ApJ, 293, L11
11. Lacy, M., Storrie-Lombardi, L.J., Sajina, A., Appleton, P.N., Armus, L., et al., 2004, ApJS, 154, 166
12. Stern, D., Eisenhardt, P., Gorjian, V., Kochanek, C.S., Caldwell, N. et al., 2005, ApJ, 631, 163
13. Cutri, R.M., Nelson, B.O., Francis, P.J., and Smith, P.S., 2002, "The 2MASS Red AGN Survey" in *AGN Surveys*, ASP Conference Proceedings 284, Astronomical Society of the Pacific, San Francisco, 2002, p. 127.
14. Borkowski, K.J., Williams, B.J., Reynolds, S.P. Blair, W.P., Ghavamian, P. et al., 2006, ApJ, 642, L141
15. Williams, B.J., Borkowski, K.J., Reynolds, S.P., Blair, W.P., Ghavamian, P. et al., 2006, ApJ, 652, L33
16. Sugerman, B.E.K., Ercolano, B., Barlow, M.J., Tielens, A.G.G.M., Clayton, G.C. et al.. 2006, Sci, 313, 196
17. Smith, N., Foley, R., and Filippenko, A.V. 2007, ApJ, submitted (astroph/0704.2249)
18. Bianchi, S., and Schneider, R. 2007, MNRAS, 378, 973

Planetary Science Goals for the Spitzer Warm Era

C.M. Lisse[*], M.V. Sykes[†], D. Trilling[**], J. Emery[‡], Y. Fernandez[§], H.B. Hammel[¶], B. Bhattacharya[||], E. Ryan[§§] and J. Stansberry[**]

[*]*Planetary Exploration Group, Space Department, Johns Hopkins University, Applied Physics Laboratory, 11100 Johns Hopkins Rd., Laurel, MD 20723, USA*
[†]*Planetary Science Institute, 1700 E. Fort Lowell Rd, Suite 106, Tucson, AZ 85719, USA*
[**] *Steward Observatory, University of Arizona, 933 N. Cherry Ave, Tucson, AZ 85721, USA*
[‡] *SETI Institute / NASA Ames Research Center, Mail Stop 245-6, Moffett Field, CA 94035, USA*
[§] *Department of Physics, University of Central Florida, 4000 Central Florida Blvd, Orlando, FL, 32816-2385, USA*
[¶] *Space* Science *Institute, 4750 Walnut Street, Suite 205, Boulder, CO 80303, USA*
[||] *Spitzer* Science *Center, California Institute of Technology, Pasadena, CA 91125, USA*
[§§]*Department of Astronomy, University of Minnesota, 116 Church St SE, Minneapolis, MN 55455, USA*

Abstract. The overarching goal of planetary astronomy is to deduce how the present collection of objects found in our Solar System were formed from the original material present in the proto-solar nebula. As over two hundred exo-planetary systems are now known, and multitudes more are expected, the Solar System represents the closest and best system which we can study, and the only one in which we can clearly resolve individual bodies other than planets. In this White Paper we demonstrate how to use Spitzer Space Telescope InfraRed Array Camera Channels 1 and 2 (3.6 and 4.5 μm) imaging photometry with large dedicated surveys to advance our knowledge of Solar System formation and evolution. There are a number of vital, key projects to be pursued using dedicated large programs that have not been pursued during the five years of Spitzer cold operations. We present a number of the largest and most important projects here; more will certainly be proposed once the warm era has begun, including important observations of newly discovered objects.

Keywords: Spitzer Space Telescope, infrared astronomical observations, Kuiper Belt objects, trans-Neptunian objects, asteroids
PACS: 95.85.Hp, 96.12.Bc, 96.30.Xa, 96.30.Ys

1. INTRODUCTION AND OVERVIEW

Condensable elements produced by stellar nucleosynthesis form icy volatile and refractory dust particles that eventually find their way into young stars and planetary systems. Dust is known to orbit a growing number of main sequence stars (Beta Pic, Vega, Epsilon Eri, and HD100546 are famous examples).

The dust in these stars is distinguished by having a lifetime that is short compared to the main-sequence lifetime of the central stars. Therefore, the dust cannot be primordial but must have been recently produced, perhaps by collisions among parent bodies in an unseen analog to the Kuiper Belt (KB) or asteroid belt, or by sublimation

CP943, *The Science Opportunities for the Warm Spitzer Mission Workshop,*
edited by L. J. Storrie-Lombardi and N. A. Silbermann

of comets. Much of the particulate material in such disks is cool, residing in the equivalent of the KB region. Strong similarities are seen between the spectra of dusty material emitted by comets, and of the dust in exo-systems (Fig. 1; Lisse *et al.* [1], [2]). The clear implication is that small bodies must exist in other planetary systems.

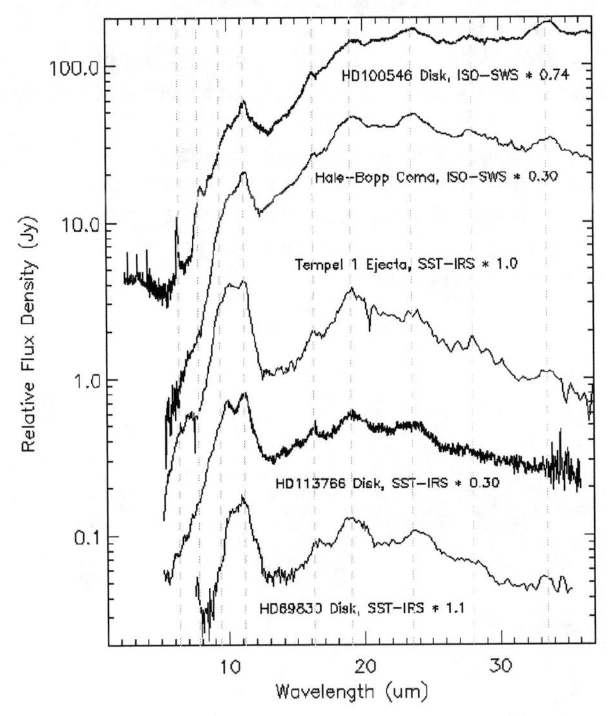

Figure 1. Similarities of mid-IR spectra of Solar System comets and dust found around other stars. Mid-IR spectra of the YSO HD100546, the young terrestrial planet building system HD113766, the mature solar-system like HD69830 system, and the comets C/1995 O1 (Hale-Bopp) and 9P/Tempel 1, showing the gross similarities and differences in the emission for the sources. Most of the obvious differences between the HD100546 and Hale-Bopp and the Tempel 1, HD113766, and HD69830 spectra are due to temperature - the latter three sources are much warmer on the average, and thus have relatively more emission in the 5-10 μm region. The other obvious difference is the very strong PAH emissions for HD100546 at 2 - 9 μm, and the superabundance of olivine in HD69830, as discerned from the pattern of the strong 9 - 12, 16 - 20, and 23 - 25 μm emission features.

The current paradigm of planetary system formation invokes the condensation of refractory materials (this term can be understood roughly as "rocky" or "solid" materials) in the thick proto-solar disk, surrounded by a gaseous envelope. Eventually the solid grains settle into a thin disk, whereupon aggregation into larger bodies begins. Dust grains grow to become pebbles, then boulders, then kilometer-sized and hundred-kilometer-sized planetesimals - comets and asteroids - the building blocks of solid planets. Planet formation has led to the clearing of the sun's disk inside about ~30AU. Many planetesimals in the inner Solar System were incorporated into the present day terrestrial planets, but many persist as 1- to 1000-km asteroids.

Planetesimals that formed in the region of the gas giants were either incorporated into the planets or rapidly ejected. Most escaped the Solar System entirely, but a large population was captured by stellar and galactic perturbations into the Oort cloud. Oort comets may be gravitationally scattered back into the inner Solar System to appear as dynamically new long-period comets, some of which are eventually captured into Halley-family short-period orbits.

The heavyweights of our Solar System are the giant gaseous planets. These formed from the planetesimals, like the terrestrial planets, but were able to also incorporate nebular material directly from the protoplanetary disk into their makeup, and were thus able to retain even the most volatile of molecular species found in the proto-solar nebula, like CO, CH_4, H, and He. With more than 200 known Jupiter-to-Neptune mass planets known to be orbiting other stars, this process of giant-planet formation is clearly ubiquitous. Jupiter and its kin hold clues to understand the processes that govern planets in other planetary systems.

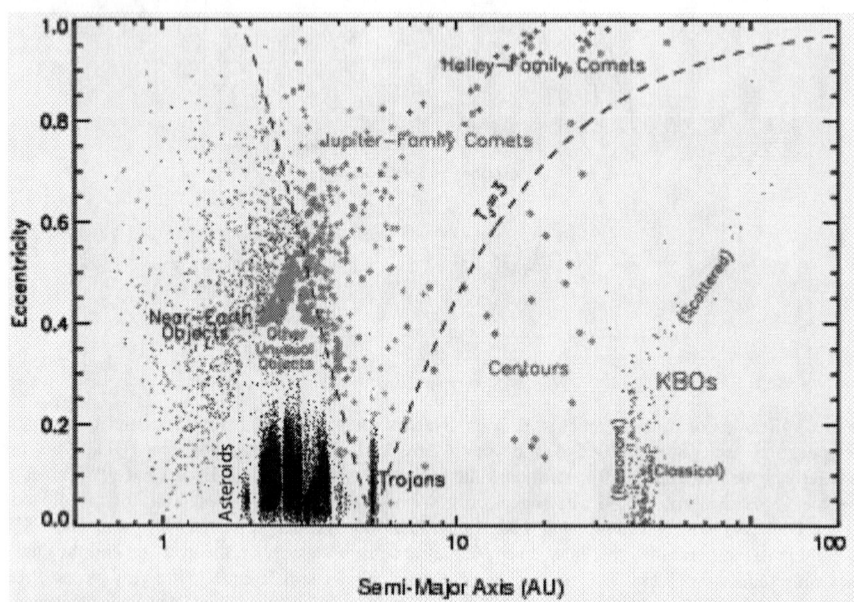

Figure 2. The major families of Solar System objects, as revealed by their dynamical structure. The Solar System's small bodies are shown plotted as a function of orbital semimajor axis (in AU) and orbital eccentricity. On the scale of this figure, the Sun and planets lie at approximately zero eccentricity. For clarity we reduced the eccentricity value of the asteroids by a factor of 2. The boundaries are not impermeable; for example there are inactivated asteroids in the nominal Jupiter-family comet region, and there are active comets in the near-Earth object region. This overlap hints at the evolutionary relationship among the groupings.

In the cool outer disk (beyond Jupiter's orbit), the planetesimals incorporated frozen volatiles as well as refractory material. The trans-Neptunian region is occupied by numerous bodies with sizes up to several thousand km (Pluto and other dwarf planets). Collisions in this population generate fragments, some of which were scattered inwards by dynamical chaos to become intra-planet wanders. These wanderers have short dynamical lifetimes, and are ejected from the Solar System entirely through gravitational interactions, or scatter further inward, where they begin to sublimate and are re-classified as first Centaurs in the giant planet region, and then Jupiter-family short-period comets in the inner Solar System. Comets that lose all their volatiles from prolonged solar heating appear as asteroids having unusual, comet-like orbits. A number of these dead comets are known among near-Earth asteroids, and are also prime targets for the Warm Era (Fig. 2).

In this scenario, there is a clear evolutionary flow-down of protoplanetary material, through planetesimals (KBOs, Centaurs, comets and asteroids) to planets. Many key questions in the details of this scenario of planetary system formation -- for our system, and others -- remain unsolved. We briefly list here six of the most relevant questions addressable in the Warm Spitzer Era (in rough priority order of importance), and describe each more completely below.

(1) **What is the dynamical history of the Solar System, and what role has giant-planet migration played?** There is substantial theoretical work suggesting that giant planets (in our Solar System and in others) undergo significant radial migration before the system settles into its final configuration, but there is still little direct evidence for this hypothesis. A Warm Spitzer can provide observational tests of specific predictions made by those theories by observing various small bodies in the Solar System. The results of those tests have strong implications for the dynamical evolution of our Solar System. [Trojans, Outer Main Belt Asteroids (MBAs), Kuiper Belt Objects (KBOs)]

(2) **What is the distribution of water and organic components in the Solar System?** The dynamical evolution of our Solar System has direct implications for the habitability of the Earth. It has been proposed that water and organic material was delivered to the Earth by impacting comets and asteroids. With a Warm Spitzer, we can map the distributions of water and organic material in the asteroid belt and constrain the consequent flux rate of this life-giving material to the early Earth. [Main Belt Asteroids]

(3) **What are the physical properties of potential Earth-impacting asteroids?** Survival on the Earth might be an "easy come, easy go" game -- an impact 65 million years ago eliminated the dinosaurs and allowed mammals to rise to prominence. Equally, there are thousands of asteroids known in near-Earth space. With a Warm Spitzer, we can understand the physical properties of these bodies and help evaluate the potential impact hazard onto the Earth. This information will potentially aid preparations for mitigating the impact hazard through direct manipulation of those asteroids. [Near-Earth Asteroids]

(4) **What is the nature of the refractory and carbonaceous material incorporated into comets from the proto-solar nebula?** Carbon is critical to the beginnings and evolution of life, as well as being an important component of the Earth's crust. Comets -- as leftover planetesimals from the era of Solar System formation -- contain important details of the mechanism of aggregation of ISM dust and gas into planets. The major reservoir of carbon in comets is in the gaseous volatile organic and CO/CO_2 ices, not in the refractory state, and is poorly studied. [Comets]

(5) **Are the structures we see in mature exo-disks explainable by the erosional and collisional processes supporting the Solar System zodiacal cloud?** For example, the major source of mass loss from active comets is in large (supermicron),

dark trail particles that are an important source for our Solar System's interplanetary dust cloud. Can the model timescales for these processes (10 - 10^6 yrs for cometary emission processes, 10^6 - 10^7 yrs for asteroid fragmentation events, 10^5 yrs for dust-dust collisions, 10^3 - 10^6 yrs for PR drag, $< 10^6$ years for radiation pressure blowout) be reconciled with zody cloud studies? [Zodiacal Cloud]

(6) **Is the major energy source driving the dynamics of the icy giant planets Uranus and Neptune gravitational contraction, insolation, or dynamical friction?** Answering this question is naturally crucial for understanding the energy budgets of exoplanet atmospheres. Current observations of Uranus suggest that seasonal insolation changes are more significant than predicted by models, and that Neptune's atmosphere may be more sensitive to insolation than expected. By extending observations of Uranus and Neptune over the five years of the Warm Era -- through major events in the seasonal cycles of the two planets -- we can determine which effects are paramount. [Ice Giants]

In sum, Solar System astronomy is unique among all fields of astronomy because of our ground truth: we can and do visit many of these primordial bodies with spacecraft missions to asteroids, comets, and planets, and we obtain telescope data of a quality and scope not possible for any other planetary system. (The ground truth also comes to us naturally, in the form of meteorites!) The Spitzer observations described below serve as a bridge between our rare but detailed ground truth data and understanding the formation and evolution of planetary systems throughout the galaxy.

2. SMALL OUTER SOLAR SYSTEM BODIES: TNOS, CENTAURS, JOVIAN TROJANS, AND OUTER MAIN BELT ASTEROIDS

Spitzer warm era observations are perfectly suited to determination of surface compositions of Trans-Neptunian Objects (TNOs), Centaurs, and Trojan and outer Main Belt asteroids through measurements of broadband reflected fluxes with IRAC. The value of IRAC is that its measurements of reflectance will 1) provide a far more sensitive search for ices than is possible at shorter wavelengths and 2) readily distinguish between candidates for the poorly understood "dark material," which is also not possible at shorter wavelengths. These capabilities critically address several longstanding questions in planetary science related to the nature of the nearly ubiquitous "dark material" and the distribution of volatiles, and more broadly address the following key question: What is the dynamical (and chemical) history of the Solar System, and what role has giant planet migration played?

2.1 Dynamical Evolution of the Outer Solar System

Two general views of the origin and dynamical evolution of the Solar System are currently in circulation. One, built up from a wealth of data and modeling in the mid to latter part of the last century, posits a relatively dynamically quiescent system after accretion. Starting in the 1990s, contemporaneous discoveries (and dynamical characterization) of Kuiper Belt objects in the Solar System and giant planets in exotic orbits around other stars have lead to increased consideration of giant planet migration and the dynamical eruptions such migrations induce. A second leading hypothesis for the Solar System has emerged in which all four giant planets have undergone migration, with severe implications for the minor body populations. In either scenario, the groups of primitive bodies we propose to observe are inherently interesting because of the information they hold on the compositional make-up and chemical evolution in critical regions of the nebula as well as the current distributions of astrobiologically and cosmochemically important materials (i.e., H_2O and organics). Furthermore, their importance increases significantly by the fact that their origins, and therefore compositions, provide the opportunity to distinguish between the two hypotheses for the dynamical evolution of the Solar System.

The Jupiter Trojans are particularly relevant for distinguishing these hypotheses. Orbiting the Sun at ~5.2 AU (trapped in Jupiter's stable Lagrange points), Trojans have traditionally been thought to have formed near their present location and to have been trapped in the Lagrange points later, perhaps as Jupiter rapidly grew (Marzari and Scholl [3]). In this case, they would represent material from the middle part of the solar nebula where ice first began to condense, a region not sampled by any other class of primitive body. Recently, however, Morbidelli *et al.* [4] proposed a scenario in which the Trojans formed in the Kuiper Belt, were scattered inward and were trapped in the Lagrange points as Jupiter and Saturn crossed their 2:1 mean-motion resonance. In this case, the Trojan asteroids would represent the most readily accessible depository of Kuiper belt material. But more importantly, *the Trojans offer a critical test of the planetary migration model of Morbidelli et al., which has implications for not only the Trojans, but for the dynamical evolution of the Kuiper belt and the Solar System as a whole.*

Trans-Neptunian objects (TNOs; perihelia beyond Neptune's orbit) also hold a wealth of information about the dynamical evolution of the Solar System. Several dynamical sub-classification systems have been proposed within this group, all of which include a cold classical population that has not suffered much scattering, resonant objects that are trapped in mean-motion resonances with Neptune (e.g., Pluto), and a scattered population. Evolution models call upon various mechanisms to explain different aspects of the current dynamical state, including migration of all the giant planets, several now lost (or hidden at very large distances) Neptune-class planetesimals, galactic tides, and close encounters with another star, among others (e.g., Chiang *et al.* [5] and references therein). Physical studies offer the best means to discriminate among these mechanisms. Currently, the only correlation between physical and dynamical properties is that the cold classical objects have systematically redder colors than other objects. Otherwise, the full range of spectral types (red, neutral, featureless, volatile-bearing) is equally well represented in the dynamical classes. As knowledge of the compositional types improves (by observations like those suggested here), it should be possible to identify source regions for the different compositional types. Then evolution can be run in reverse, and the different proposed dynamical mechanisms tested. Indeed, results from a program to observe a small number of TNOs using IRAC (Spitzer program 20769) hint at groupings not previously seen (see Fig. 3). We expect that insights from this new result will lead to eventual unraveling of the evolutionary history of the Kuiper Belt as discussed above, but observations of a much larger sample of objects are needed to flesh out and understand these groupings.

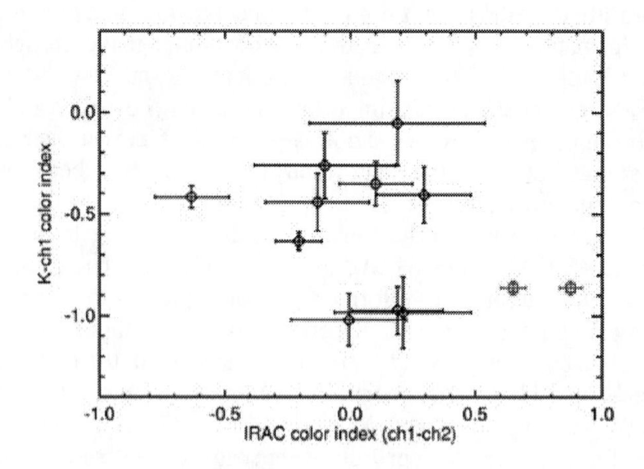

Figure 3. IRAC data of TNOs from program 20769 shows a bimodal distribution. The two IRAC observations made of each object were averaged together for this plot. The blue squares on the right of the plot are Pluto data at two different longitudes.

2.2 Composition of Dark Material in the Outer Solar System

Most TNOs, Centaurs, Trojans, and outer-belt asteroids have low visible albedos. The nature of this "dark material" that lowers the albedos is almost completely unknown. The reddest Centaurs and TNOs require organics to explain the steep spectral slopes in the visible and NIR (Cruikshank and Dalle Ore [6]). The analog species that have been put forth a) all have very strong absorptions in the wavelength range of the IRAC bands, and b) have spectra that often differ widely from one another in the region covered by the IRAC bands (compare the green dotted and red solid curves in Fig. 4). Surface compositions of the less red objects are even more uncertain. The neutral to moderately red spectra can be explained equally well with silicates or a variety of organics (e.g., Cruikshank et al. [7], Emery and Brown [8]). Moderately red-sloped objects (including KBOs, Centaurs, and asteroids) are common in the Solar System. Distinguishing their surface compositions is crucial for understanding the distributions of organics, which is an issue that bears significantly on ideas of the chemistry and processes in the solar nebula. Silicates and organics have very different spectral behavior at $\lambda > 3$ μm.

IRAC observations therefore offer the best means to constrain the class of material that causes both the ultra red spectral slopes and the more moderate slopes. We note that the IRAC data of the ultra-red Centaur Pholus and the scattered disk object Sedna (Fig. 4a), which were observed by IRAC as part of program 20769, are both best fit with nitrogen-rich tholins as opposed to more oxygen- or C-H rich organics. Similarly, IRAC data of the moderately red Centaur Asbolus suggest N-rich tholins rather than either C-H rich organics or silicates (Fig. 4b). Since there is such a wide range of possible compositions for these low-abedo, red objects, it is particularly important to observe a large sample and look for correlations between dynamical and compositional properties.

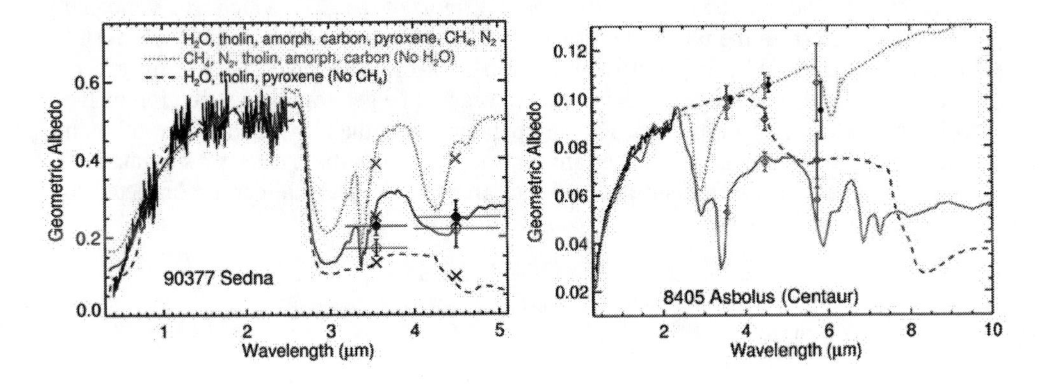

Figure 4. (a; left) IRAC data for the TNO Sedna (filled and open circles) along with the vis-NIR spectrum and three spectral models. (b; right) IRAC data of the Centaur 8405 Asbolus (black circles) compared with three compositional models that all match the shorter wavelength data.

2.3 Volatile Inventory of the Outer Solar System

H_2O ice has been detected on several TNOs and Centaurs through absorptions at 1.5 and 2.0 μm, but others do not exhibit these bands. Trojan and outer-belt asteroids may have H_2O ice in their interiors, but none has yet been detected. If these asteroids originated in the same region as TNOs, they should have icy interiors, and we would expect impacts to uncover some of these ices, particularly among dynamical family members. Because H_2O contains very strong absorptions at $\lambda > 2.7$ μm, the 3.6 μm IRAC channel provides a sensitive test. IRAC data points are also sensitive to the presence of other ices, such as CH_4 and CH_3OH (see below). The greater sensitivity to ices that IRAC offers will more robustly identify correlations between the presence of volatiles and any other dynamical or physical properties, if they exist.

Ten of the 11 KBOs observed so far from program 20769 show strong absorption at $\lambda > 3\ \mu m$ in the IRAC data (Emery *et al.* [9]). This includes the first detection of ices on four objects. In addition, Emery *et al.* [10] have used IRAC data to identify the first reported detection of H_2O on Sedna (Fig. 4a), along with confirmation of the CH_4 reported from shorter wavelength data (Barucci *et al.* [11]).

2.4 Pluto and Kin

The dwarf planets Pluto, Eris, Sedna, 2005 FY9, and 2003 EL61 are particularly interesting because of their potential for atmospheric and geologic activity. Near-IR spectra reveal the presence of CH_4 on all of these objects except 2003 EL61. Some spectra also suggest the presence of N_2 ice on Eris, and Pluto has abundant N_2.

Photolysis and cosmic ray radiolysis are expected to quickly convert CH_4 to higher-order carbon molecules, with resultant darkening of surface layers. However, Spitzer MIPS radiometry and HST imaging show that the albedos of these objects are quite high. This is probably the result of seasonal transport of the CH_4, which will segregate the volatile ice from the photolytic products. The seasonally active layer should result in changes of the albedo patterns on the surfaces of these objects, and any others which may be discovered over the next few years. The geographic distribution of the volatile ices directly influences their temperature, and thereby the pressure of their vapor in the atmosphere. As the ices are transported across the surface by sublimation, both the appearance of the surface and the atmospheric pressure can be expected to undergo seasonal changes.

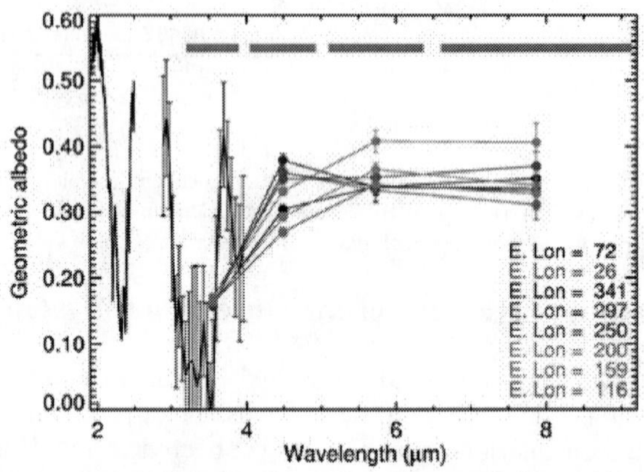

Figure 5. IRAC measurements of Pluto from GTO program 70 along with a ground-based NIR (1.9 - 4.0 μm) spectrum. The colored filled circles are IRAC data at eight different Pluto longitudes.

Detailed IRAC observations of these kinds of objects are an example of a small warm-mission program that could reveal very interesting seasonal behavior, and also significantly deepen our understanding of the compositions of their surfaces. For example, the IRAC data for Sedna (Fig 4a) clearly provide a unique test for constraining the composition in ways that cannot be achieved using current ground-based capabilities. IRAC data for Pluto (Fig. 5), to which Eris and 2005 FY9 are best compared, revealed the expected CH_4 absorption in the 3.6 μm channel. The 4.5 μm reflectance, however, has a rotational variation that is distinct from the other three bands, the visible albedo, or any of the NIR spectral features (Grundy and Buie [12]). This manifests itself as two distinct spectral units on Pluto in the IRAC channels. This may be due to organic materials or previously undetected ices. Rotationally resolved IRAC light curves for other objects in this class could also reveal unique compositional and geographic information that cannot be obtained using any other existing capability.

3. MAIN BELT ASTEROIDS

There are more than 350,000 known Main Belt asteroids. Many of them contain volatiles -- most notably water, in various forms -- and also organic material, both of which are retained from the era of planet formation 4.5 billion years ago. The traditional paradigm of Solar System formation suggests a condensation sequence from inner to outer Solar System in which high temperature silicates condensed in the inner Main Belt (E-type asteroids), followed by moderate temperature silicates (S-type asteroids) in the inner to mid Main Belt, low temperature silicates with a small fraction of organics in the mid-outer Main Belt (C-type asteroids), and increasing organic material further out (P- and D-type asteroids). Most mid-belt C-type asteroids contain hydrated silicate spectral features, which have been attributed to a post-accretion heating event - possibly magnetic induction heating - whose strength was a function of heliocentric distance. A new model proposes large-scale dynamical upheaval throughout the outer Solar System, which may have affected the mid to outer Main Belt. In this new scenario, the outer Main Belt asteroids in particular may have formed much further out in the Solar System. Furthermore, important non-equilibrium chemical pathways have been identified that can account for the formation of hydrated silicates in the nebula rather than on asteroid parent bodies themselves. Whereas in some cases meteorite petrology strongly indicates in situ formation of the hydrated silicates, in other cases formation on the body is not certain. Additionally, there has to date been no direct evidence (e.g. spectroscopic) to support the hypothesis that the low albedos and red spectral slopes are due to organic materials. Mapping these components in the main belt allows us to understand the distribution of volatiles and organic material in the early, nascent Solar System.

We therefore ask the following key critical question: *What is the distribution of water and organic compounds in the Solar System?*

Water ice, hydrated minerals, and organic materials all exhibit very strong spectral signatures in the 3 to 5 *μm* range. Spitzer IRAC observations of outer main belt asteroids, with location beyond 3 AU, provide a clear measure of the thermal emission from the asteroid surface, and therefore offer the means to perform very sensitive, systematic searches for these diagnostic materials on a large sample. The two IRAC bands are not sufficient to determine mineralogy, or in most cases even to distinguish among H_2O, hydrated minerals, and organic materials. However, identification and classification of absorptions using IRAC provides an important search that can be used to refine future, more detailed spectral studies with, for example, JWST.

We tested the feasibility of such a project using spectral models. A model fit that used hydrated minerals and organic materials but ***no*** H_2O (solid black dot in Fig. 6) was computed for a typical outer-belt D-type asteroid, then varying amounts of the absorbing material were added to investigate the effects on the K-3.6μm and 3.6μm-4.5μm color ratios. S/N ~10 will be sufficient to detect even just a few percent of these materials on the surface. Although only H_2O and hydrated minerals are shown on this plot, the effects are similar with macromolecular organics (e.g. tholins). Measurements of K-3.6μm and 3.6μm -4.5μm colors (using 2MASS with Warm Spitzer) can be used to pull out those compositions.

There are ~100,000 asteroids with orbital semi-major axes greater than 3 AU, of which ~10,000 are in the 2MASS database. To meet out science goals, a sufficient number of target asteroids can easily be selected from the 2MASS subset, choosing objects with favorable apparitions and a range of sizes, as appropriate.

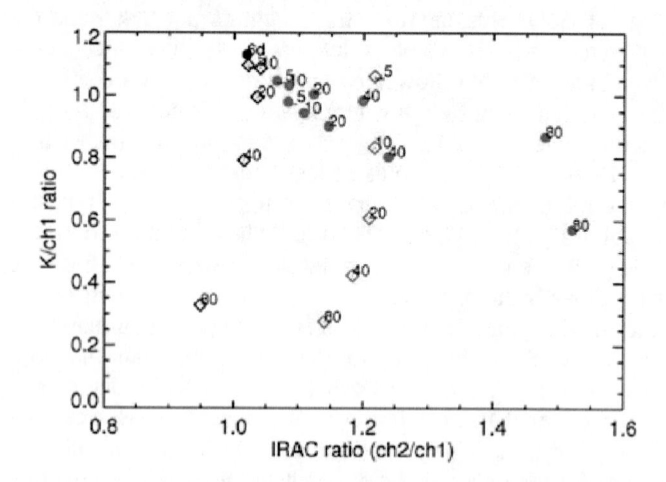

FIGURE 6. Band ratios for various fractions of H_2O ice (black diamonds - 100 μm grains, blue diamonds - 12.5 μm grains) and hydrated minerals (red filled circles - chlorite, green filled circles - serpentine). The black filled circle represents the volatile-free spectrum. The numbers adjacent to each symbol indicate the percentage of the absorber that is included in the mixture.

4. NEAR EARTH OBJECTS

65 million years ago, a 10 km body hit the Earth and caused the extinction of the dinosaurs. Asteroids and comets are known to hit the Earth periodically, with a frequency that is inversely proportional to their size; discussion of the next impact onto the Earth is a question of "if" not "when." The U.S. Congress has established a requirement that 90% of the potential Earth-impacting asteroids larger than 140 meters be identified by the year 2020. In this way, potential hazards can be identified, and mitigation plans defined, if necessary. These potentially threatening asteroids have been named Potentially Hazardous Asteroids (PHAs), and generally fall within the dynamical category of Near Earth Object (NEOs). PHAs are defined as bodies larger than 140 meters that will pass within 0.05 AU of the Earth's orbit, and NEOs are all bodies whose orbits pass within a few tenths of an AU of the Earth's orbit. As of this writing, there are around 4000 NEOs, and 850 PHAs. The Pan-STARRS program is likely to increase the number of known NEOs to ~10,000 by 2013.

These bodies are of specific interest to humans. Characterizing potential threats to life on Earth carries an almost unimaginably high value. Additionally, these near-Earth objects are the most easily reached by spacecraft, enabling our exploration of the Solar System. These nearest neighbors can be explored with Spitzer during the Warm Era.

Consequently, we ask the following key question: *What are the physical properties of potential Earth-impacting asteroids?*

4.1 NEO Sizes and Densities

NEOs have relatively hot surface temperatures (>250 K). Fluxes in IRAC 3.6μm and 4.5μm will therefore be dominated by thermal flux (Fig. 7). We can derive the sizes and albedos of NEOs by combining thermal measurements with reflected light data (i.e., visible magnitudes). At sizes <1 km, these bodies will be the smallest bodies observed by Spitzer. Deriving the sizes of hundreds of NEOs also will allow us to establish a true size distribution, which reflects the dynamical origin of this population.

Approximately 10% of NEOs are binaries. Using Keplerian dynamics, thermal measurements of NEOs that turn out to be binaries will yield densities, which in turn will suggest compositions and internal strengths. The intrinsic strengths of these bodies will be particularly important in mitigating potential hazards to the Earth -- disrupting a solid body is very different than disrupting an unconsolidated rubble pile.

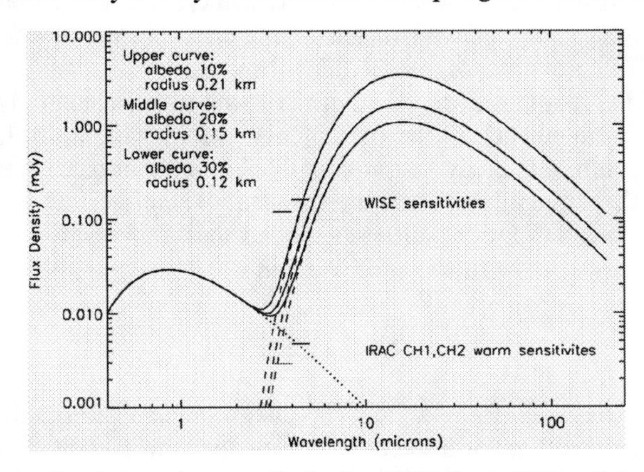

FIGURE 7. A predicted spectral energy distribution (SED) for an NEO with heliocentric distance 1.3 AU and Spitzer-centric distance 1.0 AU. Three different albedos are shown (and their corresponding diameters). The lower horizontal lines are Warm Spitzer sensitivity limits, 5σ, 500 seconds (high background). The upper horizontal lines are WISE sensitivities for the ecliptic pole, 5σ, in sky survey mode.

4.2 Regoliths and Thermal Inertia

Most airless bodies in the Solar System are covered to some degree with regolith, layers of pulverized rock that are produced over time by collisions with both large and small particles. It has recently been suggested, based on indirect evidence, that bodies smaller than ~5 km may be nearly devoid of regolith (Binzel *et al.* [13], Cheng [14]). This prediction affects the most numerous population of objects in the Solar System (sub-km asteroids) and our nearest neighbors in the Solar System.

The amount of regolith on a surface is reflected in the body's thermal inertia -- the degree to which the body is in instantaneous thermal equilibrium with solar radiation.

With thermal measurements of a sufficiently large number of hundreds of NEOs, the average thermal inertia can be derived, and therefore the typical regolith amount on these small bodies. Understanding the regolith properties of these small bodies will help us understand their dynamical ages in near-Earth space, with consequent implications for the flux of potential bodies into and across the Earth's orbit.

Asteroids that have high thermal inertias emit significant thermal flux on their night sides. In these cases, asteroid orbits can change through the Yarkovsky effect, in which anisotropic radiation of thermal inertia produces a force through which semi-major axes can change. Consequently, understanding - and potentially mitigating - the Earth impact hazard requires detailed analysis of the thermal inertia and consequent Yarkovsky effect on NEOs.

4.3 Origins and Compositions

The majority of NEOs are believed to be S class asteroids, a rocky and relatively volatile poor asteroid type. The majority of main belt asteroids are C type asteroids, more volatile and organic rich. The source and dynamical path and evolution of main belt asteroids to NEOs are not well understood. Measuring the albedos of a large number of NEOs will probe the degree to which there is a C class "tail" of NEOs, with implications for the compositions of Earth-crossing (and potentially impacting) asteroids. Although an impact in the present day would have a strongly negative impact on life on Earth, it is likely that volatiles and organic material were brought to a pre-biotic Earth through just this mechanism. Therefore, understanding the compositions of potential Earth impacting bodies has important implications for understanding the past *and* the future of life on Earth.

5. COMETS

Comets are planetesimals formed from the circumstellar material after the proto-solar nebula has collapsed to form the nascent Sun and surrounding disk of material. Comets were formed throughout the disk, from the ice line (where water ice is stable) out to its edges. The comets that formed in the giant planet region were either incorporated into the planets or scattered into the Oort cloud as the gas and ice giants cleared their local regions of space during their accretional growth phase, while the comets that formed at the edges remained in place or were scattered as the ice giants migrated outwards, forming the collection of objects now known as the Kuiper Belt. In the current paradigm for the formation and evolution of our Solar System, most of the Short Period comets dwelling mainly in the inner Solar System are collisional fragments of KBOs that have migrated inward, after having been scattered back and forth among the giant planets. The Long Period comets derive from the Oort Cloud, where they have remained until perturbations from passing stars, molecular clouds, and the galactic tide sent them into the inner Solar System. Both families of comets have been subjected to various degrees of evolution---virtually none over the last >4 Gy for those in the Oort Cloud, extensive collisional evolution for those derived from the Kuiper Belt, and extensive sublimation (and possibly fragmentation and collisional) losses for the SP comets.

Current theories predict that cometesimals decoupling from nebular gas drag should have a characteristic radius of 100m, but the known cometary nuclei and KBOs are substantially larger than this. Thus, an additional poorly understood process of aggregation determines the ultimate form of these bodies. Other physical properties may also be affected by the processes of formation, e.g. collisional fragments may vary more in composition, as revealed by the color of their reflected and thermally emitted light. Until there is an accurate observational database on the physical properties of cometary nuclei to guide the development of theory, we will be stymied in our efforts to explain the formation and evolution of the Solar System outside the ice line. This motivates the key question: What clues do the physical and compositional properties of comets give us about planetesimal accretion?

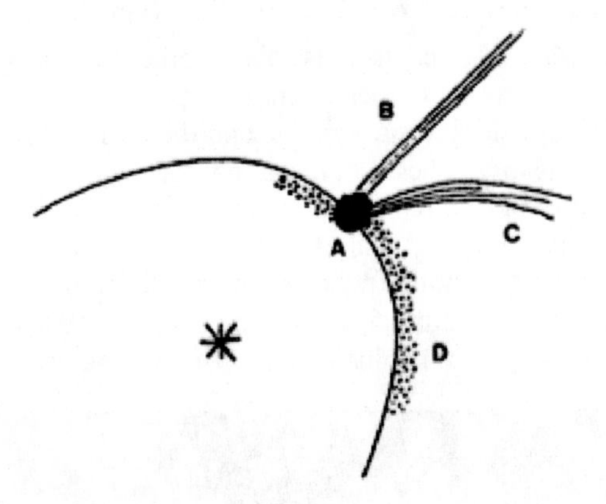

FIGURE 8. Schematic of comet morphological structures. (A) Coma, (B) Ion Tail, (C) Dust Tail, (D) Dust Trail (adapted from Sykes and Walker [15]).

5.1 Comet Dust Trails

Comets have two fundamental components: refractories and volatiles. The latter are evidenced by gaseous emissions of the coma and ion tail, while the former historically was evidenced by prominent dust tails, consisting primarily of micron-sized particles sensitive to radiation pressure (Fig. 8).

The primary emission of refractory material, by mass, is into the cometary trail, not the oft-studied coma. These dust trail particles were first detected by the Infrared Astronomical Satellite (IRAS) (Eaton et al [16], Sykes et al [17]) and were inferred to be a common property of all short-period comets in a survey of IRAS data (Sykes and Walker [15]). The large trail particles (typically mm-cm in size) are ejected at low velocities relative to the comet orbital speed, which is why they drift away from the comet along orbits close to that of the parent comet, giving rise to the appearance of 'trails' (similar in appearance to fresh airplane contrails). Particles on orbits that come within 3 AU of the Sun (the vast majority of Jupiter family comets) are quickly devolatilized with the result that trail particles allow us to study the refractory component of comets in isolation of volatiles. Trail particles have low-albedos (Sykes *et al.* [18], Sykes *et al.* [19]), making them extremely difficult to study from the ground (e.g., Ishiguro *et al.* [20]).

Spitzer has revolutionized the study of trails using the MIPS 24 μm channel (e.g., Fig. 9, Reach et. al [21]); 34 comets were observed, compared with the 8 trails observed by IRAS. Debris trails due to mm-sized or larger particles were found along the orbits of 27 comets; 4 comets had small-particle dust tails and a viewing geometry that made debris trails impossible to distinguish; and only 3 had no debris trail despite favorable observing conditions. There are now 30 Jupiter-family comets with known debris trails. The detection rate is >80%, indicating that debris trails are a generic feature of short-period comets. The mass-loss rate in trail particles is comparable to that inferred from OH production rates and larger than that inferred from visible-light scattering in comae.

Even during its post-cryo phase, Spitzer remains the preeminent facility with pointing capability for studying trails and gaining insights into an important class of Solar System objects, also inferred to exist about other stars. Spitzer allows us to obtain accurate measurements of cometary mass loss rates, details of recent comet emission history, comet dust to gas mass ratios, and the correlation of these properties with dynamical age and history. These studies allow for the probing of the diversity of comet formation locations and source populations. Understanding the cumulative mass loss of comets allows for a better understanding of their role as a source of the zodiacal dust cloud and the question of whether the existence of equivalent exo-clouds depend upon the transport of such bodies from the outer to the inner Solar System or whether such exo-clouds arise from collisional activity among asteroids.

FIGURE 9. The P/Schwassmann-Wachmann 3 dust trail and fragmented nucleus form a number of small comets, obtained at 24 μm by Spitzer MIPS (Vaubaillon and Reach [22]). Here the comae are the bright white regions surrounding each fragment, the dust tail the yellowish to reddish extensions directed away from the Sun to the right, and the comet trail is the thin red line running from upper left to lower right, connecting all the fragments. Image courtesy of NASA/JPL-Caltech/W. Reach (SSC/Caltech).

The Wide-field Infrared Survey Explorer (WISE) is scheduled for launch in 2009. WISE is a survey instrument with a fixed scanning campaign that does not allow for trail observations at optimal viewing geometries and times (within the solar elongation constraints common to both spacecraft). So, Spitzer will be able to use its more flexible pointing capabilities to observe trails at times and geometries unavailable to WISE, as well as conduct follow-on observations of trails observed by WISE. The two systems will be directly complementary.

5.2 Carbonaceous Species in Comets

The most important volatile cometary species, water, is often studied in comets using 'hotband' near-IR emission lines that avoid water absorption by the cold Earth's atmosphere. The astrophysically important carbon-bearing molecular species, CO and CO_2, are however much harder to observe from the ground and are only well measured in the brightest (most active) of comets. Studies of such comets indicate that these two species are present at the 5 to 20% level versus water in cometary nuclei, making them the second and third most abundant volatile species in comets. CO and CO_2 are the major reservoirs for carbon in comets; by comparison, carbon in refractories is roughly only one-fourth as abundant (Lisse *et al.* [2]). We need to detect and measure the dominant carbon volatile species if we are to understand the composition of the planetesimals from which the planets formed.

The budget of CO and CO_2 in comets has been unmeasured except for (as stated) the brightest – and therefore most atypical and unusual -- comets. There are almost no detections of these species in any Short Period (inner Solar System) comets, or in any "typical" Long Period comet. It is likely that we are completely missing as much as 25% or more of the volatile content of the majority of comets. Clearly this is a conundrum that needs to be investigated if we are to understand cometary composition. Our hypothesis is that the limiting factor is magnitude; comets are often simply not bright enough for infrared or millimeter spectroscopy to detect CO, and CO_2 has no feasible transitions in atmospheric windows. We expect that CO and CO_2 are there in the fainter comets, but there can be almost no ground-based measurements to quantify this. Another factor for the lack of detection could be that much of the CO and CO_2 in the topmost centimeters of the surface has sublimated away due to surface evolution.

A method of detecting CO and CO_2, using IRAC, has been found serendipitously by W. T. Reach and M. S. Kelley (priv. comm.) and provides an unprecedented and unique way to directly test the hypothesis of CO and CO_2 abundance in comets. Observations of Comet 2P/Encke in IRAC 3.6μm and 4.5μm indicate a distinct difference in the morphology of the coma that is stronger than would be expected if the coma were simply dust. The shape of the dust coma at 3.6 and 4.5 μm should be virtually identical, since size-dependent scattering and thermal emission should not change much in such a short interval of wavelength. However CO and CO_2 have emission bands in 4.5μm, while 3.6μm is sensitive only to dust (and not H_2O, for which the broad emission feature at 2.7 - 3.1 μm barely misses the onset of the 3.6μm pass band). The radial profile of the CO and CO_2 gases in the coma is significantly different than that of the dust because the gas species feel photo-dissociative effects as they travel outward from the nucleus. Thus the strength of the emission lines changes as a function of cometocentric distance differently from the dust. This means that the detection and quantification of these species will be straightforward with sufficiently deep imaging of the active comets' comae (Figure 10).

FIGURE 10. Comet 2P/Encke as observed in IRAC 3.6μm and 4.5μm in June 2004 by PID 119 (and extracted from SST archive). (A) 88-arcsecond wide subframe of the 3.6μm image centered on the comet. (B) Same as A, but at 4.5μm. (C) Difference of the two images. The coma is dramatically brighter at 4.5μm and it is morphologically distinct.

Furthermore, one can investigate the *interior* of the cometary nuclei by observing the comae at certain critical heliocentric distances. Carbon monoxide sublimates at effectively all distances, but CO_2 sublimates only within 13 AU of the Sun, and H_2O does so only within about 3 AU. Furthermore water's phase change from an amorphous to a crystalline phase occurs within 8 AU. The emission of CO and CO_2 will change at these heliocentric distances depending on how the various volatile species are mixed within the nucleus.

6. ZODIACAL CLOUD

Stars are born surrounded by circumstellar disks of dust and gas. Over timescales of millions of years, dust and gas are removed through radiation forces. However, a few percent of stars older than 400 Myr also have circumstellar dust disks (Habing et al [23]). These older debris disks require sources replenishing system dust over time. Current processes in the Solar System indicate that these sources are comet emissions and asteroidal collisional activity. Stochastic collisional activity can result in large variations in the surface area of dust (Wyatt *et al.* [24]), suggesting that the dust densities and flux levels seen in exo-disks vary strongly with time. This also implies that the Solar System is currently "relaxing" from a time of much denser interplanetary dust created by the collisions that (e.g.) formed the Veritas and Karin asteroid families (Nesvorny et al [25]) -- and with them the zodiacal dust bands -- 5-8 Myr ago. This leads to a key question we wish to address: What does our Solar System's interplanetary dust environment and our zodiacal cloud tell us about the nature of the dusty disks seen around other stars?

The zodiacal cloud results from highly structured dust formations of asteroidal and cometary origin, and is detectable in both thermal emission (Fig. 11) and scattered sunlight (Fig. 12). Both catastrophic collisions (e.g., the formation events for the Karin and Veritas families, that occurred a few Myr ago) and ongoing cometary emission

FIGURE 11. The Solar System zodiacal dust cloud at 10 μm, as it would appear 1 pc from the Sun with 0.25 AU resolution, 30° above the ecliptic plane. The stretch is logarithmic, and the density profile used for the image is derived from the IRAS model of Good et al. [26]. Jupiter can be seen at the upper left, and Saturn at the lower right.

and asteroidal grinding are thought to be source functions for the Interplanetary Dust (IPD). These sources must be continually delivering ~10^4 kg/sec of material in order to balance the dust accreted by the Sun and planets, or driven out of the Solar System by solar radiation pressure (Lisse [27]). Exo-zodiacal clouds have been detected by Spitzer, with a frequency of a few percent for mature stars, e.g. the pronounced IR excess found around the 2-10 Gyr K0 star HD69830 (Beichman *et al.* [28]). These exo-clouds are estimated to be 100x or larger the flux emitted by our zodiacal cloud, and thus represent a flux-limited sample, most likely created by recent, stochastic catastrophic asteroid fragmentation events (Lisse *et al.* [1], Wyatt *et al.* [24]).

Solar system zodiacal cloud structures are located from ~0.9 to a few AU, with temperature ranges that make them ideal for mid-infrared detection. Initial characterization of IPD by the COBE DIRBE instrument (Reach, *et al.* [21]) and IRAS (Dermott, *et al.* [29]) has identified the presence of several components, including a smooth cloud, a circumsolar ring, leading and trailing blobs, and dust bands which are associated with asteroid families (e.g., Spiesman *et al.* [30]). Spitzer is uniquely qualified to further characterize the zodiacal background, (e.g. search for dust in resonant orbits around the planets) and since launch, several observing programs have been executed to observe the zody. These projects, however, focus primarily on longer wavelength observations at 24, 70, and 160 μm.

Spitzer's warm mission using the 3.6 and 4.5 *μm* detectors provides an important opportunity to examine the sources contributing to zodiacal background, including the asteroidal dust bands. Reach *et al.* [31], have found that the location of these bands varies with temperature, as observations at different wavelengths sample different parts of the blackbody spectrum. At shorter wavelengths, on the sharply rising Wien side of the SED, warmer dust that lies closer to the sun is detected. The line of sight to this dust appears to be directed further away from the ecliptic plane. At longer wavelengths, such as MIPS 70 and 160 μm, the SED is flat and represents colder, more distant IPD. Observations at these wavelengths are along lines of sight that are directed further away from the sun and appear closer to the ecliptic plane. The above-mentioned Spitzer programs can be supplemented during the warm mission by observing the zody at shorter wavelengths.

Steps in this direction have serendipitously been taken as part of IRAC's nominal calibration plan. During each campaign, IRAC routinely observes the zodiacal background to in the ecliptic plane and at the North Ecliptic Pole to obtain flat fields and sky darks. Figure 12 shows plane and pole data taken during the early part of the

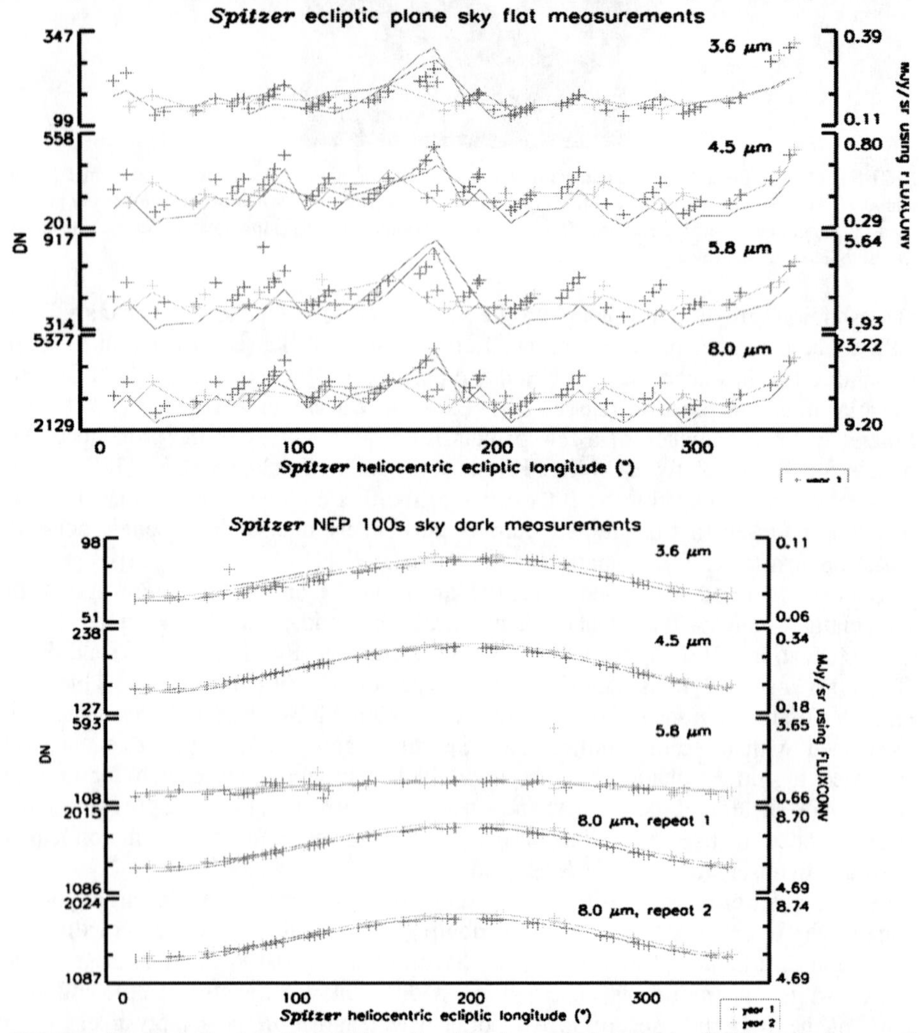

FIGURE 12. (Top) North Ecliptic Pole zodiacal background observed by IRAC. Note the annual variability due to tilt of the circumsolar dust ring relative to Spitzer's orbital plane. (Bottom) Ecliptic plane zodiacal background observed by IRAC. Note the biannual variability as Spitzer's line of sight passes through the galactic plane.

mission. The annual variability of the zody is apparent, as Spitzer moves through the circumsolar ring. IPD emission is readily detected, even at the shorter wavelengths.

During the post-cryo mission, these data that will provide a information on the scale height of the circumsolar ring, as well as other structures cane be enhanced by latitudinal scans that will provide a information on the scale height of the circumsolar ring, as well as other structures.

7. ICE GIANTS

Just as the study of the Sun plays critical role in understanding other stars, so do studies of the outer planets in our Solar System yield insight into processes and conditions that occur among the diverse sample of giant planets around other stars. The growing number of extrasolar planetary systems, including many Uranus- and Neptune-sized planets (e.g., Mayor and Queloz [32], Marcy and Butler [33], Rivera *et al.* [34], Beaulieu *et al.* [35], Lovis *et al.* [36]) elevates the importance and significance of understanding the Ice Giants within our own system: Uranus and Neptune. Although most of the giants in exoplanetary systems reside very close to their host stars, it is only a matter of time before such giants are found at larger distances. The results of Warm Spitzer Uranus and Neptune studies will aid studies of extra-solar planets in this size range by elucidating the radiation balance in our local ice-giant atmospheres. Clues to their formation and evolution lie hidden in their chemistry and atmospheric dynamics. We therefore ask the key question: What is the energy budget of the giant planets and what can this tell us about the nature of extrasolar planets?

These are special seasonal times for both Uranus (equinox in 2007) and Neptune (solstice in 2005). HST, Keck, NASA's IRTF, the VLT and many other ground-based observatories have been focused on these planets, often with surprising results as described below. The value of Warm Spitzer data for both planets is that they will provide: (1) a direct comparison with Prime-Mission observations during a time of rapid atmospheric change, and (2) a set of complementary data on two similar yet distinct large bodies, contemporaneous observations with the other seasonal observations.

Theoretical models, grounded primarily in Voyager-era data, are in some cases not adequate for explaining modern observations. What powers their winds? How deep does zonal structure go? What is their atmospheric composition as a function of altitude? How do they interact with incident radiation from the Sun? Of particular interest are the relative roles of dynamics and insolation in controlling atmospheric properties, along with the timescales and phase lags associated with significant seasonal change.

7.1 Uranus

Obliquity plays an important role in climate change on Earth (e.g., Zachos *et al.* [37]) and Mars (e.g., Nakamura and Tajiki [38]). Within the field of planetary atmospheres, Uranus' atmosphere is uniquely important because of its extreme obliquity (98 deg). The markedly low value of its internal heat source relative to each of the other giant planets also enhances its seasonal extremes from both radiative and dynamical perspectives.

During—and in the decade following—the 1986 Voyager flyby, when Uranus was near solstice, its atmosphere was quiescent and nearly featureless. Most atmospheric models of Uranus are based from data taken in this epoch or earlier. In late 2007, a rare opportunity is permitting us to study Uranus in detail: the Earth and the Sun are crossing the planet's equatorial plane (Fig. 13). This geometry takes place just twice every Uranus year, i.e., every 42 Earth years.

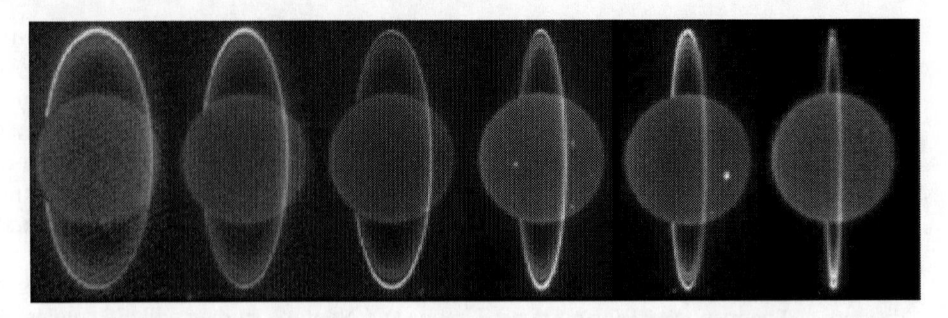

FIGURE 13. Annual Keck images of Uranus from 2001 to 2006 showing the impending 2007 equinox. This time sequence highlights the changing viewing aspect as Uranus approaches the equinox, or Ring-Plane crossing (RPX). The images were obtained with adaptive optics at 2.2 μm, where methane and hydrogen absorb sunlight in the planet's atmosphere. The white spots in the last three years are high-altitude cloud activity, possibly triggered by seasonal insolation changes. [Images from de Pater et al. [39], updated with 2006 data by HBH.]

Some Uranus atmospheric models predicted a radiative seasonal response (e.g., Wallace [40]; Friedson and Ingersoll [41]). However, those models predicted that the planet's long radiative time constant at all levels below the upper stratosphere would permit only small physical seasonal changes within the visible atmosphere, with phase lags of order 10-20 years (Friedson and Ingersoll [41]). In contrast, observers have been seeing notable atmospheric change on Uranus as equinox approaches, apparently with no phase lag. Hubble Space Telescope (HST) images of Uranus showed the south polar region darkened as it began to receive less direct sunlight (Rages *et al.* [42]). Pre-equinox images from HST and from the Keck 10-meter telescope in Hawaii (Hammel *et al.* [43], [44]; Sromovsky and Fry [45]) showed tremendous tropospheric cloud activity across the planet (Fig. 14), including changes on time-scales as short as days (Hammel *et al.* [43]).

At equinox, solar forcing is symmetric over the two hemispheres. Hemispheric asymmetries at equinox may be a measure of a seasonal radiative phase lag or may alternatively be the signature of a much more rapidly evolving dynamical instability forced by the reversal of insolation between warm and cold hemispheres. At present, the two hemispheres differ strongly in the albedo of tropospheric cloud bands and in the dynamical activity producing localized bright and dark cloud features (Fig. 14).

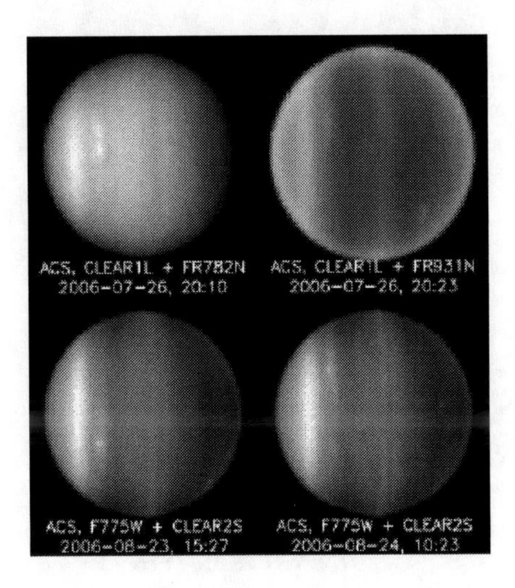

FIGURE 14. Uranus images from HST. These images from the Advanced Camera for Surveys on the Hubble Space Telescope illustrate the significant asymmetry in the troposphere of Uranus. [Images courtesy L. Sromovsky (University of Wisconsin).]

One manifestation of enhanced dynamical activity could be increased generation of upwardly propagating gravity waves and hence an increased eddy diffusion coefficient in the stratosphere. Methane would get carried up to higher altitudes, and photochemical products like acetylene and ethane would be produced at a higher rate. As a consequence, hydrocarbon species may be more readily detectable. Encrenaz *et al.* [46], using 1996 observations with the Infrared Space Observatory (ISO), derived an acetylene abundance on Uranus that was 30 times greater than was seen during the 1986 Voyager encounter (Bishop *et al.* [47]). The high-SNR disk-averaged Spitzer IRS spectra of Uranus (Fig. 15) confirm the detection of ethane, as well as other hydrocarbon species. More importantly, the spectra also show evidence for seasonal, or at least temporal, change between ground-based data in the Voyager era and the newer Spitzer data.

Realistic models of the atmosphere of Uranus are needed to explain the plethora of time-variable modern observations described above. Data from the Warm Spitzer era - together with additional data from the Spitzer Prime Mission as well as SMA, VLA, VLT, and many other facilities - can be used to create realistic physical models of the planet's changing infrared emission. The 3.6- and 4.5-μm data taken during the Warm Era can be directly compared with Prime-Mission measurements at those wavelengths. Since the atmosphere of Uranus is demonstrably changing right now, a cadence of observations of order once a year, or once every six months, can capture time-scales of change.

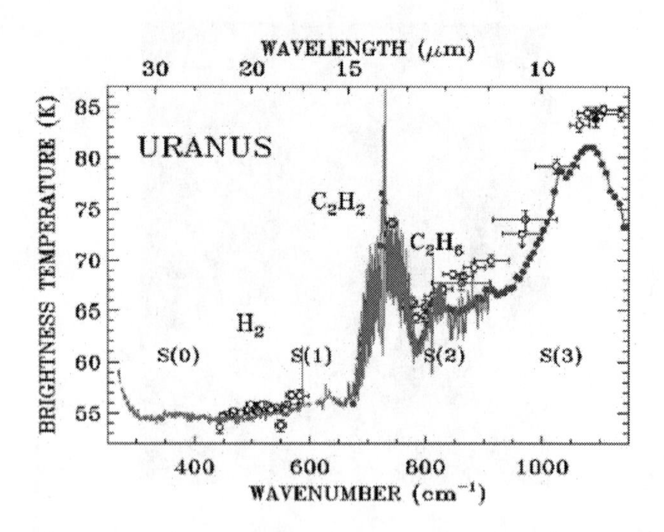

FIGURE 15. - Uranus changes as detected by Spitzer. This proprietary figure, provided by G. Orton, illustrates the change in the mid-infrared spectrum of Uranus during the past two decades (Orton et al. [48]). The open symbols are ground-based mid-infrared data from the mid-1980s, compared with current-era Spitzer data (lines and small closed symbols). Currently Uranus has lower brightness temperatures throughout this spectral region.

7.2 Neptune

The more distant ice giant, Neptune, has also been exhibiting significant atmospheric change in recent decades, though the driving mechanisms are less well understood that for seasonally-dominated Uranus.

At optical wavelengths, Neptune in 2003 reached its brightest level in nearly 30 years of photometric monitoring (Lockwood and Jerzykiewicz [49]; Hammel and Lockwood [50]). The planet has exhibited pronounced cloud activity throughout this time, possibly undergoing a transformation from one cloud distribution to a new configuration (Hammel and Lockwood [50] and references therein). Hammel and Lockwood [50] predicted that the stratospheric temperature may have increased on Neptune since the mid 1980s, because stratospheric temperatures on Uranus (Young et al. [51]) rose and fell in lockstep with that planet's visible wavelength reflectivity (Hammel and Lockwood [49]).

In the mid-infrared, early 7-13 μm IRTF observations of Neptune (Orton *et al.* [52], [53]; Hammel et al. [54]) hinted at variation in the 12.2-μm ethane emission feature. Data taken a decade later showed significantly increased ethane and methane emission (Hammel *et al.* [55]). Radiative-transfer models, discussed in detail in Hammel *et al.* [55], suggested that increasing stratospheric temperature has been driving the long-term mid-infrared change, an interpretation supported by independent observations at sub-millimeter wavelengths (Marten et al. [56]) and also by the visible-wavelength photometry discussed above. Recent images of Neptune at methane (7.7-um) and ethane (12-um) emission bands with the MICHELLE instrument on Gemini (Hammel *et al.* [57]) showed enhanced south polar emission (Fig. 16, upper panel), similar to that of Saturn (Orton and Yanamandra-Fisher [58]). Both enhancements probably result from long periods of continuous exposure to sunlight (i.e., are seasonal in nature).

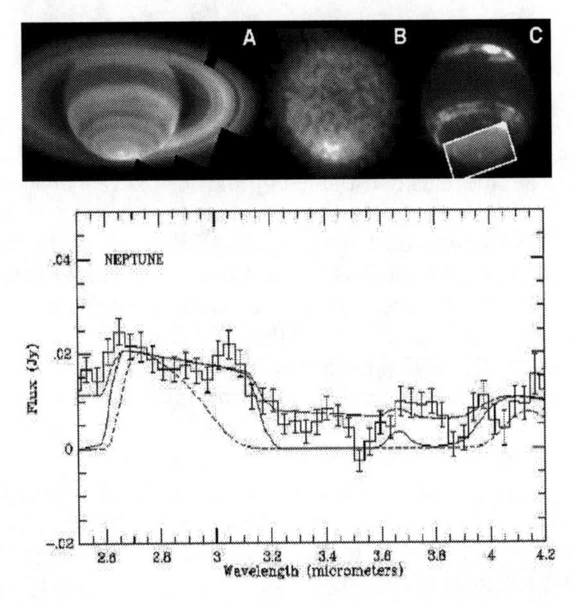

FIGURE 16. Neptune images and spectra. (Upper panel) Neptune compared with Saturn (Hammel et al. [57]). Image A shows Saturn at 8.0 μm (Orton and Yanamandra-Fisher [58]) with strong methane emission from its southern pole. Image B shows Neptune at 7.7 μm taken with the Gemini telescope; as with Saturn, the methane emission arises from the south polar region. (Note that Saturn and Neptune are not displayed to the same scale.) In Image C, Neptune -- taken at 1.6 μm with the Keck Adaptive Optics system -- reveals zonal circulation that is as tightly confined in the polar region as that of Saturn. (Lower panel) Neptune spectrum from 2.5 to 4.2 μm (Encrenaz et al. [59]). The ISOPHOT-S spectrum (data with error bars) is compared with atmospheric models (lines) using the parameters of Fink and Larson [60] with differing column densities of methane. Most other wavelengths exhibit temporal variation (see text), thus Warm Spitzer observations will likely also differ from earlier Spitzer and ISO observations. Comparisons of Warm Spitzer with data and models as shown here will permit methane column densities to be determined as a function of time.

The Warm Spitzer IRAC data are sensitive to intermediate wavelengths, and likely also bear evidence for atmospheric change, since these regions are acutely sensitive to methane (Fig. 16, lower panel). The most important contribution Warm Spitzer will make for Neptune is in the comparison with Prime-Mission observations, and in obtaining adequate temporal samples to characterize the time-scales of atmospheric change. Yearly samples are a minimum sample rate; twice per year would provide a more robust sampling since short-term variations are seen at wavelengths ranging from the optical to the mid-infrared (Hammel et al. [54], [55]).

8. OBSERVING TIME ESTIMATES AND STRATEGIES

8.1 Small Point Sources

The limiting fluxes for the small body point-source programs are all estimated in a very similar manner, so they are summarized together here. The limiting sensitivity for each project/group of objects is constrained by the desired S/N and the longest desired integration time. For illustrating potential observing time to perform key Solar System observations, we have constrained the longest AOR times to ~1 and 2 hours. The required S/N for each group of objects is listed in the table below, and summarized in the individual science sections above. We have assumed the Warm Era sensitivities for IRAC 3.6μm and 4.5μm are the same as in the Cold Era.

Reflected fluxes in IRAC 3.6μm and 4.5μm are estimated from known visible magnitudes. For the 30 TNOs, Centaurs, and Trojans observed in program 20769, the mean color indices are V-3.6μm ~ V-4.5μm = 1.6. For the calculations here, we used a slightly more pessimistic estimate of V-3.6μm = V-4.5μm = 1.5. Visible fluxes were calculated (using the IAU Minor Planet Center ephemeris generator) for all TNOs and Centaurs at three month intervals for the first three years of the Spitzer warm mission (March 2009 - March 2012). The above color indices were then applied to estimate the fluxes in the IRAC channels of each object.

The table below summarizes the estimated sensitivities expected for each channel and the corresponding number of observable objects (i.e. those with fluxes greater than or equal to the limiting sensitivities). The 3σ confusion limit in clean, low-background fields is ~1.8 μJy. Because we have little choice of where to observe our targets, we must assume that confusion will be significantly worse than 1.8 μJy. The magnitude distribution is a power-law, with many more faint objects than bright. As a simplification, we assume below that all observable targets will benefit from being observed twice at slightly different epochs, with the image pairs being used to subtract away confusing background sources (see *e.g.* Stansberry *et al.* [61]). The sky-subtracted image pairs can also be shifted and co-added to increase the SNR of the detections by about 1.4.

For observations of reflected sunlight (all objects except NEOs), 4.5μm provides the more stringent constraint on observability. For detecting thermal flux from NEOs, 3.6μm provides the more stringent constraint. Almost all presently known PHAs have maximum IRAC 3.6μm/4.5μm fluxes during the Spitzer era greater than 1 μJy. Furthermore, around half of all known PHAs have maximum IRAC 3.6μm /4.5μm fluxes during the Spitzer era greater than 1 mJy. Consequently, observing these objects should be quite easy. (The only complication is that, because these objects have orbits ~1 AU, their fluxes as observed by Spitzer vary by enormous factors during the Spitzer era as the Sun-asteroid-Spitzer geometry evolves. These factors are calculable, and have been taken into consideration for the technical summary above.)

A base level program for each type of object would observe, using both channels, at least 100 - 200 objects that are brighter than the limiting flux. Many objects are significantly brighter than the limiting flux, so would require shorter observing times.

We have grouped objects into coarse brightness bins to estimate the total time required to observe all the observable objects.

TABLE 1. Observeable Objects

Objects	Channel	S/N	Sensitivity μJy	T_{exp} seconds	dithers	AOR time	# of Potential Targets
TNO/Centaurs	2	3	1.1	100	32	~1hr	140
TNO/Centaurs	2	3	.8	100	64	~2h	213
Trojans	2	10	3.7	100	32	~1hr	1436
Trojans	2	10	2.6	100	64	~2hr	1694
Main Belt	2	10	3.7	100	32	~1hr	~10,000
Main Belt	2	10	2.6	100	64	~2hr	~10,000
NEOs	1	10	3.7	100	32	~1hr	~2000
NEOs	1	10	2.6	100	64	~2hr	~2000
Inactive comets	1	10	3.7	100	32	~1hr	~100
Inactive comets	2	10	2.6	100	64	~2 hr	~150
Active comets	1	10	3.7	100	32	Variable	~500
Active comets	2	10	2.6	100	64	Variable	~500

8.2 KBO/Centaur/Trojan/Outer MBA Observing Strategy

One potential strategy is to make two observations of each object, separated by enough time for the object to move at least 30", but remain in the 5.2' x 5.2' FOV. There are four advantages to this strategy: (1) certain identification of the object by its motion in the field, (2) background subtraction (this technique is effectively a shadow observation, alleviating the negative effects of confusion/high background near the ecliptic, in which we collect data in both frames rather than just sky in one), (3) if our moving object happens to pass in front of a background source in one observations, we will still collect data from the other (this avoids very detailed timing constraints), 4) for objects with known rotation periods, the second observations would be phased with the object's rotation so that opposite hemispheres are observed in order to search for rotational variability. Dithered observations will allow removal of cosmic rays and stray light correction. Assuming the flux from these objects is spread over four pixels, then fluxes of some of them are comparable to the "high" background flux. The follow-on "shadow" observations are therefore crucial and will allow accurate removal of the background.

8.3 Comet Trail Observing Strategies

Figure 17 below presents estimates of comet trail surface brightness assuming ~$5x10^{-9}$ and particle albedos of 0.05 for the scattered light component. Within 2 AU, with some effort, trails will be detectable in IRAC 3.6μm and/or 4.5μm. More than 140 comets come within this distance to the sun and some fraction of them will be available for study during the Spitzer Warm Cycle. Using SENS-PET integration times of less than an hour allows for 1s detection of 0.002 and 0.003 MJy/sr in IRAC 3.6μm and 4.5μm, respectively, with a medium background. Because trails are extended in their orbital direction, increase S/N can be obtained by integrating in that direction.

Eight comet dust trails were detected by IRAS. This was expanded to more than 30 trails during the Spitzer cryo-phase. This may be more than tripled during the Spitzer warm phase.

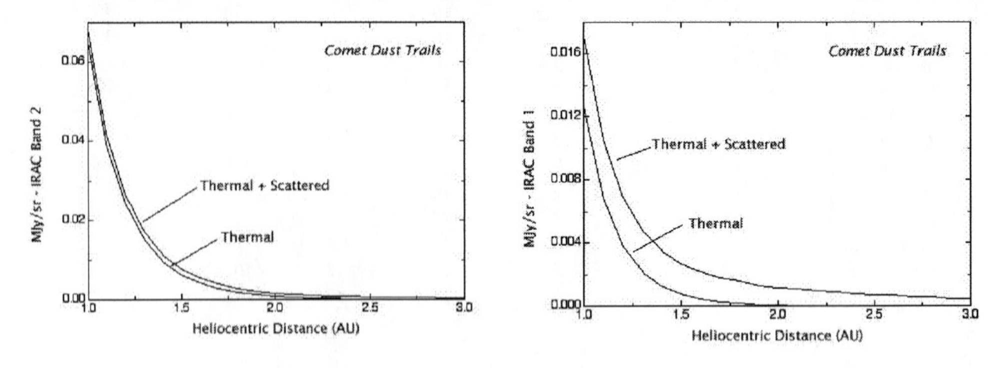

FIGURE 17. Trail surface brightness in the functioning IRAC bands during the Spitzer Warm Phase are shown assuming a near-infrared reflectivity of 0.05 and color temperature of 310K at 1 AU (Sykes et al. [19]) and median optical depth of $5x10^{-9}$ (After Reach et al. [21]).

8.4 Observations of Active Comets

Predicting surface brightnesses of active comets is famously difficult (cf. Kohoutek in 1973), but we can make a rough estimate. We note that in the Cold Era, for 3200 seconds of integration (100 second frame time and 32 dithers), one can achieve S/N=10 in a pixel in 3.6μm with a surface brightness of 0.97 μJy/pixel and in 4.5μm with 1.22 μJy/pixel. A typical case is the IRAC observation of comet 2P/Encke by Gehrz (PID 119) in June 2004 when the comet was 2.6 AU from the Sun. In only 120 seconds of integration at 3.6μm and 4.5 μm, the S/N in the inner coma was sufficiently high to detect differences in the coma morphology (as described in the Carbonaceous Species in Comets section). Comet Encke has a "typical" infrared coma and so we can conservatively estimate that ~50% of the ~350 Short Period comets will have observable comae when they are within 3 AU of the Sun. Thus at the very minimum it will be possible to perform a survey for CO/CO_2 with a sample encompassing a large fraction of the known population.

Predicting the activity of comets out to 13 AU is more difficult however, since observations of such comets have so far been scarce. As a point of comparison, we can use some of the preliminary results obtained by one of us (YRF) in a current Spitzer survey of "distant" (4-5 AU) Short Period comets (PID 30908). Approximately 15% of the sample shows activity, revealing itself as thermal emission of dust in either the IRS PU bands or the MIPS 24 μm band. This gives us confidence that there are a significant number of Short Period comets that are active beyond the nominal water-sublimation distance. Naturally the fraction of comets that are active diminishes as heliocentric distance increases, but we can use the Cold Era surveys and observations as guides to specifically target comets known to be active while far from the Sun.

9. REQUIRED SSC RESOURCES AND TOTAL SPITZER WARM ERA OBSERVING TIME

Observing all available objects is not necessary nor advocated to address the key questions identified in this proposal. The times listed in the above table represent the maximum time required to exhaustively survey the identified targets. In practice, many subsets of these times will likely be requested to address a range of topics, ranging from investigations of individual objects, to studies of small groups, and on up to larger surveys.

*Overall, the application of Spitzer's IRAC 3.6*μm *and 4.5* μm *in the warm era to the important planetary science questions presented here will require a number of resources to be in place at the SSC. These include the following:*

- A large, current database of Solar System orbital elements.
- Staff members knowledgeable in moving target observations.
- The ability to ingest similar AORs for a large list of diverse objects.
- ~1000 hours of observing time with IRAC 3.6 and 4.5 μm to perform the useful science presented herein.

REFERENCES

1. Lisse, C. M., *et al.* 2007a, *Astrophys. J* **658**, 584
2. Lisse, C.M., et al. 2007b, *Icarus* **187**, 69
3. Marzari, F. and Sholl, H. 1998, *Icarus* **131**, 41.
4. Morbidelli, A., *et al.* 2005, *Nature* **435**, 462.
5. Chiang, E., *et al.*, 2007, in Protostars and Planets V, edited by Reipurth, Jewitt, and Keil, Univ. Arizona Press, Tucson, p. 895.
6. Cruikshank, D.P. and Dalle Ore, C.M., 2003, *Earth, Moon, Plan.*, **92**, 315
7. Cruikshank, D.P., Dalle Ore, C.M., Roush, T.L., *et al.*, 2001, *Icarus* **153**, 348
8. Emery, J.P. and Brown, R.H., 2004. *Icarus* **170**, 131
9. Emery, J.P., Cruikshank, D.P., Van Cleve, J., 2006. *Icarus* **182**, 496
10. Emery, J. *et al.* 2007, *A&A*, **466**, 395
11. Barucci, A. *et al.* 2005, *A&A*, **439**, L1.
12. Grundy, W. and Buie, M., 2001, *Icarus* **153**, 248.
13. Binzel, R.P., *et al.*, 2004, *Icarus*, **170**, 259
14. Cheng, A., 2004, *Icarus*, **169**, 357
15. Sykes, M. and Walker, R., 1992, *Icarus*, **95**, 180
16. Eaton, N., Davies, J.K., and Green, S.F., 1984, *MNRAS*, **211**, 15P
17. Sykes, M.V., Lebofsky, L.A., Hunten, D.M., and Low, F., 1986, *Science* **232**, 1115
18. Sykes, M., Walker, R., and Lien, D., 1990, *Icarus* **86**, 236
19. Sykes, M.V., Gruen, E., Reach, W.T., and Jenniskens, P., The interplanetary dust complex and comets. In Comets II, edited by M. Festou and H. Weaver, Univ. Arizona Press, Tuscon, 2004, pp. 677-693.
20. Ishiguro, M., Watanabe, J., Usui, F., Tanigawa, T., Kinoshita, D., Suzuki, J., Nakamura, R., Ueno, M., and Mukai, T., 2002, *Astrophys. J*, **572**, L117
21. Reach, W.T., Kelley, M., and Sykes , M.V., 2007, *Icarus*, accepted, astro-ph/07042253.
22. Vaubaillon. J.J. and Reach B., 2006, *BAAS*, **38**, 490
23. Habing, H.J., *et al.* 1999. *Nature*, **401**, 456
24. Wyatt, M. C., *et al.* 2007, *Astrophys. J*, **658**, 572
25. Nesvorny, D. et al, 2003. *Astrophys. J*, **591**, 486

26. Good, J. C., et al. 1986. Adv. Sp. Res. **6**, 83.
27. Lisse, C.M., 2002. *Earth, Moon, and Planets,* **90**, 497
28. Beichman, C. A. *et al.* 2005, *Astrophys. J,* **626**, 1061
29. Dermott, S.F., Jayaraman, S., Xu, Y.L., Gustafson, B.A., and Liou J.C, 1994. *Nature,* **369**, 719
30. Spiesman, W.J., *et al.* 1995. *Astrophys. J,* **442**, 662
31. Reach, W.T., Morris, P., Boulanger, F., and Okumura, K., 2003. *Icarus* **164**, 384
32. Mayor, M., and D. Queloz 1995. *Nature,* **378**, 355.
33. Marcy, G. and R. Butler 1996. *Astrophys. J,* **464**, L153.
34. Rivera, E. J. *et al.* 2005. *Astrophys. J,* **634**, 625.
35. Beaulieu, J.-P. *et al.* 2006. *Nature,* **439**, 437.
36. Lovis, C., Mayor, M., Pepe, F. *et al.*, 2006, *Nature,* **441**, 305
37. Zachos, J. C. *et al.* 2001. *Science,* **292**, 274
38. Nakamura, T., and Tajiki, E., 2002, LPSC XXXIII, Abs. No. 1057.
39. de Pater, I. *et al.* 2006a. *Icarus,* **180**, 186
40. Wallace, L., 1983. *Icarus,* **54**, 110
41. Friedson, A.J., Ingersoll, A.P., 1987, *Icarus,* **69**, 135–156.
42. Rages, K.A., Hammel, H.B., Friedson, A.J., 2004, *Icarus* **172**, 548
43. Hammel, H.B., *et al.* 2005, *Icarus* **175**, 284
44. Hammel, H.B., *et al.* 2005,. *Icarus* **175**, 534
45. Sromovsky, L. and Fry, P., 2005. *Icarus* **179**, 459
46. Encrenaz, T., *et al.* 1998, *Astron. & Astrophys.* **333**, L43-L46.
47. Bishop, J., *et al.* 1990. *Icarus,* **88**, 448
48. Orton, G. S., *et al.* 2005, *BAAS,* **37**, 662
49. Lockwood and Jerzkiewicz, 2006 , *Icarus*, 180, 442
50. Hammel, H. B. and Lockwood, G. W., 2006,. *Icarus* **186**, 291
51. Young et al. 2001, *Icarus* **153**, 236
52. Orton, G. S., et al. 1987, *Icarus* **70**, 1
53. Orton, G. S., et al. 1990, *Icarus* **85**, 257
54. Hammel, H. B., *et al.* 1992, *Icarus* **99**, 347
55. Hammel, H.B., *et al.* 2006. *Astrophys. J* **644**, 1326
56. Marten, A., *et al.* 2005. *Astron. & Astrophys.* **429**, 1097
57. Hammel *et al.* 2007, *Astron. J,* **134**, 637
58. Orton, G. S., and Yanamandra-Fisher, P. 2005, *Science* **307**, 696
59. Encrenaz, T., et al., 2000, *Astron. & Astrophys.* **358**, L83
60. Fink and Larson, 1979, Astrophys. *J.* **233**, 1021
61. Stansberry, Grundy, W., Brown, M., Cruikshank, D., Spencer, J., Trilling, D., and Margot, J.-L., "Physical Properties of Kuiper Belt and Centaur Objects: Constraints from Spitzer Space Telescope," astro-ph/0702538, to appear in *Kuiper Belt*, edited by M.A. Barruccietal, University of Arizona Press, 2007.

5. – CONTRIBUTED WHITE PAPERS

Microlens Parallax Measurements With A Warm Spitzer

Andrew Gould

Department of Astronomy, The Ohio State University, 140 West 18th Avenue, Columbus, OH 43210, USA

Abstract. Because the Spitzer Space Telescope is in an Earth-trailing orbit, losing about 0.1 AU/yr, it is excellently located to perform microlens parallax observations toward the Magellanic Clouds (LMC/SMC) and the Galactic bulge. These yield the so-called "projected velocity" of the lens, which can distinguish statistically among different populations. A few such measurements toward the LMC/SMC would reveal the nature of the lenses being detected in this direction (dark halo objects, or ordinary LMC/SMC stars). Cool Spitzer has already made one such measurement of a (rare) bright red-clump source, but warm (presumably less oversubscribed) Spitzer could devote the extra time required to obtain microlens parallaxes for the more common, but fainter, turnoff sources. Warm Spitzer could observe bulge microlenses for 38 days per year, which would permit up to 24 microlens parallaxes per year. This would yield interesting information on the disk mass function, particularly old brown dwarfs, which at present are inaccessible by other techniques. Target-of-Opportunity (TOO) observations should be divided into RTOO/DTOO, i.e., "regular" and "disruptive" TOOs, as pioneered by the Space Interferometry Mission (SIM). LMC/SMC parallax measurements would be DTOO, but bulge measurements would be RTOO, i.e., they could be scheduled in advance, without knowing exactly which star was to be observed.

Keywords: Spitzer Space Telescope, infrared astronomical observations, dark matter, masses, parallaxes, distances
PACS: 95.85.Hp, 95.35.+d,97.10.Nf,97.10.Vm

1. INTRODUCTION

Microlens parallaxes measure a vector quantity, the "projected velocity" $\tilde{\mathbf{v}}$, which is the projection of the lens-source relative velocity on the plane of the observer. Another way of writing this quantity is

$$\tilde{\mathbf{v}} = \mathrm{AU}\frac{\mu_{\rm rel}}{\pi_{\rm rel}}. \tag{1}$$

where $\pi_{\rm rel}$ and $\mu_{\rm rel}$ are the lens-source relative parallax and proper motion. It is a useful quantity to measure because it depends only on the kinematic properties of the lens and source, and is independent of the mass. Once it is measured, one can also determine the "projected Einstein radius",

$$\tilde{r}_{\rm E} = \tilde{v}t_{\rm E} = \sqrt{\frac{\kappa M}{\pi_{\rm rel}}}, \tag{2}$$

where $t_{\rm E}$ is the "Einstein timescale" (which is almost always well-measured) and $\kappa \equiv 4G/c^2\mathrm{AU} = 8.1\,\mathrm{mas}/M_\odot$.

Hence, an ensemble microlens parallax measurements can distinguish statistically between different kinematic populations.

CP943, *The Science Opportunities for the Warm Spitzer Mission Workshop*,
edited by L. J. Storrie-Lombardi and N. A. Silbermann

2. MICROLENS PARALLAX SCIENCE WITH A WARM SPITZER

Microlensing parallax measurements are feasible with Spitzer toward two classes of targets, the Large and Small Magellanic Clouds (LMC and SMC) and the Galactic bulge. The science cases and technical challenges are substantially different for the two classes.

2.1. Science Toward the LMC/SMC

The nature of the microlensing events detected toward the Magellanic Clouds by the MACHO (Alcock et al. [1]) and EROS (Tisserand et al. [2]) collaborations is unknown. They are definitely too infrequent to make up most of the dark matter, but could make up of order 20%. The basic problem is that, while a few of the lenses have special properties that allow their distances to be determined, for most lenses it is not known whether they are in the Milky Way (MW) halo, the MW disk, or in the Clouds themselves. While it would be tempting to use the lenses with known locations to address this question, actually the very properties that allow their distances to be measured predispose them to be in the Clouds or the MW, so they do not constitute a fair sample. It would be better to find a way to choose "typical" events, not selected by any characteristic, and determine their position.

Spitzer has already been used to measure the microlens parallax of one lens. Dong et al. [3] found that its projected velocity was $\tilde{v} \sim 230\,\mathrm{km\,s^{-1}}$, which is typical of halo lenses, but much bigger than expected for MW disk lenses and much smaller than expected for SMC lenses. However, they also found that (with no priors on the location or properties of the lens) that it could be in the SMC with 5% probability. Thus, to draw rigorous scientific conclusions, it will be necessary to obtain parallaxes for at least 3, and more comfortably 5, such events.

2.2. Science Toward the Galactic Bulge

Han and Gould [4] conducted a systematic analysis of what could be learned about lenses observed toward the bulge by obtaining parallaxes for an ensemble of them. They found that such parallaxes greatly enhance microlensing as a probe of the stellar mass function. For example, comparing the precision of mass estimates (for lenses assumed to be in the disk in both cases), they found that without parallaxes the masses could be individually "measured" with a $1\,\sigma$ precision of 0.6 dex (so, essentially no measurement) whereas with parallaxes, the uncertainty was reduced to 0.2 dex.

Of particular interest would be to measure the frequency of disk brown dwarfs (BDs). Of course, BDs can be directly detected, but only while they are young, so that inferences about their global frequency depend sensitively on both their history of formation and their assumed cooling properties. It would be nice to constrain their frequency based solely on their mass.

The main characteristic of disk BDs from a parallax standpoint is that they have small $\tilde{r}_E$ (see eq. [2]). For example, a BD with $M = 0.03M_\odot$ and distance $D_l = 3\,\text{kpc}$ would have $\tilde{r}_E = 1.1\,\text{AU}$. In order for an M dwarf with mass $M = 0.08M_\odot$ to produce the same $\tilde{r}_E$ would require a distance $D_l = 1.5\,\text{kpc}$. This is much less likely, but of course not impossible. So an ensemble of measurements would be needed to estimate to estimate the BD frequency. In fact, as I will discuss in the § 3.2, parallax measurements are actually easier for lenses with smaller $\tilde{r}_E$ (up to a point). So Spitzer is well-placed to probe this key regime.

3. TECHNICAL CHALLENGES

The technical challenges in the two directions differ significantly, although they are related.

3.1. Technical Challenges Toward the Clouds

As mentioned, there is already one Spitzer microlens parallax toward the Clouds, which is a proof of concept. However, the source was an $I = 18$ clump giant (i.e., quite cool, so large $I - L$), which is what permitted excellent photometry with just 2 hours of integration time in each of the four measurements required. Such microlensing sources toward the Clouds are extremely rare (this was the only one detected so far during the entire Spitzer mission). To obtain the 3–5 events that are needed for the science goal, requires acceptance of more typical turnoff stars, which are both fainter and bluer, and hence demand integration times that are 10–20 times longer. Such integrations were prohibitive for cold Spitzer but might be possible for warm Spitzer when the observatory is under less severe demand. If we imagine an average of 60 hours per event for 4 events, this would take about 250 hours over 4 years. Note that the individual exposures cannot be made any longer than the 27 second exposures for the clump giant because of artifacts from very bright stars.

3.2. Technical Challenges Toward the Bulge

Here there are many challenges. First, because the bulge is near the ecliptic, it can only be observed for two 38-day periods per year (in late spring and late autumn). Moreover, the autumn window is almost useless for microlensing because no events can be discovered then from the ground. Second, in contrast to the Clouds, a large number of events must be monitored to obtain scientifically interesting results, close to 50 and preferably 100. Third, there are technical challenges related to reconciling ground-based and Spitzer photometry. This is a classic problem first discussed by Gould [5]. Basically, unlike trigonometric parallax, microlens parallax is actually a vector, π_E, which is related to observations by

$$\pi_{\rm E} = \frac{\rm AU}{d_{\rm proj}}\left(\frac{\Delta t_0}{t_{\rm E}}, \Delta u_0\right) \tag{3}$$

where Δt_0 is the measured difference in times of peak of the event as seen from Spitzer and the ground, Δu_0 is the measured difference in the impact parameter, and $d_{\rm proj}$ is the magnitude of the Earth-Spitzer separation vector projected onto the plane of the sky. The problem is that while Δt_0 can be robustly measured (because the time of the peak can be read directly from the lightcurve), u_0 from each observatory is much harder to measure because it is a fit parameter that is partially degenerate with the amount of blended light. Gould [5] showed this degeneracy could be strongly constrained by arranging for the space and ground observatories to have identical cameras (and so identical blending) but that is obviously impossible for Spitzer observations. Dong et al. [3] evaded this problem by making Spitzer responsible only for measuring the first component of $\pi_{\rm E}$: the events toward the Clouds are long enough that the acceleration of the Earth toward the Sun (roughly perpendicular to Spitzer) allowed ground-based measurement of the other component. However, most events toward the Galactic bulge are too short to make this trick work.

In order to understand whether the first two challenges can be overcome, I plotted (not shown), all 90 OGLE events that peaked during a random 38-day period in Spring 2007, with events plotted in red before they were alerted and in black afterward. Events must be alerted before peak or it is impossible to measure t_0 from Spitzer. The diagram proved to be quite a mess, but much of this mess is caused by 24 events that were alerted after peak because they are so faint. The remaining 66 events are shown in Figure 1. Based on experience with Spitzer observations of the SMC event, it should be possible to obtain better than 1% photometry in one hour exposures of events that reach $I < 17$. (The bulge sources are intrinsically bluer than the SMC clump giant, but the fields are heavily reddened, so the *observed* $I-L$ colors are similar.) There are 24 such events. Hence, over 5 years, a sample of 120 events could yield measured parallaxes.

Before addressing the challenge of aligning the photometry of the two observatories, one should take note of the bottom panel in Figure 1. It shows the Earth-Spitzer separation projected onto the bulge, during each of the 5 38-day intervals of the Warm Spitzer Mission, red for the first year and blue for the last. The average value is about 0.6 AU, about 2.5 times the leverage available for the SMC event (Dong et al. [3]). This is important because, until the projected separation gets to be the same order as $\tilde{r}_{\rm E}$, the signal-to-noise ratio of the measurement scales directly as the projected separation.

A possible approach to aligning the ground and Spitzer photometry would be to obtain L band images during the event from the ground with a high-resolution, large aperture telescope. Such measurements would yield an accurate $I-L_{\rm ground}$ color, independent of blending. While $L_{\rm ground}$ is certainly not identical to Spitzer $3.6\,\mu$m, it should be possible to use this color, together with the colors of other stars in the image, to align the two photometry systems based on stars of similar color. Only empirical testing will determine whether this is a practical method. One potential problem with this approach is that there may not be enough bright bluish stars to perform the alignment for the (typically bluish turnoff) stars that dominate the event rate. In this case, one might be

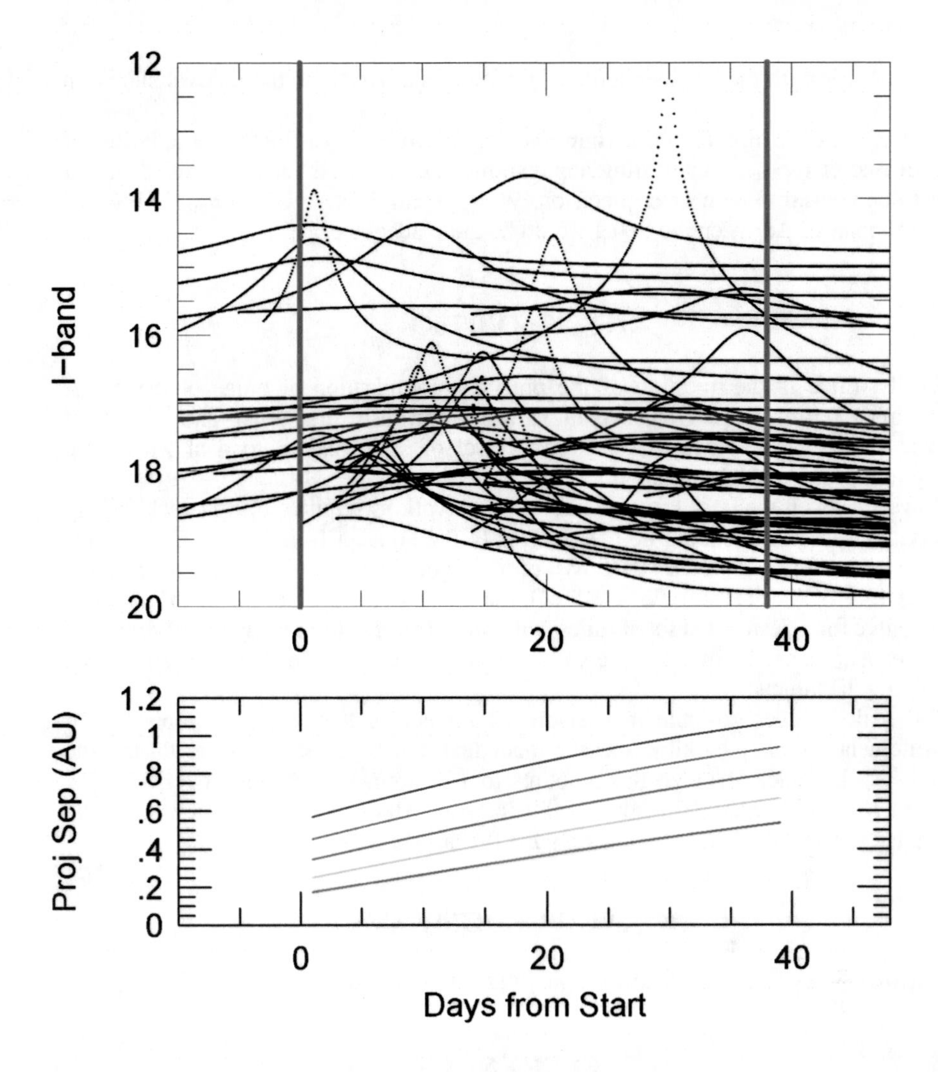

FIGURE 1. Upper Panel: Light curve trajectories of all 66 events discovered before peak by the OGLE collaboration (Udalski et al. [6]) during an arbitrarily chosen 38-day period beginning 1 April 2007. Thirty-eight day interval is delineated by the vertical lines. Light curves are in red prior to discovery of event and in black afterward. Lower Panel: Earth-Spitzer separation (projected onto the plane of the sky toward the Galactic bulge) as a function of time for each of the five 38-day intervals that warm Spitzer will be able to observe the bulge. Red to blue corresponds to 2009 to 2013.

forced to stick with the brighter stars, which comprise 7 stars in the interval shown in Fig. 1.

Even if it proves difficult to measure Δu_0 very accurately, it is important to point out that for one of the most interesting applications, the BD mass function, Δu_0 does not have to be measured with great precision. With typical $\tilde{r}_E < 2\,\mathrm{AU}$ and $d_{\mathrm{proj}} \sim 0.6\,\mathrm{AU}$, measurements of Δu_0 accurate to 0.1 would be quite adequate.

3.3. RToO/DToO

Finally, there is the question of the practical organization of bulge observations. Target-of-Opportunity (ToO) observations are notoriously disruptive, and the Warm Spitzer Mission is likely to have a lower level of staff support to deal with these disruptions than the cold mission does.

However, given the very large number of observations required during each 38-day observing "season", it should be possible apply the concept [originally developed for the Space Interferometry Mission (SIM)] of a "Regular ToO" or RToO (as opposed to a "Disruptive ToO" or DToO). For an RToO, one plans in advance to make observations of the bulge for a designated set of times, but without specifying exactly what the targets are. Then a day or so before the observations, instructions are uploaded to the spacecraft specifying the targets.

If it really proves possible to measure 24 targets, and each one requires 7 1-hr measurements (plus 1 baseline measurement that can be done at leisure during the autumn 38-day interval), then this amounts to 168 hours of observing over 38 days, or about 18% of observing time during this interval. All of these observations could be scheduled in advance, without knowing what the targets were.

ACKNOWLEDGMENTS

This work was supported by NASA grant 1277721 issued by JPL/Caltech.

REFERENCES

1. Alcock, C. et al. 2000, ApJ, 542, 281
2. Tisserand, P. et al. 2007, A&A, 469, 387
3. Dong, S. 2007, ApJ, 664, 862
4. Han, C. and Gould, A. 1995 ApJ, 447, 53
5. Gould, A. 1995, ApJ, 441, L21
6. Udalski, A., et al. 1994, Acta Astron., 44, 227

IRAC Deep Survey Of COSMOS

Nick Scoville*, Peter Capak*, Mauro Giavalisco†, Dave Sanders**, Lin Yan‡, Herve Aussel§, Olivier Ilbert**, Mara Salvato*, Bahram Mobasher¶ and Emeric LeFloc'h**

*California Institute of Technology, MC 105-24, Pasadena, CA 91125, USA
†Department of Astronomy, University of Massachusetts, LGRT-B 520, 710 N. Pleasant St., Amherst, MA 01003, USA
**Institute for Astronomy, University of Hawaii, 2680 Woodlawn Dr., Honolulu, HI 96822, USA
‡Spitzer Science Center, Caltech MC 220-6, Pasadena, CA 91125, USA
§CNRS, CEA/Saclay, Bat. 709, Gif-sur-Yvette, 91191, France
¶Department of Physics and Astronomy, UC Riverside, Riverside, CA, 92521, USA

Abstract. Over the last four years, we have developed the COSMOS survey field with complete multi-wavelength coverage from radio to X-ray, including a total of 600 hours of Spitzer Legacy time (166 hours IRAC, 460 hours MIPS). Here we propose to deepen the IRAC 3.6 μm and 4.5 μm coverage with 3000 hours over 2.3 deg^2 area included in deep Subaru imaging. This extended mission deep survey will increase the sensitivity by a factor of 3-5. The most important impact will be that the COSMOS survey will then provide extremely sensitive photometric redshifts and stellar mass estimates for approximately a million galaxies out to z $\sim$ 6. We expect these data to detect approximately 1000 objects at z = 6 to 10. The data will also provide excellent temporal coverage for variability studies on timescales from days to the length of the extended mission.

Keywords: Spitzer Space Telescope, infrared astronomical observations, origin and formation of universe, external galaxies, distances, large scale structure
PACS: 95.85.Hp, 98.62.Py, 98.62.Ve, 98.80.Bp

1. COSMOS OVERVIEW

The COSMOS survey probes the coevolution of galaxies, AGN and cosmic large scale structures (LSS). The 2 square degree field samples scales of 30–180 Mpc at z = 0.2–4 – including a million galaxies in a volume of over 10^7 Mpc^3, similar to SDSS in the local universe (Scoville et al. [1]). Star formation, galactic stellar masses and morphologies and AGN are sampled with greatly reduced cosmic variance as a function of redshift over the full range of environments at each epoch.

2. THE UNIQUE COSMOS DATASET

The Cosmic Evolution Survey (COSMOS) was the largest project ever undertaken on the Hubble telescope. During the years 2003 and 2004, 10% of the Hubble's time was devoted to imaging a 2 degree area of the sky on the celestial equator with ACS. The COSMOS Spitzer survey also is one of the two largest Spitzer Legacy programs with 650 hours over two cycles (Sanders et al. [2]). A major CXO survey (Elvis et al. in prep.) with 1.8 million seconds (Ms) is being done this year, building upon the 1.6 Ms

CP943, *The Science Opportunities for the Warm Spitzer Mission Workshop*,
edited by L. J. Storrie-Lombardi and N. A. Silbermann

project already completed with XMM (Hasinger et al. [3]). Major investments have also been made by virtually every large ground-based telescope– most importantly 35 nights with Subaru for broad band imaging to yield photometric redshifts (Taniguchi et al. [4], Mobasher et al. [5]) and 540 hours with the VLT for spectroscopy (Lilly et al. [6]). These comprehensive datasets coordinated and made public through the COSMOS team are enabling major studies of evolution for both the luminous and dark matter at $z > 0.5$ to 5. Over a million galaxies are seen in these data – by far the largest sample ever studied in the early universe.

An important feature of COSMOS is the development of complete, very sensitive, multi-wavelength datasets. Deep optical/infrared imaging (Subaru, CFHT, and NOAO) provides photometric redshifts (photo-z with $\sigma_z/(1+z) \simeq 0.02$, see Fig. 1) and SEDs for ~700,000 galaxies at $I_{AB} < 26$ mag, enabling the tracing of LSS in the galaxies. XMM and CXO probe the densest clusters in diffuse X-ray emission (and also detect 3000 AGN point-sources). Weak lensing analysis of the HST-ACS imaging reveals the dark matter LSS. SFRs and AGN measures are derived independently from the IR (Spitzer), UV (GALEX), radio (VLA) and X-ray (XMM/CXO) observations. Initial results including source counts and sensitivities for the S-COSMOS Legacy programs in IRAC and MIPS are presented in Sanders et al. [2].

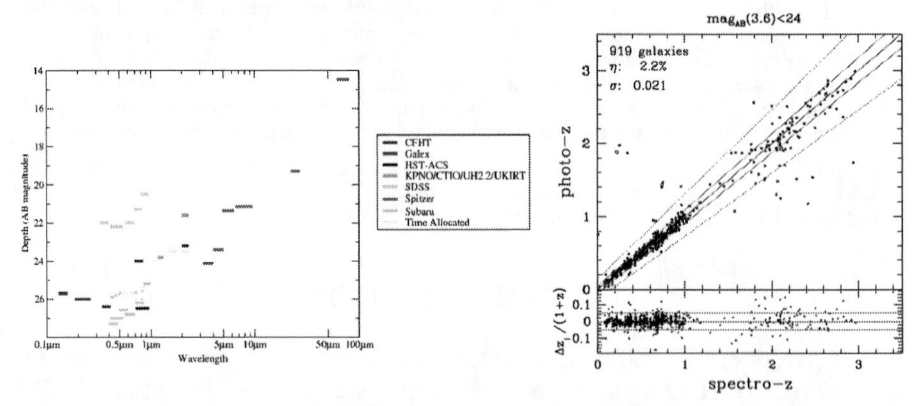

FIGURE 1. Left – The 5-σ sensitivities are shown for the UV-optical and IR bands in COSMOS (3 arcsec apertures except ACS 0.15 arcsec) Capak et al. [7]. Right – Comparison of the newly derived COSMOS photo-z's using 20 bands (Ilbert et al. in prep.) with the spectroscopic redshifts (Lilly et al. [6]) for 2100 galaxies with 3.6μm < 24 mag$_{AB}$ (some with I_{AB} down to 26 mag) indicates $\sigma_z/(1+z) < 0.02$.

3. LARGE SCALE STRUCTURE IN COSMOS

A major goal of COSMOS has recently been realized and was a highlight of the Jan'07 AAS meeting and a Nature cover (see Fig. 2) – the first imaging of large scale structures (38 LSS on scales of 3–30 Mpc at z = 0.2 - 1.1) in both the dark matter from weak lensing (Massey et al. [8]) and in the baryons from galaxy overdensities (Scoville et al. [9], Guzzo et al. [10]) and diffuse X-ray emission (Hasinger et al. [3], Finoguenov et al. [11]). The dependence of morphology and galactic spectral energy distribution (SED) on environment has been followed in these structures and into the field population (Scoville et al. [9], Capak et al. [12]). Large structures have also been seen in COSMOS at z $\sim$ 3 – 4.5 in Lyman Break Galaxies (LBG: Lee et al. in prep.) and at z = 5.7 in Lyα emitting sources Murayama et al. [22]).

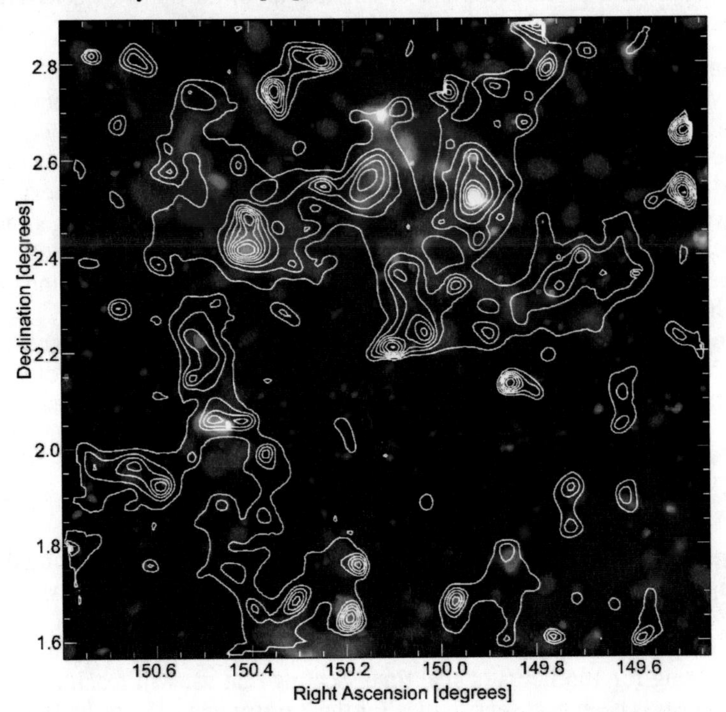

FIGURE 2. The large structures of dark matter (contours, Massey et al. [8]) and baryons (colored background, Scoville et al. [9]) are shown for the COSMOS area of the sky. The blue color indicates the concentrations of galaxies and the red shows the hot X-ray emitting gas (Finoguenov et al. [11]) at the centers of the most dense galaxy clusters. The largest filaments can be tracked over ~30 Mpc in the dark matter and the galaxies.

4. COSMOS ULTRA-DEEP IRAC

Ultra-deep IRAC observations as proposed here will be a very significant advance over earlier IRAC COSMOS data, enabling:

- detection and sampling of the galaxy mass function down to few$\times 10^{10}$ $M_{\odot}$ (less than the Milky Way) out to z $\sim$ 7;
- very deep IRAC data for the large COSMOS field will provide an unparalleled sample of high redshift galaxies (z > 0.5) with masses and SED types for precision galaxy evolution studies with respect to both environment and redshift;

- photometric redshifts at $1.5 < z < 5$ with very high accuracy ($\sigma_z/(1+z) < 0.02$). The present optical photometry catalog detects over a million objects at ($I_{AB} < 26$) while the present IRAC depth detects only 300,000. The greater depth proposed here should yield IRAC detections of virtually all of the optically detected objects – greatly improving their photo-z's and stellar mass estimates.

Precision redshifts are essential for the definition and tracing of large scale structure – in COSMOS the photometric redshifts are of sufficiently high accuracy to enable large scale structure studies at z < 1.1. (The large scale structures shown in Fig. 2 were at z < 1.1.) A major goal of the deep COSMOS IRAC observations is to provide sufficiently deep and uniform near infrared data that the photo-z's can be extended out to z $\sim$ 5 with sufficient accuracy to enable large scale structure investigations at the earlier epochs spanning the peak of galaxy and AGN evolution. The COSMOS team has proven experience in the derivation of very high accuracy photo-z, combining the IRAC data with the COSMOS ancillary imaging.

Compared with other survey fields, COSMOS offers much more extensive panchromatic ancillary data and the very large COSMOS field cover the full range of large scale structure environment and has greatly reduced cosmic variance at each redshift.

4.1. Sensitivities

There are two 38.9 day windows a year when COSMOS is observable and the moon is far enough away from the field. If we assume a 2 year mission that works out to 3734 hours. If we assume 2/3 of the time with low zody background we can observe COSMOS for 2500 hours. In the present data we get 1200 seconds of integration time per pointing for 166 hours of real time. Scaling that we will reach 2.5 hours per pointing per year. The expect sensitivities for the 3.6 μm and 4.5 μm IRAC exposures are given in Table 1, using the backgrounds actually encountered in previous COMOS IRAC observations. For comparison, we also include in this table the depths obtained for GOOD IRAC. The last column assumes that the extended mission might extend to 2014!

TABLE 1. IRAC Sensitivities – 5σ (AB)

$\lambda_{\mu m}$	**pres. (0.3 hr/pt)**	**2.5 hr/pt = 1 yr**	**7.5 hr/pt = 3 yr**	**12.5 hr/pt = 5 yr**
3.6	24.0	25.2	25.7	26
4.5	23.3	24.4	25.0	25.4
total time	166 hours	1250 hours	4750 hours	6250 hours

$\lambda_{\mu m}$	GOODS deep	GOODS ultra-deep
3.6	26.4	27.2
4.5	25.6	26.4
total time	23 hours	100 hours

4.2. The Necessity For Deep Rest-frame NIR

The major science drivers for deeper IRAC coverage of the COSMOS field are :

- **Understanding the buildup of stellar mass in galaxies during the critical epoch of galaxy assembly and peak star formation at z = 1 to 3.** The IRAC 3.6 μm and 4.5 μm coverage probes the restframe near infrared to visible – the ultra deep coverage proposed here will yield detections and photometrically-derived mass estimates for galaxies down to a few $\times 10^{10}$ $M_{\odot}$ in this redshift range, well below the knee in the mass function. And at z < 1.5, the greatest evolution for the early type galaxies is in the lower mass systems; these will be now detected in the deeper IRAC data.
- **Much higher reliability photometric redshifts for galaxies at z > 1** where the limited sensitivity of the existing COSMOS near infrared data compromises the accuracy. Higher sensitivity near infrared data will help break degeneracies in the photo-z fitting. As noted earlier, the deeper IRAC data should enable detections of nearly a million objects, compared with 300,000 detected now.
- **Detection of very large samples of objects at z >** 6 **to 10.** Although few if any objects have yet been detected and verified at z > 6, recent studies (e.g. Haiman and Loeb [13], Yahata et al. [14], Yan et al. [15]) indicate that over 1000 objects could be detected at these redshifts with the improved sensitivity over the large COSMOS field.
- **Characterization of the population of high-z IR-luminous sources** invisible from the optical and near-IR windows. IRAC will be the most efficient (and unique way) to determine which are massive starbursts, which are power-law dominated AGNs, as well as to pin down their accurate localization (in preparation of their follow-ups with NIRCAM and MIRI on JWST and with ALMA).
- These data obtained on repetition timescales spanning from days to the maximum extent of the Spitzer mission (as great as 9 years) will enable **major investigations of variable sources such as AGN and stars.**
- Lastly, we note that the additional IRAC data with many more exposures on each pointing can be programmed with a dither pattern to **optimally reach the theoretical Spitzer 3.6 μm PSF** FWHM = 1.44 arcsec). This could be extremely helpful for sources which are somewhat blended in the existing IRAC images.

Very little is known about high redshift ($z > 3$) galaxies because they are faint and relatively rare. Also, the region of the galaxy Spectral Energy Distribution (SED) containing mass and age information is redshifted to the near and mid infrared where it is more difficult to measure. Deep IRAC observations of COSMOS complement the extremely deep observations in GOODS and UDF by probing the bright end and knee of the luminosity and mass function at $z > 3$ as well as probing a large enough volume to reduce cosmic variance. Importantly, the objects in COSMOS are bright enough to allow for detailed studies, not possible with the fainter samples. Furthermore, the VLT spectroscopic program will provide redshifts many of the Spitzer-detected objects between $1.5 < z < 4$.

The COSMOS data is amongst the best ever collected for studying large scale structure at high redshift. Over 60,000 galaxies are found at $z > 3$ in COSMOS. These data represent a factor of 10 improvement in our knowledge of the $z > 3$ universe. The largest high redshift survey to date (Steidel et al. [16]) provides 2347 galaxies brighter than $I < 25.5$ at $2.5 < z < 3.5$, of which 940 are spectroscopically confirmed. In contrast, COSMOS contains ~34,000 galaxies brighter than $I < 25.5$ at $2.5 < z < 3.5$ of which over 9,000 will be spectroscopically confirmed. The deep 3.6 and 4.5μm imaging of COSMOS provided by the warm mission will probe a magnitude below L_* at $z \sim 3$ (Steidel et al. [17]) and to L_* at $z = 6$ (Iwata et al. [18]), providing an unprecedented census of the high redshift universe.

At $z > 5$ the universe is approximately a billion years old, providing very little time for galaxies to form and evolve. Yet, several studies show large structures were already visible (Hu et al. [19, 20], Ouchi et al. [21], Murayama et al. [22]) and some massive, ($10^{10} M_\odot$) relatively old (400Myr) galaxies were already in place at these redshifts (Mobasher et al. [23], Eyles et al. [24, 25], Yan et al. [26, 15], Lai et al. [27]). Detailed studies of these first galaxies are key to understanding galaxy and structure formation in the universe.

The space density and physical properties of massive galaxies provide direct tests of galaxy formation scenarios in the context of Cold Dark Matter (CDM) structure formation models. These models predict galaxies assemble through mergers of dark matter haloes over time (Springel et al. [28], Cole et al. [29], Kauffmann et al. [30]). Since the universe is not old enough at $z \simeq 5$ for many mergers to occur, the mass in large galaxies is closely related to the mass in its initial dark matter halo. The distribution of initial halo masses is strongly constrained by the Cosmic Microwave Background (CMB) measurements (Bennett et al. [31]). So, the mass function of very massive galaxies at $z \simeq 5$ is also strongly constrained. As a result, the mass function and clustering of the LAEs and LBGs at $z \simeq 5$ provide a key test of the CDM models.

The present data on the $z \simeq 5$ mass function suggest more high mass galaxies at $z \simeq 5$ than predicted by most variants of CDM models (Eyles et al. [24, 25], Yan et al. [26, 15]). We will dramatically improve these measurements by constraining the density of bright, and hence very high mass objects (those with magnitudes of $z' < 25.5$), increasing the present sample size by a factor of 100. The environmental dependence of the galaxy mass function at $z = 5.7$ and $z = 4.95$ will also constrain CDM models, which predict the largest galaxies should fall in high density regions (Springel et al. [28]). The LAEs and LBGs in the COSMOS field are the only sample allowing such studies.

Time resolved 3.6 and 4.5 μm observations of the about 3000 AGN contained in the COSMOS field will allow to probe variability on time scales of minutes to years. The existence of such variability is already indicated by ground-based observations; however, only for a very small sample of sources. Systematic studies will lead to an increasing insight to the physical processes responsible for this variability (e.g., inhomogeneities in the medium along the line of sight, encountered by relativistic outflow or small scale instabilities in the accretion disk).

4.3. Photometric Measurement of Physical Properties of Galaxies

The stellar mass of a galaxy is traced by long lived red stars, which dominate the integrated light of galaxies at rest-frame optical/near-infrared wavelengths. While active star formation is dominated by short lived blue stars, which dominate the integrated light at rest frame ultraviolet (UV) wavelengths. As a result, a sensitive indicator of galaxy age and integrated stellar mass is the strength of the Balmer break at 4000Å. The size of the Balmer break is directly proportional to the age of the stellar population (i.e. starburst *vs.* post-starburst) (Yan et al. [26], Bruzual and Charlot [32]) . Hence, three critical pieces of information are needed to study the mass and age of galaxies. These are: redshifts to constrain the rest frame position of the 4000Å break, a rest frame optical flux which is related to the stellar mass, and the rest frame ultraviolet to optical flux ratio, that indicates the age of the stellar population.

At the redshifts of the galaxies in this study, the Balmer break lies between 2.2μm and 3μm. Hence, the proposed IRAC data along with ground based z', J, and K_s data (and NICMOS F160W data in some cases) provide the rest-frame ultraviolet fluxes needed to constrain ages, dust content and star formation rates. The existing S-COSMOS (Sanders et al. [2]) Spitzer/IRAC data are not deep enough to probe the knee of the luminosity function which lies at $z' \simeq 25$ (Iwata et al. [18]). The proposed data will allow us to probe 1.5 magnitudes farther down the luminosity function than the S-COSMOS data to $z' = 25.5$. This is 0.5 magnitudes fainter than L_* at $z = 5$. A stacking analysis of the existing IRAC observations indicated galaxies at $z' \simeq 25.5$ have an IRAC flux of $\sim 0.3\mu$Jy ($m_{3.6} \simeq 25.2$ AB mag), which is a factor of 3 deeper than the existing IRAC observations (Sanders et al. [2]).

The COSMOS team has extensive knowledge of the multi-wavelength data which is critical to the maximal effectiveness of the deep IRAC imaging – both the IRAC observations planning, image reduction and mosaicing, PSF matching and source extraction and photometry. Extremely critical also, is the extensive knowledge of the the other COSMOS photometric datasets and the estimation and validation of photometric redshifts and galaxy properties from these diverse data.

5. THE LEGACY OF COSMOS

The COSMOS field is the largest contiguous area ever imaged by HST; Spitzer and Chandra have also devoted some of their largest ever allocations to this survey. Being equatorial and accessible to all telescopes (in both hemispheres), it is likely to remain the most thoroughly observed extragalactic field well into the JWST era. As such, it is destined to represent the reference field for future studies of observational cosmology, attracting massive time investments by every new facility coming on line, e.g., Herschel, ALMA and JWST. (It is already a prime target for Hershel GTO.)

COSMOS builds on the earlier, pioneering cosmic evolution studies (HDF, GOODS, GEMS, AEGIS and UDF) by probing environmental dependences on all scales of cosmic large-scale structure at the epochs of maximum activity (with a large sample volume to reduce cosmic variance). Over 2 million galaxies and AGN are detected in the 2 square degree field. Our most recent photometric redshifts derived from 20 bands (Fig. 1) and with deeper IR photometry are now yielding $\sigma_z/(1+z) < 0.02$ out to $z \sim 2.5$ with a sample of ~800,000 galaxies (see Fig. 1)! Photometric masses are also obtained for over a million galaxies from the ground-based and Spitzer-IRAC imaging. The growth of galaxies, AGN and dark matter can be traced by COSMOS over ~75% of the age of the universe.

The COSMOS (and the deeper but smaller GOODS) surveys are the backbones to planning of observational programs for future space missions such as JWST. A critical feature of these surveys is the combination of data from many facilities, across the electromagnetic spectrum for a comprehensive investigation of cosmic evolution. By doing so on such a highly statistically significant sample (by virtue of their size and homogeneous data quality), investigators can test theories on a much stricter quantitative basis.

REFERENCES

1. Scoville, N. et al., 2007, ApJS, 172, 1
2. Sanders, D. et al., 2007, ApJS, 172, 86
3. Hasinger, G. et al., 2007, ApJS 172, 29
4. Taniguchi, Y. et al., 2007, ApJS, 172, 9
5. Mobasher et al., 2007, ApJS, 172, 17
6. Lilly, S. 2007 et al., 2007, ApJS, 172, 70
7. Capak, P. et al., 2007, ApJS, 172, 99
8. Massey, R et al., 2007, Nature, 445,286
9. Scoville, N. et al., 2007, ApJS, 172, 150
10. Guzzo, L. et al., 2007, ApJS, 172, 254
11. Finoguenov, A. et al., 2007, ApJS, 172, 182
12. Capak, P. et al., 2007, ApJS, 172, 284
13. Haiman, Z. and Loeb, A., 1999, ApJ, 519, 479
14. Yahata et al.,, 2000, ApJ, 538, 493
15. Yan et al., 2006, ApJ, 651,24
16. Steidel, C et al., 2003 ApJ, 592, 728
17. Steidel, C et al., 1999 ApJ, 519, 1
18. Iwata, I. et al., 2007, MNRAS, 376 , 1557
19. Hu, E., and Cowie, L., 2006, Nature, 440, 1145
20. Hu, E et al., 2004, AJ, 127, 563
21. Ouchi, M. et al., 2005, ApJL, 620, L1
22. Murayama, T., 2007, ApJS, 172, 523
23. Mobasher, B. et al., 2005, ApJ, 635, 832
24. Eyles, L. et al., 2005, MNRAS, 364, 443
25. Eyles, L. et al., 2007, MNRAS, 374, 910
26. Yan, H et al., 2005, ApJ, 634, 109
27. Lai, K. et al., 2007, ApJ, 655, 704
28. Springel, V., 2005, Nature, 435, 629
29. Cole, D. et al., 2000, MNRAS, 319, 168
30. Kauffmann, G et al., 1993, MNRAS, 264, 201
31. Bennett, C., 2003, ApJS, 148, 97
32. Bruzual, G. and Charlot, S., 2003, MNRAS, 344, 1000

A Spitzer Warm Mission Ultra-Wide Survey As A Target Finder For The James Webb Space Telescope

Jonathan P. Gardner*, Xiaohui Fan†, Gillian Wilson** and Massimo Stiavelli‡

*NASA Goddard Space Flight Center, Code 665, Greenbelt, MD 20771, USA
†Steward Observatory, University of Arizona, 933 N. Cherry Ave, Tucson, AZ 85721, USA
**Spitzer Science Center, Caltech MS 220-6, Pasadena, CA 91125, USA
‡Space Telescope Science Institute, 3700 San Martin Dr., Baltimore, MD 21218, USA

Abstract. The James Webb Space Telescope (JWST) is the successor to the Hubble and Spitzer Space Telescopes. It has a broad scientific mission which includes spectroscopic studies of the epoch of reionization through observations of $z > 8$ quasars. The Spitzer warm mission provides a unique opportunity to conduct an infrared survey of several hundred square degrees to a depth of several micro-Janskys, capable of finding quasars out to $z = 10$. Deep JWST continuum spectroscopy of these quasars will establish the epoch and history of the Universe through detection of the Gunn-Peterson trough and/or Lyman-α damping wings. The statistics and luminosity function of high-z quasars will reveal the early history of accretion in the most extreme systems, providing insights in the role of black holes in galaxy evolution. Data obtained from an ultra-wide warm Spitzer survey will also be useful for other science, including studies of high-redshift galaxy clusters.

Keywords: Spitzer Space Telescope, infrared astronomical observations, extragalactic objects, origin and formation of the universe
PACS: 95.85.Hp, 98.62.Py, 98.62.Ai, 98.80.Bp

1. INTRODUCTION

The James Webb Space Telescope (JWST; Gardner et al. [1]) will be a large infrared-optimized space telescope launched in 2013. It is designed to address four scientific goals: The End of the Dark Ages: First Light and Reionization; The Assembly of Galaxies; The Birth of Stars and Proto-planetary Systems; and Planetary Systems and the Origins of Life. The science goals require high sensitivity in both imaging and spectroscopy in the near and mid infrared. JWST studies of the first galaxies, and their evolution until the present day, will be done with deep and deep-wide imaging and spectroscopic surveys. Studies of stellar and planetary systems will primarily be done with targets in star-forming regions that have already been identified.

JWST studies of reionization, however, will require the identification of bright sources that can be used for absorption spectroscopy. These sources would be quasars at $z > 8$, with brightnesses of a few micro-Jy. It is unlikely that JWST can find suitable targets by itself, as they are expected to be no more common than one per 10s to 100s of square degrees; JWST will not be able to survey an area this large. The JWST imaging field of view is comparable to Hubble's at 10 arcmin^2, and while the observing and survey efficiency is higher than Hubble's, it will not be orders of magnitude better. Spitzer, with

CP943, *The Science Opportunities for the Warm Spitzer Mission Workshop*, edited by L. J. Storrie-Lombardi and N. A. Silbermann

a 25 arcmin^2 imaging field of view, and comparatively rapid re-pointing capability, is a much more efficient surveyor than either Hubble or JWST; the largest extra-galactic Spitzer survey is 50 deg^2, while the largest Hubble survey is 2 deg^2. Finding $z > 8$ quasars will require a survey of hundreds of square degrees to micro-Jy depths in the infrared, with optical support for identification. The Spitzer warm mission will provide a unique opportunity to conduct such a survey.

2. JWST STUDIES OF REIONIZATION

Neutral Hydrogen atoms first formed in the early universe when primordial protons and electrons joined together with the release of the cosmic microwave background radiation, about 380,000 years after the Big Bang. About 400 million years later, the first stars and galaxies formed out of the very-low metallicity primordial gas. The first stars were very massive, 30 to 100 $M_\odot$ or more, and lived only a few million years. They produced large amounts of ultraviolet (UV) radiation. The UV radiation reionized their neighborhoods, blowing bubbles in the surrounding neutral gas. As more galaxies formed, these bubbles expanded and become more numerous, joined by bubbles of ionized gas surrounding the build-up of black holes in the centers of the galaxies. Eventually, the bubbles of ionized gas joined together, and the universe was reionized.

Soon after the first light sources appear, both high-mass stars and accretion onto black holes become viable sources of ionizing radiation. We do not know which was primarily responsible for reionizing hydrogen in the surrounding intergalactic medium (IGM). Active galactic nuclei (AGN) produce a highly energetic synchrotron power spectrum, and would reionize helium as well as hydrogen. Because observational evidence reveals that helium was reionized at a much later time, hydrogen was probably reionized by starlight at earlier epochs. However, it is possible that helium recombined after being reionized for the first time together with hydrogen. A second Helium reionization would occur during the epoch when quasar activity peaks. Recently, a combination of observations by the WMAP of the cosmic microwave background polarization (Kogut et al. [2], Page et al. [3], Spergel et al. [4]) with spectra of $z > 6$ quasars found by the Sloan Digital Sky Survey (SDSS) (Fan et al. [5, 6]) has revealed the possibility that reionization was an extended process. In some models, the completion of the reionization epoch that is seen at $z = 6$ was proceeded by an earlier partial reionization beginning in the first light epoch (Cen [7, 8]). Although the observations allow for other possibilities, there is evidence that the reionization history of the universe was complex (e.g., Gnedin [9]).

The most direct observational evidence of re-ionization is the detection of a Gunn-Peterson trough (Gunn and Peterson [10]) in the spectrum of high redshift quasars. Neutral hydrogen clouds along the line of sight (the Lyman-α forest) produce increasing absorption as the redshift increases. At $z \sim 5$, some signal is detected shortward of the Lyman-α line, suggesting that the universe is fully ionized at $z = 5$ and that re-ionization was completed at still higher redshifts.

Fan et al. [11, 12, 13]) detected high redshift quasars using the Sloan Digital Sky Survey, including some at $z > 6$, showing a drop in continuum flux just shortward of Lyma- α as much as a factor 150 (see Fig. 1). This is evidence that a Gunn-Peterson trough has been detected in these objects (Becker et al. [14], Fan et al. [6]). Other QSOs

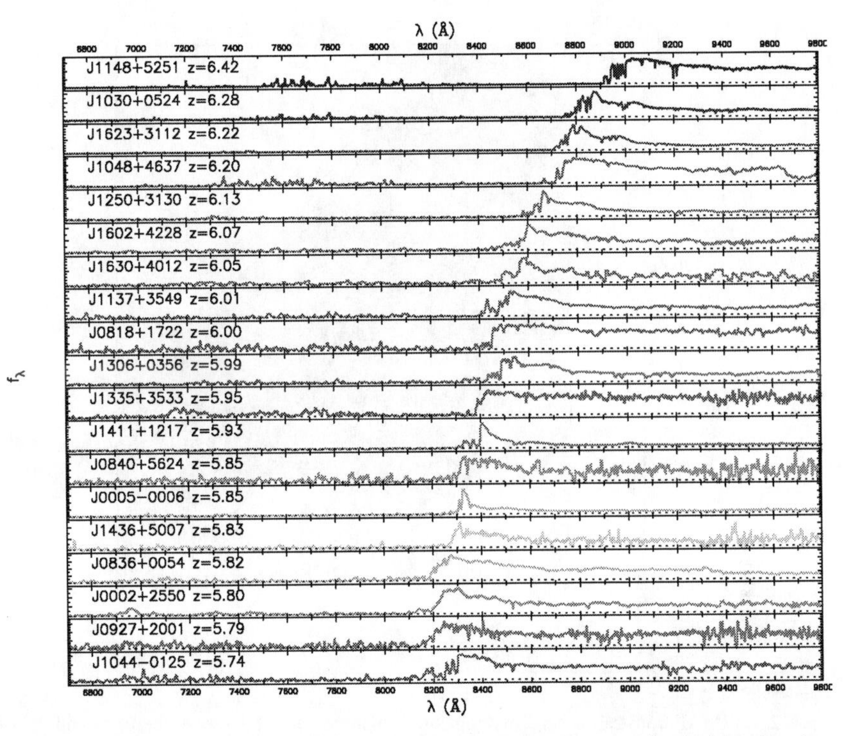

FIGURE 1. Spectrum of 19 SDSS quasars at $5.74 < z < 6.42$. The absence of flux in the region shortward of the Lyman-α line is a possible indication of a Gunn-Peterson trough in the highest redshift quasars, indicating that the fraction of neutral hydrogen has increased substantially over this redshift range, and that the universe is approaching the epoch of complete reionization at $z \sim 6$ (From Fan et al. [15]).

at slightly lower redshift show a much smaller continuum drop. Variation in the QSO properties indicates that the reionization did not occur abruptly at the same time throughout the universe. Haiman and Holder [16] argue for an extended 'percolation' period of reionization. We cannot conclude that the reionization epoch has been determined on the basis of these few objects, particularly since even a very modest local neutral hydrogen column density could produce the observed Gunn-Peterson troughs. However, these detections open up the possibility that re-ionization was completed by the relatively low redshift of $z \sim 6$.

The correlations between the cosmic microwave background temperature and polarization, as measured by WMAP, support a much earlier reionization of hydrogen, giving $z_{reion} = 10.9^{+2.7}_{-2.3}$, under the assumption of a single epoch of full reionization (Fig. 2; Spergel et al. [4]). This may be an indication that hydrogen at least partially recombined after the first epoch of reionization, only to be reionized again at a lower redshift. In contrast to the epoch of hydrogen reionization, the epoch of helium reionization has been firmly identified at $z \sim 3$ through the detection of a Gunn-Peterson trough in quasar spectra (Jakobsen et al. [17], Davidsen et al. [18], Heap et al. [19]).

It is possible to compute the minimum surface brightness required to reionize the

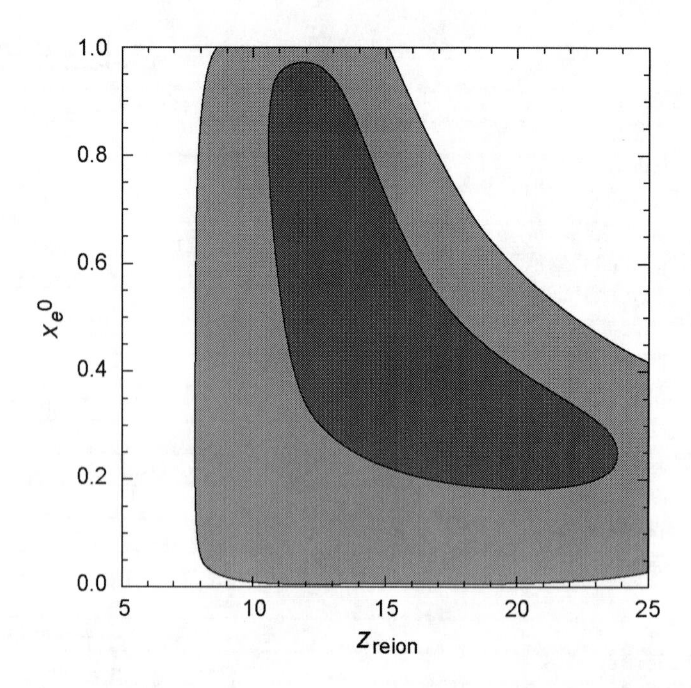

FIGURE 2. WMAP constraints on the reionization history. The plot shows the 68% and 95% joint two-dimensional marginalized confidence level contours for a model in which the Universe is partially reionized with an ionization fraction x_e^0 at z_{reion}, and then fully reionized at z = 7. The WMAP data are inconsistent with a single epoch of reionization at $z \sim 6$, and argue for a complex reionization history. (From Spergel et al. [4]).

universe, under the assumptions that the universe was reionized by hot population III stars, and that all UV photons can escape the system. This minimum surface brightness of ionizing sources at $z > 6$ is about AB = 29 mag arcmin^{-2} in a redshifted $\lambda = 1400$Å band (Stiavelli et al. [20]), when counted as the typical ionizing flux seen per unit area. For a luminosity function similar to that of z = 3 Lyman break galaxies and with M* not fainter than −15 mag, this implies a few sources per square arcmin with AB = 31 or brighter.

Even though one often refers to the epoch of reionization as if it were a sudden transition, the time elapsed between the epochs when 10% and 90% of hydrogen was reionized can last a significant fraction of the age of the universe. The WMAP detection of a significant Compton opacity is evidence of either an extended reionization process, or of two distinct reionization epochs (Cen [7, 8], Haiman and Holder [16], Holder et al. [21], Stiavelli et al. [20]). Regardless of the specifics of the reionization process, inhomogeneities along the line of sight may create dispersion in optical depth shortwards of Lyman-α. Moreover, only a very low residual fraction of neutral hydrogen is needed to produce a Gunn-Peterson trough in the spectra of high redshift quasars. In addition, the opacity near Lyman-α would be modified in the neighborhood of ionizing sources (Miralda-Escudé and Rees [22]), in analogy to the proximity effect in QSOs (Møller and Kjaergaard [23]).

While models differ significantly in the details of how the reionization was started by these various possible first light populations at $15 < z < 25$, they all converge to produce roughly the same cosmic star-formation history of population II stars in the mini halos of dwarf galaxies at $6 < z < 10$. This is simply the consequence of the need to fit the nearly complete Gunn-Peterson troughs now seen in the spectra of at least four SDSS quasars in the range $6.05 < z < 6.43$ (Fan et al. [12]). While these indicate non-zero flux shortward of 0.8 μm, there is essentially zero flux longwards of 0.810 μm. Hence, there was significant HI in front of these quasars at $z > 5.7$, although the HI-fraction at $z = 6$ was still very small (of order 10^{-4} to 10^{-5}).

In most models, the conclusion of this reionization epoch is modeled by dwarf galaxies producing an increasing number of population II stars at $6 < z < 11$. Most models are rather similar in their predictions of the cosmic star formation history at $6 < z < 8$, in order to match the SDSS Gunn-Peterson troughs seen at $z = 6$. For example, the Cen [7] models predict an increase in the cosmic star-formation history of a full factor of 10 over $16 > z > 11$ and another factor of 10 increase over $11 > z > 6$. In other words, most of the population II stars that we see today were born in dwarf galaxies, but most were not born until about $z = 8$ (consistent with the oldest ages of population II measured today of 12.8 Gyr), and it was likely the high-mass end of those population II stars that completed the epoch of reionization by $z = 6$. In WMAP cosmology (Spergel et al. [24]) there was only 300 Myr between at $6 < z < 8$ and another 170 Myr at $8 < z < 10$, so the stellar population that was formed in those galaxies, and whose O, B and A stars helped complete the reionization of the universe by $z = 6$, is still visible as the low-mass population II stars seen today.

3. JWST OBSERVATIONS

The epoch of reionization is revealed through signatures in the Lyman-α forest: a black Lyman-α Gunn-Peterson trough, islands in the Lyman-α forest, and appearance of a Lyman-α damping wing. In addition to these techniques, the epoch of reionization can be identified as the redshift at which there is fast evolution of the properties of Lyman-α emitters. However, a sharp transition in the Lyman-α luminosity function can be suppressed if, for instance, a relatively long reionization onset is coupled to a smooth increase in metal content. Sources at higher redshifts will have increasingly more absorbed Lyman-α but also increasingly stronger intrinsic equivalent widths because of the lower metallicity. It is easy to build models where the two effects cancel out. Alternative methods, not sensitive to this limitation, are the study of the evolution of the ratio between Lyman-α and Balmer lines.

JWST will make deep spectroscopic observations of $z > 8$ quasars to study the Lyman-α forest. High signal-to-noise, $R \sim 1000$, near-infrared spectra will reveal the presence of a Gunn-Peterson trough or of a Lyman-α damping wing. High S/N is needed to discriminate between optical depths τ of a few and $\tau \gg 10$. A damping wing should be present for a few million years, before the ionizing radiation is sufficient to create a large Strömgren sphere around each ionizing source. Failure to detect a damping wing does not necessarily imply that the universe is ionized. R $\sim$ 100 spectra will be able to determine the presence of a Lyman-β 'island'. This is relevant if reionization occurs relatively abruptly. In this case, objects at redshifts between the redshift of reionization, z_{reion}, and $z = (\lambda_\alpha/\lambda_\beta)(1+z_{reion}) - 1$, will show an island of normal, finite, forest absorption between the Lyman-α and the Lyman-β forests. Here, λ_α and λ_β are the rest frame wavelengths of Lyman-α and Lyman-β, respectively.

If there are indeed two distinct reionization epochs, the (possibly partial) recombination following the first reionization may be detectable in continuum spectra of high redshift objects as an absorption signature in the region shortward of Lyman-α.

4. HIGH REDSHIFT QUASARS

Discoveries of luminous quasars at $z \gtrsim 6$ (Fan et al. [5, 12, 13], Fan, Carilli, and Keating [25], Mahabal et al. [26], Cool et al. [27], McGreer et al. [28]) show that active, massive black holes of order a billion solar masses existed a few hundred million years after the first star formation in the Universe. Their absorption spectra reveal the rapid evolution of the IGM at the end of reionization.

4.1. The Quasar Luminosity Function: Black Hole Growth at z > 6

Evolution of the density and bright-end luminosity function of quasars have been well-studied using the SDSS sample at $0 < z < 6$ (Fan et al. [11, 13], Richards et al. [29], Hopkins et al. [30]). At $z > 3$, the density declines exponentially towards high redshift: for quasars at $M_B < -27$, the density at $z \sim 6$ is a factor of about 40 lower than where it peaks at $z \sim 2.5$ (see Fig. 3). Meanwhile, the slope of quasar luminosity also appears to be a strong function of redshift: while it is relatively flat $\Phi(L) \propto L^{-2.5}$ at $z \sim 4$, it appears to have considerably steepened at $z \sim 5$ to $z \sim 6$, with $\Phi(L) \propto L^{-3.1}$, suggesting a large population of faint quasars at high-redshift (see Fig. 4). These faint quasars could also be a significant contributor of the reionization photon budget.

The existence of such objects is surprise. The SDSS $z \sim 6$ quasars are among the most luminous quasars at any redshift; their apparent magnitudes are about 19 even at $z > 6$! Their black hole masses are in the range 10^9 to 10^{10} $M_\odot$, and they are likely to reside in dark matter halos of 10^{13} $M_\odot$, comparable to the most massive black holes and galaxies found locally. The black hole masses are close to the theoretical maximum allowed by Eddington-limited accretion for black hole growth. It is a challenge to understand how the universe managed to assemble these systems so fast and so efficiently; this rapid growth places one of the strongest constraints on the theory of supermassive black holes.

The evolution of the high-redshift quasar luminosity function provides direct constraints on the accretion mode of the highest redshift quasars. The standard supermassive black hole model assumes that it grows from a stellar mass seed black hole through radiatively efficient, Eddington-limited accretion and black hole-black hole mergers. Assuming a radiative efficiency of 0.1 for black hole accretion, the Salpeter time, defined as the e-folding time for black hole growth in Eddington-limited accretion, is about 4 $\times$

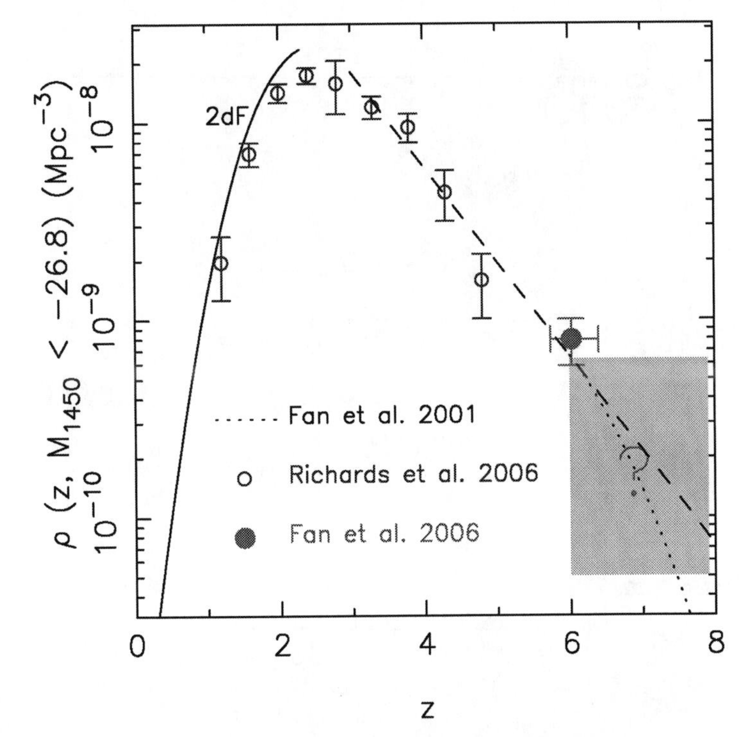

FIGURE 3. Evolution of the quasar number density. The quasar number density at $z \sim 6$ is a factor of 40 times lower than the peak at $z \sim 2$. It is unknown whether the slope continues to higher redshift, or whether the evolution is even steeper.

10^7 yr. Even if the first supermassive star formed as early as $z \sim 30$, there would only be about 15 Salpeter times available for black hole growth until $z \sim 7$. If quasars with $10^9\ M_\odot$ existed at $z \sim 7$, even non-stop Eddington accretion requires the existence of seed black holes with $M > 300\ M_\odot$, probably collapsed from massive Pop III stars (e.g., Heger et al. [31], Abel et al. [32], Tumlinson et al. [33]).

A number of negative feedback effects would further slow down black hole growth. Using semi-analytic modeling, Volonteri and Rees [34] recently modeled black hole growth through accretion and mergers, including feedback effects such as gravitational recoil and growth timescale changes due to black hole spin-up. They conclude that unless super-Eddington accretion or direct formation of intermediate-mass black holes with 10^3 to $10^6\ M_\odot$ is important, high-redshift black holes have to be limited to moderate spin (with low radiative efficiency) and the formation needed to be highly selective at early epochs. They predict that there will be a rapid decline of quasar density and a rapid steepening of the quasar LF at $z > 6$, as the growth of the most massive black holes is limited in that case by the number of e-folding times available. The steepening LF at $z \sim 6$ discovered in the SDSS, if confirmed towards higher redshift, would strongly suggest that the Salpeter timescale constraint is becoming the limiting factor for black

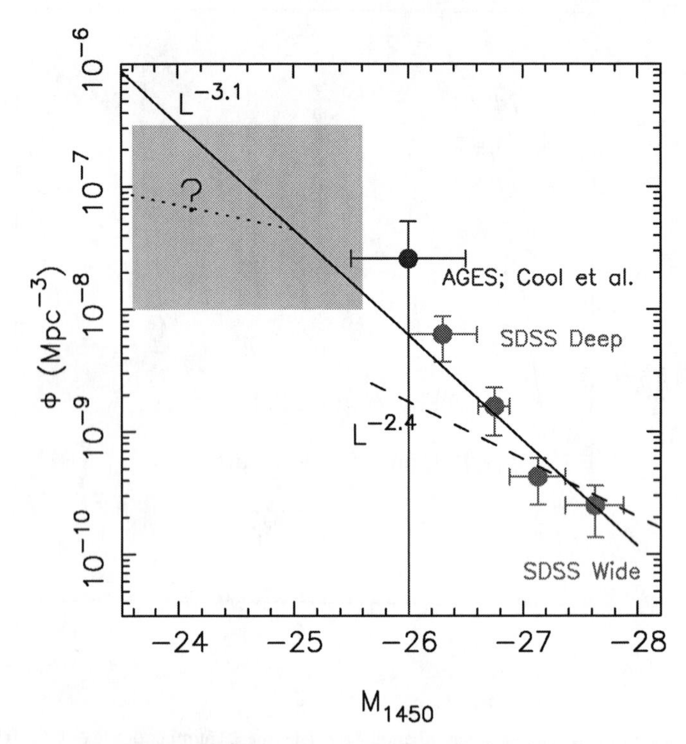

FIGURE 4. Quasar luminosity function at z ~ 6. The luminosity function of quasars at z ~ 6 detected by the SDSS and the Spitzer AGES survey shows a steeper slope ($L^{-3.1}$) than at lower redshift. A break is expected at faint magnitudes, but is currently undetected.

hole growth. New quasar samples will enable the determination of the quasar LF down to $M_{1450} \sim -23.5$, possibly reaching the expected break of the LF at $z > 6$.

Discovery of any extremely luminous quasar at $z > 7$ or beyond will present a major challenge to the theory of black hole formation. To overcome the timescale constraint, a number of models (e.g. Madau [35], Begelman et al. [36], King and Pringle [37]) suggest rapid early black hole growth either through super-Eddington accretion, or direct formation of intermediate-mass black holes. Recently, Li et al. [38] used hydrodynamic simulations that include both hierarchical galaxy formation and black hole feedback effects to study the growth of $z > 6$ quasars. They show that in a large simulated volume, for Eddington-limited accretion, only the most biased peaks in the most extreme galaxy formation environment could have reproduced observed properties of luminous quasars at $z \sim 6$ towards a small fraction of selected viewing angles. Detailed comparison between the simulated and observed quasar LF at $6 < z < 8$ (Fig. 3 & 4) will test whether the earliest supermassive black holes can still grow by Eddington accretion, or alternative accretion modes are truly needed for their rapid growth.

The discovery of such quasars at $z > 7$ would answer several key questions: Will their properties be very different from those at lower redshift? Will they pose a serious challenge to our current models of Eddington-limited black hole growth and hierarchical galaxy formation?

4.2. Evolution of Quasar Properties at High Redshift

One of the somewhat surprising results from the studies of SDSS quasars is the similarity in the quasar spectral energy distributions (SEDs) between high and low-z quasars. This is especially evident in UV and X-ray observations, which trace the emission from the broad line region and accretion disk. Fan et al. [13] presented the average spectrum of a sample of $z \sim 6$ quasars in the UV; it is virtually identical to the standard quasar template constructed from low-redshift quasars, in terms of both emission line strengths and continuum shape. Using emission line ratios, they showed that quasar broad emission line regions have roughly solar or even higher metallicities at $z \sim 6$ (e.g., Pentericci et al. [39], Freudling et al. [40], Dietrich et al. [41], Venkatesan et al. [42]). This lack of evolution demonstrates that the emission line region and accretion disk can form on a short timescale and are decoupled from the cosmological environment.

However, there is now some evidence for the possible evolution in properties of the AGN dust torus and host galaxy that are further away from the central engine. Most low-redshift dust is generated in AGB star envelopes; at $z > 6$, there is not yet enough time for intermediate mass stars to evolve. This timescale constraint suggests possible evolution in the dust origin and properties. Spitzer observations reveal two $z \sim 6$ quasars that do not show any sign of hot dust emission in mid-IR wavelengths, while no such quasar has ever been found at lower redshift (Jiang et al. [43]). Some SDSS $z \sim 6$ quasars also exhibit dust extinction inconsistent with any standard extinction curve, but can be fit by extinction generated from the much smaller grains that are produced in short-lived Type II supernova envelopes (Maiolino et al. [44]). It will be highly exciting if the emission line properties of quasars also show strong evolution when we reach $z > 7$.

Luminous high-redshift quasars are likely to be located in the densest environments in the early Universe. However, direct imaging of the stellar light of host galaxies is extremely difficult with current technology. At observed far-IR to mm wavelengths, radio-quiet quasars are dominated by the reprocessed radiation from cool dust in the host galaxy. Roughly 40% of the radio-quiet quasars are detected at the 1 mJy level at 1 mm wavelength (e.g. Carilli et al. [45], Priddey et al. [46], Wang et al. [47]). Their FIR luminosity ($\sim 10^{13}$ $L_{\odot}$) implies a dust mass of order 10^8 $M_{\odot}$. If the dust heating came from starburst activity, that would suggest star formation rates as high as 10^2 to 10^3 $M_{\odot}$ yr^{-1}. Molecular gas emission from these starforming host galaxies provides the only method so far to constrain their dark matter halo masses at $z \sim 6$. Walter et al. [48] obtained high-resolution VLA observations of the CO in the $z = 6.42$ quasar SDSS J1148+5251. The gas has a line-width of 280 km/s, with the size of the CO emitting region ~ 2 kpc. This is the first time a high-redshift quasar host is resolved both spatially and kinematically. Assuming the gas is bound, the estimated dynamical mass is about 10^{10} $M_{\odot}$ within the 2 kpc radius. The total bulge mass is about one order of magnitude lower than one would expect from the local black hole mass - velocity dispersion relation. This result, as well as CO measurements in other quasars at $z \sim 4$ (Shields et al. [49]), indicate that luminous quasars at high redshift do not live in a fully developed galaxies with massive central bulges. Does this suggest that the black hole formed before galaxy assembly? Pushing such studies to higher redshift will require JWST and ALMA.

4.3. Spitzer Warm Mission Survey for Quasars at z > 7

Given the uncertainties in black hole accretion and feedback models and dust obscurations, it is currently still difficult to make accurate theoretical calculations of the density of quasars at $z > 7$. However, Volonteri and Rees [34] predict a relatively steep luminosity function at $z \sim 6$, consistent with the new SDSS observations. They also find relatively smooth evolution in the black hole mass function at high redshift. Meanwhile, Li et al. [38] show that the progenitors of the $z \sim 6$ quasars are roughly one and two orders of magnitude fainter in bolometric luminosity at $z \sim 8$ and $z \sim 10$, respectively, although the energy output of AGN light in the rest-frame optical and UV is highly uncertain. These calculations suggest that a simple extrapolation of the SDSS quasar luminosity function from $z \sim 6$ is possible.

A major consideration in survey design is the issue of depths vs. area. A steep quasar luminosity function suggests a higher efficiency with deeper coverage; however, the quasar luminosity function will have an *unknown* break in the slope at fainter luminosities. At $z \sim 6$, this break occurs at $M_B > -24$. The same luminosity corresponds to $f_{3.6} = 1.5\mu$Jy at $z = 8$, and $f_{3.6} = 1\mu$Jy at $z = 10$. We estimate the number of quasars to be detected with a Spitzer warm mission in two surveys. A shallow survey of 500 deg^2 down to ~ 3 μJy (SWIRE depth) will detect about 130 quasars at $z > 6$, 22 quasars at $z > 8$ and 0.8 quasars at $z > 10$. Because the steepness of the luminosity function approximately cancels the effects of $\sqrt{t}$, a deeper survey of 50 deg^2 observed to ~ 1 μJy will give approximately the same number of quasars: about 120 quasars at $z > 6$, 20 quasars at $z > 8$ and 0.7 quasars at $z > 10$. Although these two straw-man surveys would find about the same number of quasars, the quasars in the deeper survey would be fainter. As JWST requires high signal-to-noise continuum spectra for its reionization studies, the wider survey would be better suited.

While a Spitzer survey can detect these high-z quasars, candidate selection and identification observations are not trivial. These quasars have $J_{AB} \sim 23$ to 24. Even with VISTA, there is no ground-based JHK survey reaching this depth in such wide area. So the candidate selection will likely rely on combining Spitzer and deep optical (i, z, and Y) data. For example, the planned PANSTARRS medium deep survey reaches $z_{AB} = 24.7$ and $Y_{AB} = 23.9$ over 1200 deg^2.

The main contaminants are cool dwarfs and dusty, red galaxies at intermediate redshift. Cool dwarfs are much brighter in the 3.6 and 4.5 micron bands, and have much redder [3.6]-[4.5] colors. Most of the low-redshift galaxies can also be separated out by combining optical and Spitzer colors (e.g. Stern et al. [50]). More detailed modeling is required to refine the color selection for dusty galaxies and whether they will present a major challenge to the color selection. Once the candidate list has been reduced to a few hundred sources, then ground-based near-infrared follow-up of individual candidates will be possible. Spectroscopic confirmation may require JWST.

5. OTHER SCIENCE FROM AN ULTRA-WIDE SPITZER SURVEY

A wide-area survey is ideal for finding rare objects. The power of Spitzer lies in its ability to rapidly survey large areas of the sky to depths of a few μJy in the infrared, reaching normal galaxies at redshifts of $1 < z < 3$. The most massive objects in the Universe are clusters of galaxies, and their existence at high redshift places strong constraints on both cosmology and galaxy evolution. Crucial stages in cluster galaxy evolution, where vigorously star-forming proto-cluster regions at $2 < z < 6$ (Kurk et al. [51], Steidel et al. [52]) transform into the quiescent early-type population seen at $z \sim 1$ (Blakeslee et al. [53], Holden et al. [54]) appear to occur in the hard-to-access redshift range $1 < z < 2$, the current 'cluster desert,' where clusters have been difficult to find and identify. Recent results from Spitzer (Stanford et al. [55], Wilson et al. [56], Brodwin et al. [57]) have shown that it is possible to detect clusters at $z > 1$ using IRAC's 3.6 and/or 4.5 micron channels in combination with optical data. The largest of the IRAC cluster surveys is being carried out by the SpARCS collaboration in the 50 square degree SWIRE fields (Muzzin et al. [58], Wilson et al. [59]) and uses an infrared adaptation of the two-filter Cluster Red-Sequence technique (Gladders and Yee [60], Gilbank [61], Gladders and Yee [62]). The SpARCS survey is now about two-thirds complete, and using a combination of z'-band and IRAC 3.6 micron imaging, is on track to discover a total of about 200 new clusters at $1 < z < 2$ by the survey's end in late 2007. Several of these clusters are awaiting spectroscopic confirmation and are candidates for the most massive distant objects yet discovered. In the coming years, the number density of these new massive clusters in the $z > 1$ redshift range has the potential to place important independent constraints on cosmological parameters (Gladders et al. [63]), and in particular, on the equation of state of the dark energy.

In addition to studies of high-z quasars and clusters, an ultra-wide survey would extend all of the science results from the short-wavelength SWIRE survey data with a factor of 10 greater statistics. Results would include luminosity functions, galaxy-galaxy correlation functions, weak lensing dark matter tomography and low-mass field stars.

6. CONCLUSION

High-redshift quasars are needed as background sources for JWST studies of reionization, and are interesting in their own right. A 500 deg^2 survey to the SWIRE depth offers the best chance for finding bright quasars out to $z \sim 10$.

Deep JWST continuum spectroscopy of bright $z > 8$ quasars will establish the epoch and history of reionization of the Universe through detection of the Gunn-Peterson trough and/or Lyman-α damping wings. The statistics and luminosity function of high-z quasars will reveal the early history of accretion in the most extreme systems, providing insights in the role of black holes in galaxy evolution.

The data obtained from this survey would also be useful for a wide variety of other science, including constraints on cosmology and galaxy evolution from the detection of $z > 1$ galaxy clusters.

REFERENCES

1. Gardner, J. P., et al. 2006, Space Science Reviews, 123, 485
2. Kogut, A., et al. 2003, ApJS, 148, 161
3. Page, L., et al. 2007, ApJS, 170, 335
4. Spergel, D. N., et al. 2006, ApJS, 170, 377
5. Fan, X. et al., 2001b, AJ 122, 2833
6. Fan, X. et al., 2002, AJ, 123, 1247
7. Cen, R. 2003a, ApJ, 591, 12
8. Cen, R. 2003b, ApJL, 591, L5
9. Gnedin, N. Y. 2004, ApJ, 610, 9
10. Gunn, J. E., and Peterson, B. A. 1965, ApJ, 142, 1633
11. Fan, X. et al., 2001a, AJ, 121, 54
12. Fan, X. et al., 2003, AJ, 125, 1649
13. Fan, X. et al., 2004, AJ, 128, 515
14. Becker, R.H., et al. 2001, AJ, 122, 2850
15. Fan, X. et al., 2006, AJ, 131, 1203
16. Haiman, Z., and Holder, G. P. 2003, ApJ, 595, 1
17. Jakobsen, P., et al. 1994, Nature, 370, 35
18. Davidsen, A. F., Kriss, G. A., and Wei, Z. 1996, Nature, 380, 47
19. Heap, S.R., et al. 2000, ApJ, 534, 69
20. Stiavelli, M., Fall, S. M., and Panagia, N., 2004, ApJ, 600, 508
21. Holder, G. P., Haiman, Z., Kaplinghat, M., and Knox, L. 2003 ApJ, 595, 13
22. Miralda-Escudé, J., and Rees, M. J. 1994, MNRAS, 266, 343
23. Møller, P., and Kjaergaard, P., 1992, A&A, 258, 234
24. Spergel, D. N., et al. 2003, ApJS, 148, 175
25. Fan, X., Carilli, C. L., and Keating, B. 2006, ARA&A, 44, 415
26. Mahabal, A., Stern, D., Bogosavljevic, M., Djorgovski, S. G., and Thompson, D., 2005, ApJ, 634, L9
27. Cool, R. J., et al., 2006, AJ, 132, 823
28. McGreer, I. D., Becker, R. H., Helfand, D. J., and White, R. L., 2006, ApJ, 652, 157
29. Richards, G. T., et al., 2006, AJ, 131, 2766
30. Hopkins, P. F., et al., 2007, ApJ, 654, 731
31. Heger, A., Fryer, C. L., Woosley, S. E., Langer, N., and Hartmann, D. H., 2003, ApJ, 591, 288
32. Abel, T., Bryan, G. L., and Norman, M. L., 2002, Science, 295, 93
33. Tumlinson, J., Venkatesan, A., and Shull, J. M., 2004, ApJ, 612, 602
34. Volonteri, M., and Rees, M. J., 2006, ApJ, 650, 669
35. Madau, P., 2005, Growing Black Holes: Accretion in a Cosmological Context, 3
36. Begelman, M. C., Volonteri, M., and Rees, M. J., 2006, MNRAS, 370, 289
37. King, A. R., and Pringle, J. E., 2006, MNRAS, 373, 90
38. Li, Y., et al., 2007, ApJ, 665, 187
39. Pentericci, L., et al., 2002, AJ, 123, 2151
40. Freudling, W., Corbin, M. R., and Korista, K. T., 2003, ApJL, 587, L67
41. Dietrich, M., Hamann, F., Appenzeller, I., and Vestergaard, M., 2003, ApJ, 596, 817
42. Venkatesan, A., Schneider, R., and Ferrara, A., 2004, MNRAS, 349, L43
43. Jiang, L., et al., 2006, AJ, 132, 2127
44. Maiolino, R., Schneider, R., Oliva, E., Bianchi, S., Ferrara, A., Mannucci, F., Pedani, M., and Roca Sogorb, M., 2004, Nature, 431, 533
45. Carilli, C. L., et al., 2001, ApJ, 555, 625
46. Priddey, R. S., Isaak, K. G., McMahon, R. G., Robson, E. I., and Pearson, C. P., 2003, MNRAS, 344,
47. Wang, R., et al. 2007, AJ, 134, 617

48. Walter, F., Carilli, C., Bertoldi, F., Menten, K., Cox, P., Lo, K. Y., Fan, X., and Strauss, M. A., 2004, ApJ, 615, L17
49. Shields, G. A., Menezes, K. L., Massart, C. A., and Vanden Bout, P., 2006, ApJ, 641, 683
50. Stern, D., et al. 2005 , ApJL, 631, 163
51. Kurk, J.D., et al.2004, A&A, 428, 817
52. Steidel, C.C., et al.2005, ApJ, 626, 44
53. Blakeslee, J.P. et al.2003, ApJL, 596, 143
54. Holden, B. et al., 2004, AJ, 127, 2484
55. Stanford, S. A. et al, 2005, ApJ, 634L
56. Wilson, G., et al. 2006, in proceedings 'The Spitzer Space Telescope: New View of the Cosmos', ASP Vol. 357, 238
57. Brodwin, M., et al., 2006, 651, 791
58. Muzzin, A., et al., 2006, in proceedings 'The Spitzer Space Telescope: New View of the Cosmos', ASP Vol. 357, 246
59. Wilson, G., et al. 2007, ApJ, 660, 59
60. Gladders, M. D. and Yee, H. K. C., 2000, 120, 2148
61. Gilbank, D. G., 2004, MNRAS, 348, 551
62. Gladders, M. D. and Yee, H. K. C., 2005, 157, 1
63. Gladders, M. D., et al. 2007, ApJ, 655, 128

The Porcupine Survey: A Distributed Survey and WISE Followup

Edward L. Wright*, P. R. M. Eisenhardt†, A. K. Mainzer†, J. D. Kirkpatrick** and M. Cohen‡

*UCLA Dept. of Physics & Astronomy, Los Angeles CA 90095-1547, USA
†Jet Propulsion Laboratory, Mail Stop 264-740, Pasadena, CA 91109, USA
**Infrared Processing and Analysis Center, MS 100-22, California Institute of Technology, Pasadena, CA 91125, USA
‡University of California Berkeley, Radio Astronomy Lab., 601 Campbell Hall, Berkeley, CA 94720, USA

Abstract. Spitzer post-cryogen observations to perform a moderate depth survey distributed around the sky are proposed. Field centers are chosen to be WISE brown dwarf candidates, which will typically be 160 μJy at 4.7 μm and randomly distributed around the sky. The Spitzer observations will give much higher sensitivity, higher angular resolution, and a time baseline to measure both proper motions and possibly parallaxes. The distance and velocity data obtained on the WISE brown dwarf candidates will greatly improve our knowledge of the mass and age distribution of brown dwarfs. The outer parts of the Spitzer fields surrounding the WISE positions will provide a deep survey in many narrow fields of view distributed around the sky, and the volume of this survey will contain many more distant brown dwarfs, and many extragalactic objects.

Keywords: Spitzer Space Telescope, infrared astronomical observations, brown dwarfs, Wide-field Infrared Survey Explorer (WISE)
PACS: 95.85.Hp, 97.20.Vs, 97.10.Cv

1. INTRODUCTION

The Wide-field Infrared Survey Explorer (WISE) will survey the entire sky in 4 mid-infrared bands centered at 3.3, 4.7, 12 and 23 μm, with angular resolution $6''$ ($12''$ at 23 μm), to 5σ sensitivity requirements of 120, 160, 650 and 2600 μJy. WISE will be launched into an IRAS/COBE-like Sun-synchronous nearly polar low Earth orbit in November 2009. The WISE Band 1 and Band 2 filters are optimized for the detection of methane dominated brown dwarfs. The 4.7 μm filter samples the strong peak of brown dwarf emission due to a hole in the methane absorption, and the 3.3 μm filter is in a strong methane band so brown dwarfs are faint, but this band is sensitive to emission from normal stars and can be used to veto normal stars. Very few objects are expected to mimic the extreme [3.3]-[4.7] μm colors of T (and Y) brown dwarfs. WISE will find several hundred brown dwarfs colder than 750 K, which is approximately the lowest currently known temperature. For such cold objects the optical flux is negligible, and the near-IR flux accessible by ground-based telescopes is quite weak. Since the WISE survey covers the whole sky, these objects will include the closest and brightest objects in all spectral classes. As such they will provide the best targets for further study with large telescopes such as the JWST. But it will be very important to collect astrometric

CP943, *The Science Opportunities for the Warm Spitzer Mission Workshop,*
edited by L. J. Storrie-Lombardi and N. A. Silbermann

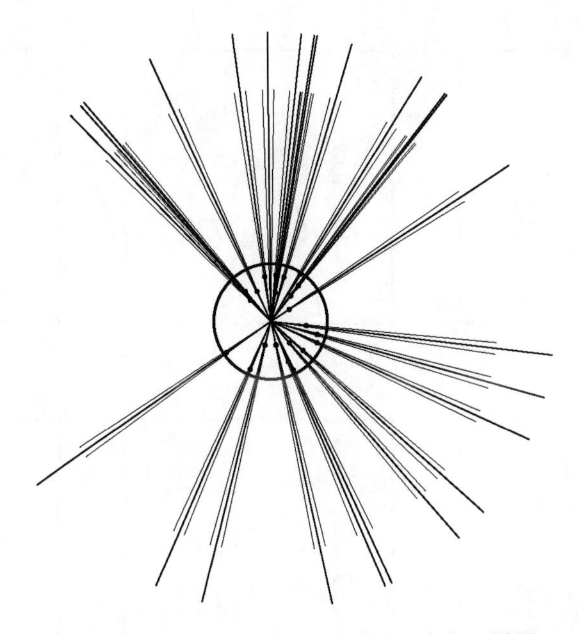

FIGURE 1. Cartoon showing the WISE survey volume (the inner circle), WISE candidates within the circle, and the Spitzer observations as rays extending out to great distance.

information on these sources in order to concentrate JWST time on the nearest and most interesting systems. For objects with small distances from the Solar System, JWST cameras will be able to resolve and separately image self-luminous Jovian planets or detect astrometric wobbles of the central stars, and JWST spectrographs will be able to obtain high SNR data yielding data about the elemental and isotopic abundances in these stars.

TABLE 1. Astrometric accuracies for positions at the mean epoch of the observations θ, the annual proper motion μ, and the annual parallax π.

Datasets	$\sigma(\theta)$	$\sigma(\mu)$	$\sigma(\pi)$
WISE baseline	0.50″	...	...
WISE extended	0.35″	1.40″/yr	...
Spitzer + WISE baseline	0.07″	0.20″/yr	0.08″
Spitzer + WISE extended	0.07″	0.14″/yr	0.07″

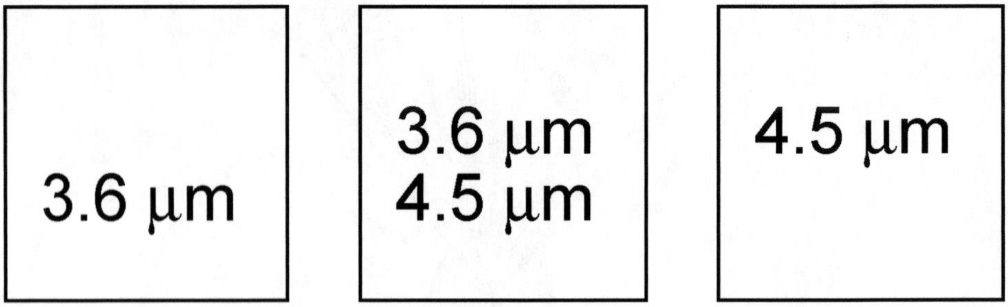

3.6 μm

3.6 μm
4.5 μm

4.5 μm

FIGURE 2. First epoch with Spitzer.

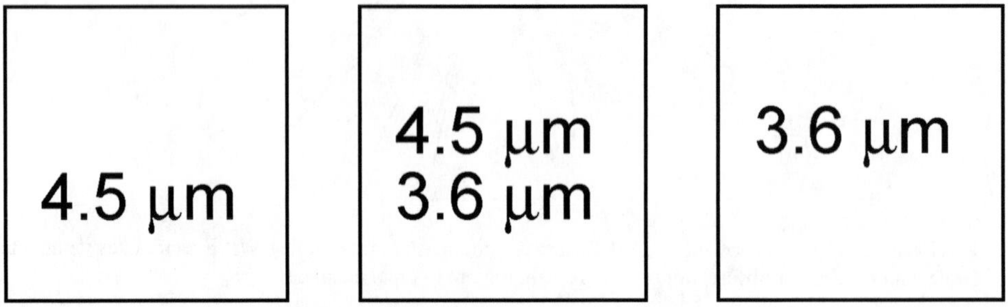

4.5 μm

4.5 μm
3.6 μm

3.6 μm

FIGURE 3. Second Spitzer epoch, 6 months later.

2. WISE CANDIDATE BROWN DWARFS

The WISE catalog will have $\mathcal{O}(10^8)$ sources at 4.7 μm, and the reliability requirement is > 0.999. The methane dominated brown dwarfs found by WISE will be brightest at 4.7 μm, and many of them will only be detectable at 4.7 μm. This means that there might be as many as $\mathcal{O}(10^5)$ false 4.7 μm only sources that could be confused with brown dwarfs.

The WISE team will make a special effort to reduce the false positive rate, both by visual inspection of the 8 or more frames covering each object, and by leveraging the photometric and astrometric information gained by cross identification of the WISE sources with other very large area surveys. Thus there might be $\mathcal{O}(10^4)$ candidates. We propose to take Spitzer 3.6 and 4.5 μm images of these fields, which will generate a fairly large area survey spread into many pencil beams or quills, as shown in Figure 1.

3. WISE SCHEDULE

WISE observations are planned to occur during calendar year 2010. In the baseline WISE mission, the sky is surveyed once in a six month interval, but all of the eight or more images of a given object are taken in a one or two day period, so there is only one epoch for each source. The WISE cryostat is currently estimated to have a lifetime greater than a year, which would allow a second pass over the sky. If this

WISE extended mission occurs, there will be two epochs separated by 6 months on every object. Astrometry from WISE is required to be better than 0.5 arc-sec relative to 2MASS, so if the extended mission occurs WISE alone will be able to estimate proper motions to an accuracy better than 1.4 arc-sec/yr if the parallax does not confuse the proper motion determination. In any case, candidates that show significant motion on a 6 month baselines deserve followup with Spitzer. The baseline WISE mission will only give a position at a single epoch and no proper motion information.

But no proper motion information does not mean there will be no proper motion. For a typical transverse velocity of 30 km/sec, the proper motion will be $(6/D)$ arc-sec/yr, where D is the distance in parsecs. This can easily confuse the cross-identification of WISE brown dwarfs with other surveys such as 2MASS, SDSS, the UKIDSS Hemisphere Survey, VISTA, and the POSS. WISE candidates that show up in these other surveys, with no position displacement and colors inconsistent with a cool brown dwarf, can be vetoed. But positive confirmation of nearby brown dwarfs using non-simultaneous data will be difficult without good proper motions.

The preliminary WISE data release will occur in early to mid 2011, with a final data release coming one year later. Thus the followup observations with Spitzer could occur in the 2012 to 2014 time frame. The observations proposed here would give two Spitzer observations with a 6 month separation, and the time base between the WISE and Spitzer observations would be 2 to 3 years. Spitzer positions should be accurate to 0.1 arc-sec, since the WISE candidate brown dwarfs will be quite high SNR in the Spitzer data.

4. UTILITY OF WARM SPITZER DATA

Both the WISE and Spitzer observations will be taken very close to 90° elongation from the Sun which means that the baseline for the parallax measurement will be the full 2 AU diameter of the Earth's orbit. Thus the Spitzer data will provide parallax accuracies very close to $\sigma(\theta)/\sqrt{2}$ as shown in Table 1.

The proposed Spitzer observations involve two epochs taken 6 months apart. At each epoch the WISE candidate is observed with both fields of the IRAC. We propose a 5 point Gaussian dither pattern with small steps in each field of view, with 30 second frame times, leading to 150 seconds of integration time in each band per epoch. A single epoch thus provides both 3.6 and 4.5 μm data on the central position, plus two flanking fields covered in only one of the bands, as shown in Figure 2. The second epoch taken 6 months later will have the orientation of the IRAC fields reversed as seen in Figure 3, so that each of the flanking fields gets covered in both bands, while the central field is covered in both bands during each epoch. Since the IRAC fields are quite close together, there is only a weak requirement on the orientation of the second epoch. We hope that 7 to 9 candidate positions can be done per hour at each epoch. Then 2500 total hours can provide followup data in two epochs for about 10,000 WISE candidate positions.

The data taken in this survey will provide a large sample of serendipitous objects. The 5σ limits in the flanking fields will be 3 μJy and 6 μJy in IRAC bands 1 and 2, if the post-cryogen performance matches the cryogenic performance. The central field will be covered twice, giving 2 and 4 μJy limits. For serendipitous brown dwarfs, each WISE candidate leads to surveying a solid angle of $2 \times 5' \times 5'$ to a distance 5 times greater than the WISE depth, and the central $5' \times 5'$ to a distance 6 times greater than WISE. Therefore the volume surveyed by Spitzer at each WISE candidate position corresponds to a volume of 3.7 sq deg surveyed to the WISE depth. Thus 2500 hours, covering 10,000 WISE candidates, would also survey a volume corresponding to 37,000 square degrees of WISE data. Thus there will be a comparable number of cool brown dwarfs discovered serendipitously in these data.

5. REQUIRED TIME

The survey plan depends on how much overhead there is for each observation. If the overhead is T and the integration time is t, the survey depth goes like $t^{-1/2}$, and the number of serendipitous sources discovered per hour goes like $t^{3/4}/(t+T)$. This discovery rate is a maximum when $0.75/t - 1/(t+T) = 0$ or $(1+T/t) = 4/3$. Thus the integration time should be 3 times larger than the overhead, as noted by Eisenhardt *et al.* (2004, ApJS, 154, 48). The actual time per AOR computed by Spot is 629 seconds, but the density of WISE candidates on the sky will be high so the slews between candidates will be a few degrees or less, and the slew penalty in Spot may be too high. Thus an overhead of 100 to 150 seconds per candidate may be reasonable. An overhead of 150 seconds gives 2500 hours of total time to do 10,000 candidates in two epochs, while the Spot time per AOR gives 3500 total hours.

6. OTHER USES

The value of the serendipitous extragalactic data will be substantial. For 2500 hours a solid angle of 208 square degrees is covered. This is 50 times the solid angle of the Spitzer Shallow Survey, covered to a greater depth. Since the WISE brown dwarf candidates will be randomly distributed over the high galactic latitude sky, the effect of cosmic variance on the statistics of extragalactic sources will be minimized. Each $5' \times 5'$ spans 2.5 Mpc for redshifts in the range 1-2 where the angular size distance is a maximum in the concordance model, and this is a large enough patch to study clustering.

A Spitzer post-cryo survey of WISE ULIRGs may also be worthwhile. ULIRGs with a $F_\nu \propto \nu^{-2}$ will be detected by WISE in its 12 and 23 μm bands but may not be seen at 3.3 and 4.7 μm. Spitzer data will extend the spectral energy distribution to shorter wavelengths and give improved positions. For ULIRGs, a survey deep enough to detect neighboring galaxies down to L_* would be useful since it would allow one to measure the clustering properties of the population. Since the ULIRGs could be at $z \approx 3$, this would require longer exposures on each field to reach L_*.

7. CONCLUSION

The Porcupine Survey proposed here will greatly enhance the value of the WISE survey, and will positively identify very nearby star systems that will be prime targets for follow-up with the JWST. The combination of WISE, Spitzer post-cryo and JWST observations could very well lead to the discovery and verification of the closest star systems to the Sun and the closest extrasolar planets.

8. FURTHER READING

See the WISE web site for astronomers at `http://www.astro.ucla.edu/~wright/WISE` or the WISE public Web site at `http://wise.astro.ucla.edu`.

ACKNOWLEDGMENTS

Spitzer and WISE are managed by the Jet Propulsion Laboratory, California Institute of Technology, under contracts with NASA.

A Spitzer Warm Mission Proposal To Survey Single White Dwarfs

Michael Jura

Department of Physics and Astronomy, University of California Los Angeles, Los Angeles, CA 90095, USA

Abstract. At least 2% of single white dwarfs with 10,000 K $< T <$ 20,000 K have an infrared excess at 3.6 μm and 4.5 μm. A Spitzer warm mission survey of 1000 white dwarfs can enable an improved characterization of the occurence of an excess as a function of white dwarf properties such as age, mass and composition. The tidal-disruption of an asteroid is the most promising model to explain an infrared excess around a cool, single white dwarf, and thus this survey can help improve our understanding of extrasolar planetary systems.

Keywords: Spitzer Space Telescope, infrared astronomical observations, white dwarfs, stellar ages, stellar masses
PACS: 95.85.Hp, 97.20Rp, 97.10.Cv, 97.10.Nf

1. INTRODUCTION

An infrared excess produced by dust orbiting the white dwarf G29-38 was discovered 20 years ago (Zuckerman and Becklin [1]). It took 18 years for the second excess, that around GD 362, to be found (Becklin et al. [2], Kilic et al. [3]). Progress during the past year has been dramatic; there are now eight white dwarfs known to have an infrared excess (Mullally et al. [4], Farihi et al. [5], Jura et al. [6], Kilic et al. [7], Kilic and Redfield [8]). The Spitzer warm mission can be used to continue to improve our understanding of dust orbiting white dwarfs.

2. BACKGROUND

The most promising model to explain an excess around a white dwarf is that a minor-body is perturbed (Debes and Sigurdsson [9]) to orbit within the tidal radius of the star where it is destroyed and a dusty disk is produced (Jura [10]). Besides producing an infrared excess, accretion from this disk can pollute the white dwarf. Gravitational settling of heavy elements is so effective in white dwarfs cooler than 20,000 K that their atmospheres are expected to be pure hydrogen or pure helium (Paquette et al. [11]), and the $\sim$ 20% of these stars which exhibit photospheric metals are thought to be externally polluted. Since the gravitational settling times in a white dwarf's atmosphere for different elements vary by less than a factor of 2 (Paquette et al. [11]), the metal composition of the white dwarf directly measures the composition of the accreted material. Therefore, the atmospheres of white dwarfs with an infrared excess may provide a unique and powerful tool for studying the bulk compositions of extrasolar

CP943, *The Science Opportunities for the Warm Spitzer Mission Workshop,*
edited by L. J. Storrie-Lombardi and N. A. Silbermann

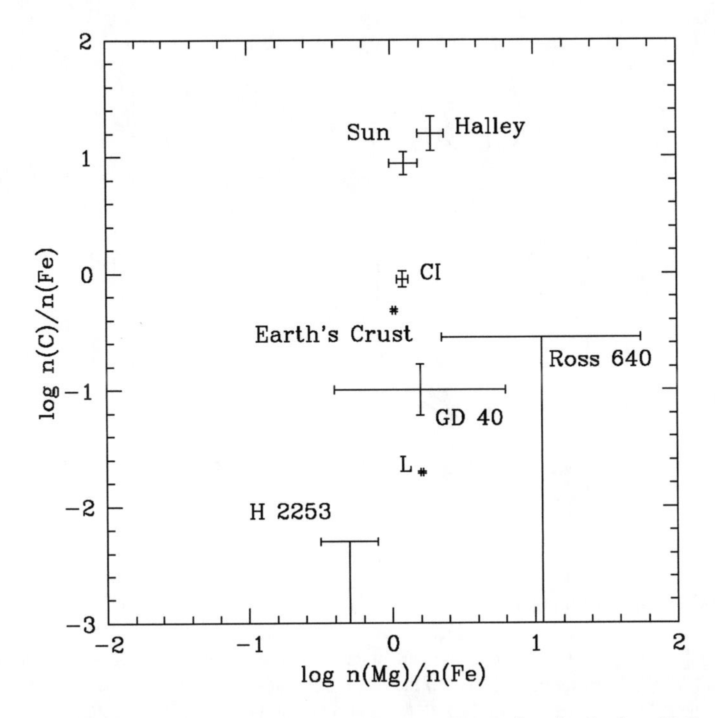

FIGURE 1. Abundance ratio of carbon to iron vs. magnesium to iron in the Sun, Halley's comet, the Earth's crust, CI chondrites, L chondrites and three externally-polluted white dwarfs (see Jura [12]). The accreted material onto white dwarfs is markedly deficient in carbon. From Jura [12]. Reproduced by permission of the AAS.

minor planets. Recent work shows that similar to both the Earth and even the most primitive meteorites, carbon is deficient relative to refractory atoms by more than a factor of 10 in the atmospheres of six white dwarfs[1].

To date, the white dwarfs with definite infrared excesses all possess dust as warm as 1000 K. A representative result showing a marked excess at both 3.5 μm and 4.6 μm is shown in Fig. 2 for GD 40, an externally polluted white dwarf with m_K = 15.6 mag and a value of n(C)/n(Fe) more than a factor of 30 below Solar.

Approximately 200 white dwarfs have been observed with the IRAC camera on Spitzer, and 8 stars are known to display infrared excesses. However, because of selection effects, the true frequency of white dwarfs with an infrared excess is unknown; a conservative bound is that at least 2% of white dwarfs with $10,000K < T_{eff} < 20,000K$ have an excess.

---

[1] n(C)/n(Fe) is shown for three stars in Figure 1. GD 362 (Zuckerman et al. [13]) also has a markedly low value of n(C)/n(Fe) while GD 61 and PHL 962 are externally-polluted carbon-deficient white dwarfs with low values of n(C)/n(Si) (Desharnais et al. in preparation).

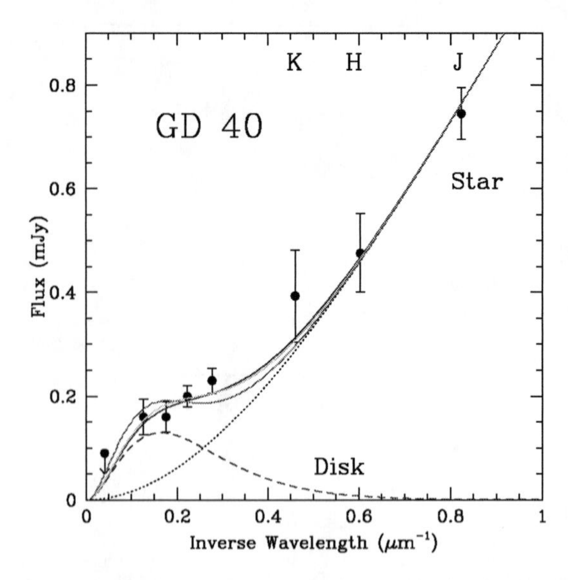

FIGURE 2. Comparison of models and data (2 errors) for GD 40. The plot shows the contribution from the stellar photosphere (dotted line), one disk model (dashed line) and totals for three models described in Jura et al. [6]. A representative model has an inner disk temperature of 1200 K and an outer disk temperature of 600 K; the dust lies well within the star's tidal radius. From Jura et al. [6]. Reproduced by permission of the AAS.

There are now enough white dwarfs with an infrared excess that it is possible to begin to discern some patterns. First, all the white dwarfs with an infrared excess also display atmospheric metals. Thus, it is highly plausible that the stars are accreting from reservoirs of circumstellar material. Second, Kilic et al. [7], Farihi et al. [5], Kilic and Redfield [8] have shown that the stars with relatively high calcium abundances also tend to display an infrared excess. Third, the stars with an infrared excess all have effective temperatures greater than $\sim$ 9500 K and white dwarf cooling ages less than $\sim$ 1 Gyr [2]. Figure 3 (taken from Jura et al. [6]) presents a comparison of $\dot{M}_{dust}$ vs. effective temperature for DAZs with IRAC photometry, distinguishing between those with and without excess emission.

There are many unknowns:

- Do stars with T < 10,000 K display infrared excesses as frequently as stars with T > 10,000 K? Cooler white dwarfs are older, and their asteroid belts may have been depleted.
- How robust is the correlation, based on a handful of stars, between accretion rate

[2] GD 362 previously was thought to be hydrogen-rich and have a relatively low luminosity and therefore a cooling age well in excess of 1 Gyr (Gianninas et al. [14]). However, the star is now known to be helium-rich and have a much larger luminosity and correspondingly shorter cooling age (Zuckerman et al. [13]).

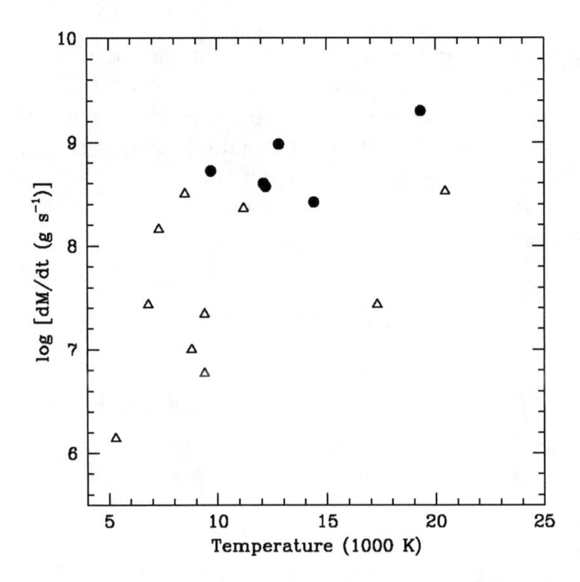

FIGURE 3. Mass accretion rates vs. effective temperature for hydrogen-rich white dwarfs with an excess (solid circles) and without an excess (open triangles) in published Spitzer data. GD 133 and PG 1015+161 (Jura et al. [6]), GD 56 (Jura et al. [6], Kilic et al. [7]), G29-38 (Zuckerman and Becklin [1]), WD 2115-560 (Mullally et al. [4], von Hippel et al. [16]), WD 1150-153 (Kilic and Redfield [8]) have an excess while EC 12043-1337, HE 1225+038, and HE 1315-1105 (Jura et al. [6]), WD 1202-232, WD 1337+705 and WD 2149+021 (Mullally et al. [4]), and WD 0208+396, WD 0243-026, WD 0245+541, and WD 1257+278 (Debes and Sigurdsson [17]) do not. Two helium-rich white dwarfs (GD 40 and GD 362) that also display an excess are not shown here. From Jura et al. [6]. Reproduced by permission of the AAS.

and the presence of an infrared excess? Why do some externally-polluted white dwarfs display an excess, but most do not? Are there different modes of accretion?

- Is there any correlation between having an infrared excess and the mass of the white dwarf?
- Is there any difference between the fraction of helium-rich and hydrogen-rich white dwarfs with an infrared excess?
- Is there any correlation between having an infrared excess and the chemical composition of the accreted material? Are all externally-polluted stars also carbon deficient? What is the fraction of white dwarfs with detectable gaseous disks (Gaensicke et al. [15]) and do these stars also have infrared excesses?

3. EXTENDED-MISSION PROPOSAL

A survey of the $\sim$ 1000 single white dwarfs that are brighter than $m_K \approx$16 mag to measure 3.6 μm and 4.5 μm fluxes of 100 μJy is possible with the Spitzer warm mission. (The IRAC data shown in Fig. 1 for GD 40 (m_K = 15.6 mag) were obtained with a total of 600 seconds of integration time.) To date, the most extensive IRAC survey of white dwarfs was by Mullally et al. [4] who observed 124 white dwarfs at 4.5 μm and 7.9 μm. A preliminary estimate is that the proposed survey requires $\sim$200 hours of telescope time. By substantially increasing the number of observed stars, it should be possible to understand better the origin and evolution of dust orbiting white dwarfs. In turn, spectroscopic studies of these stars can enable a better quantitative measure of the bulk compositions of extrasolar minor planets.

REFERENCES

1. Zuckerman, B., and Becklin, E. E. 1987, Nature, 330, 138
2. Becklin, E. E., Farihi, J., Jura, M., Song, I., Weinberger, A., and Zuckerman, B. 2005, ApJL, 632, L119
3. Kilic, M., von Hippel, T., Leggett, S. K., and Winget, D. E. 2005, ApJL, 632, L115
4. Mullally, F., Kilic, M., Reach, W. T., Kuchner, M. J., von Hippel, T., Burrows, A., and Winget, D. E. 2007, ApJS, 171, 206
5. Farihi, J., Zuckerman, B., and Becklin, E. E. 2007, ApJ, submitted
6. Jura, M., Farihi, J., and Zuckerman, B. 2007, ApJ, 663, 1285
7. Kilic, M., von Hippel, T., Leggett, S. K., and Winget, D. E. 2006, ApJ, 646, 474
8. Kilic, M., and Redfield, S. 2007, ApJ, 660, 641
9. Debes, J. H., and Sigurdsson, S. 2002, ApJ, 572, 556
10. Jura, M. 2003, ApJL, 548, L9
11. Paquette, C., Pelletier, C., Fontaine, G., and Michaud, G. 1986, ApJS, 61, 197
12. Jura, M. 2006, ApJ, 653, 613
13. Zuckerman, B. et al. 2007, ApJ, submitted
14. Gianninas, A., Dufour, P., and Bergeron, P. 2004, ApJL, 617, L57
15. Gaensicke, B. T., Marsh, T. R., Southworth, J., and Rebassa-Mansergas, A. 2006, Science, 314, 1908
16. Von Hippel, T., Kuchner, M.J., Kilic, M., Mullally, F., and Reach, W.T., 2007, ApJ, 662, 544
17. Debes, J. H., and Sigurdsson, S., 2007, AJ, 134, 1662

AUTHOR INDEX

M

N

P

R

S

T

V

W

Y